MW00586643

THE DANGEROUS
SNAKES OF AFRICA

STEPHEN SPAWLS AND BILL BRANCH

PRINCETON UNIVERSITY PRESS
Princeton and Oxford

Published in the United States and Canada in 2020 by
Princeton University Press, 41 William Street,
Princeton, New Jersey 08540
press.princeton.edu
First published in Great Britain in 2020 by
Bloomsbury Publishing Plc

Library of Congress Control Number 2019955916
ISBN 978-0-691-20792-6

2 4 6 8 10 9 7 5 3 1

Maps by Julian Baker
Design by Rod Teasdale

Printed and bound in India by Replika Press Pvt. Ltd.

This book is dedicated to
Royjan Taylor (1975–2019):
Kenyan, herpetologist
and friend.

CONTENTS

Namaqua Dwarf Adder,
Bitis schneideri
(Tyrone Ping).

PREFACE – Stephen Spawls

In 1995, Bill Branch and I published *The Dangerous Snakes of Africa*, a guide to Africa's dangerous snakes and the treatment of their bites. It was well received; many herpetologists and medical personnel remarked on how useful it was. Snakebite is a major problem in Africa. A Somali doctor wrote to us and said that, with our book, for the first time in his life he actually had in his hands concrete information about the deadly snakes of his country, where they lived, the risk they posed and how to take action against them.

Since 1995, the taxonomy of African snakes and the medical expertise around their bites has advanced. So when Johan Marais suggested to us in 2017 that our book was 'well overdue for a revision', Bill and I agreed. Our publishers showed interest, we signed a contract and we started the preliminary work. In late 2017, Bill cautiously wrote to me to tell me he was having mobility issues, and wasn't sure how this might affect our project. And in early 2018, to my shock, I heard from Bill that he had been diagnosed with motor neurone disease. But he remained full of optimism and said that he fully intended to do his bit, cheerfully pointing out that Stephen Hawking had lasted many years with the same affliction. Bill sorted out a superb set of digital pictures and started work on the snakebite and viper accounts.

But it was not to be. The disease quickly took hold. Bill could not move around and was confined to one floor of his house. He continued to work, but eventually he could no longer even hit the computer keys. Tragically, Bill died on 14 October 2018. His untimely death is a major loss to African herpetology; few herpetologists have reached both the public and their fellow scientists with as much verve and accuracy as Bill did.

So it has fallen to me to complete the book. With his wide-ranging and ever-precise knowledge of every aspect of African herpetology, Bill was one of my sternest critics. As I write, I find myself constantly asking, 'What would Bill think of this?' And so, on behalf of Bill and myself, I hope this book is going to be useful in putting into the public domain the information about Africa's dangerous snakes, where they live, and how to prevent and treat their bites. We hope it reduces the suffering of the people of Africa.

Puff Adder, *Bitis arietans*, Dodoma, Tanzania (Bill Branch).

ACKNOWLEDGEMENTS

A lot of people have helped with this book; we are profoundly grateful to them all, and we apologise if we have inadvertently missed anyone. We must start by expressing our gratitude to the herpetologists and snake handlers who risked life and limb helping us photograph dangerous snakes. Five institutions and their personnel stand out: Royjan Taylor, Sanda Ashe, Boniface Momanyi, Anthony Childs, Kyle Ray and Charles Wright at Bio-Ken Snake Farm, at Watamu on the Kenyan coast (sadly Royjan and Sanda are no longer with us); Joe Beraducci at MBT's Snake Farm and Reptile Centre at Arusha, Tanzania; Deon Naude and Lynn and Barry Bale at Meserani Snake Park, near Arusha in Tanzania; Paul Rowley, Edd Crittenden and Dr Rob Harrison at the Liverpool School of Tropical Medicine; and Patrick Malonza, Victor Wasonga, Beryl Bwong and Vincent Muchai at the Herpetology Section of the National Museum in Nairobi, Kenya. They have made a major difference. All have live collections and allowed us to freely photograph their often rare specimens. The Bio-Ken team also generously allowed access to their photographic database. In Arusha, Joe Beraducci frequently put aside his work in order to ferret out information for us. In the UK, we thank Mike Nolan and Ted Wade, who let us photograph the snakes in their private collections, and Patrick Campbell and David Gower, at the Natural History Museum in South Kensington, for their expertise and kind hospitality.

We thank also the medical professionals, those pioneering snakebite doctors, who allowed us to pick their brains. Professor David Warrell read much of the text, gently corrected the mistakes and generously allowed us to make extensive use of his major work on African snakebite. Dr Colin Tilbury wrote much of the original draft, and alerted us to (and kindly supplied) relevant literature. The mistakes that remain, of course, rest with us.

Several expert herpetologists also read our drafts. We have benefitted from the expertise of Wolfgang Wuster, who kindly sorted out the cobra accounts for us, and Tomáš Mazuch who greatly improved our treatment of the carpet vipers. Johan Marais kindly commented on the snakebite section and gave freely of his many years of expertise dealing with snakebite in southern Africa. Jean-François Trape also kindly allowed us to make use of his unpublished work on

Berg Adder, *Bitis atropos*, Storms River, South Africa (Bill Branch).

the snakes of Chad. Other friends who commented upon and improved the draft or gave willingly of their expert opinions include Zarek Cockar, Ton Steehouder, Jordan Benjamin, Bob Drewes, Aaron Bauer, Florian Finke, Werner Conradie, Luis Ceriaco, Jean-Philippe Chippaux, Harry Greene, Thomas Hakansson, Thea Litschka-Koen, Paula Kahumbu, Darrell Raw, Mark-Oliver Rödel, Barry Hughes, Bob Hansen, Kate Jackson, Mikey Mccartney, Petr Necas, Tyrone Ping, Norbert Rottcher, Washington Wachira, Sherif Baha El Din, Luca Luiselli, Laura Parker, Francois Theart, Vlada Trailin, Eli Greenbaum, Ted Wade, Ralph Braun and David Williams.

Next, the photographers. Over 60 herpetological photographers kindly allowed us to use their pictures, including top professionals who can command high fees for their pictures, yet generously allowed us to use their super images just for the reward of a copy of the book. In alphabetical order, they are: Jonas Arvidsson, Jordan Benjamin, Richard Boycott, Dayne Brayne, Don Broadley, Gary Brown, Evan Buechely, David Buttle, Juan Calvete, Luis Ceriaco, Alan Channing, Anthony Childs, Laurent Chirio, Tyler Davis, Max Dehling, Gerald Dunger, Dietmar Emmrich, Vincenzo Ferri, Florian Finke, Yannick Francioli, Paul Freed, Eli Greenbaum, Jelmer Groen, Václav Gvoždik, Harald Hinkel, Daniel Hollands, Barry Hughes, Kate Jackson, Tony Kamphorst, Chris Kelly, Luke Kemp, Sebastain Kirchof, Warren Klein, Johan Kloppers, Matthijs Kuijpers, Mark Largel, Dwight Lawson, Luke Mahler, Johan Marais, Jesse Mathews, Tomáš Mazuch, Mike McLaren, Konrad Mebert, Michele Menegon, Abubakr Mohammed, David Moyer, Deon Naude, Mark O'Shea, Laura Parker, Olivier Pauwels, Johannes Penner, Mike Perry, Tyrone Ping, Fabio Pupin, Gunther Rath, Darrell Raw, Eduardo Razzetti, Peter Riikula, Mark-Oliver Rödel, Paul Rowley, Steve Russell, Ignas Safari, Bill Roland Schroeder, Tim Spawls, Francois Theart, Colin Tilbury, Jean-François Trape, Sonya Varma, Luke Verburgt, Lorenzo Vinciguerra, Philipp Wagner, Alfred Wallner, Frank Willems, Elvira Wolfer, Wolfgang Wuster and Andre Zoersel.

Stephen would also like to thank his wife Laura for her endless unstinting support, and his old friend Jonathan Leakey, who kindly made available the field notebooks of C. J. P. Ionides, Tanzanian herpetologist; the natural history information here has proved invaluable. Thanks also to Julia Leakey and Dena Crain for getting the material to us. Stephen also thanks his many field companions, friends and those who gave hospitality and stimulating conversation, especially his sons Jonathan and Tim, Daniel Hollands, Glenn Mathews, Conrad and Linda Thorpe, Dave Morgan, Anthony and Emma Childs, Dave Brownlee, Jonathan Leakey and Dena Crain. Stephen is grateful to his fellow science lecturers at City College Norwich, Janet Cross and Ian Cummings, for stimulating daily discussion, and his principals at City College, Corienne Peasegood, Jerry White and Julia Buckland, who generously allowed him use of the college resources. At Bloomsbury, we thank Jim Martin for his enthusiastic championing of this book, Jenny Campbell, who has guided it to fruition, and the designer, Rod Teasdale. And finally, Stephen thanks James ('Jim') Ashe, who was not only responsible for Stephen's early herpetological education, but also a man who saved the lives of many Kenyans. He is remembered in the James Ashe Antivenom Trust (JAAT), whose work continues Jim's legacy.

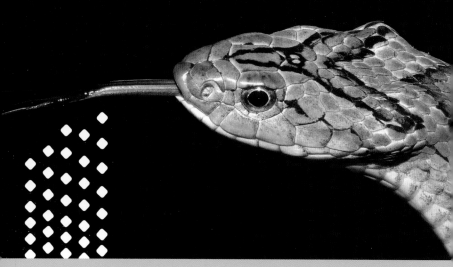

INTRODUCTION

Snakebite is best treated by use of the correct antivenom
for the correct snake, given at the correct time.
James Ashe, founder of Bio-Ken Snake Farm, Watamu, Kenya

This book is intended to meet a real need in Africa. Dangerous snakes occur virtually throughout Africa and are often common, and snakebite is an ever-present deadly hazard. The risk is exacerbated because it is disproportionately faced by the poorest members of society, who are constantly exposed to venomous snakes all of their lives and who often lack the means to protect themselves and reduce the risk of being bitten. In addition, snakebite treatment can be hideously expensive. Pharmaceutical companies can expect little profit from developing effective antivenom, since those who are bitten largely cannot afford the treatment. In addition, medical authorities in Africa have often been reluctant to take the problem seriously and devote resources to combating snakebite, although this is now improving. We have expanded on this in the section on snakebite (pp. 279–308).

The estimated number of victims of snakebite varies, due to problems gathering data in remote areas. But the figures are large. Worldwide, it is confidently believed that snakebite affects the lives of 4.5 million people yearly, seriously injuring 2.7 million people and claiming 125,000 lives. In Africa, a study of snakebite in 2008 estimated there were between 80,000 and 420,000 bites per year, and between 3,500 and 30,000 deaths per year. Whatever the true figure, this is an awful medical problem that needs attention.

And thus the major purpose of this book: comprehensive descriptions, with diagnostic photographs and maps, of all the 136 species of dangerous snakes of Africa. We have also provided photographs and descriptions of the 'look-alikes' – innocuous snakes that are likely to be confused with dangerous forms. This is followed by a section on snakebite in Africa, how to prevent it, some practical considerations, and the medical aspects of snakebite, including first aid and hospital treatment. The Appendix (pp. 309–24) lists useful resources. It is our intention that this book can be used, anywhere in Africa, to work out: (a) what dangerous snakes occur here and how they may be identified, (b) the risks involved from the bite of a local snake, (c) the means to prevent or lessen the risk of snakebite and (d) guidelines on how to treat bite victims. We hope it serves its purpose.

A brief word on semantics is necessary. We use the word 'dangerous' to describe snakes that represent a danger to humans, and that is why we have included the large pythons. We use the term 'venomous' to describe snake species that have a venom delivery system; that is, modified teeth (fangs) and toxic mouth secretions. In the past, the term 'poisonous' was often used for venomous snakes; a classic 1968 work was called *Poisonous Snakes of the World*. However, scientists working on venomous creatures nowadays use the word 'poisonous' to mean organisms that are toxic to eat, and 'venomous' to mean those that inject toxins, by biting or stabbing.

AFRICA'S SNAKES: WHICH ONES ARE DANGEROUS?

Snake classification is a formal business, based on a mixture of morphology and molecules, and we shall shortly expand upon it. However, an informal but most handy way of classifying Africa's dangerous snakes was coined by Professor David Warrell, a medical doctor with a pioneering interest in snakebite. Warrell's classification is based on medical significance: really dangerous snakes that bite a lot of people, really dangerous snakes that hardly ever bite anyone, not very dangerous species that bite a lot of people and not very dangerous species that hardly bite anyone. This is a useful division and we refer to it from time to time. For the people of Africa, it is the first group that are really bad news: the deadly animals that bite a lot of people. It is impossible (at present) to accurately put them in order of medical importance as we don't have the data, but those snakes are: the four widespread species of carpet viper (*Echis*), the Puff Adder (*Bitis arietans*) and the seven spitting cobras (*Naja nigricollis, N. ashei, N. nigricincta, N. mossambica, N. katiensis, N. pallida* and *N. nubiae*). They are followed by snakes of the Egyptian cobra (*Naja haje*) and forest cobra (*Naja melanoleuca*) complexes, and the Black Mamba (*Dendroaspis polylepis*). If you are rural dweller in Africa, those are the snakes you really want to avoid.

Worldwide, there are over 3,700 species of snake. Around 600 species are known from Africa, in nine families, as follows: Pythonidae (pythons), Boidae (boas), Colubridae (colubrid or 'ordinary' snakes), Lamprophiidae (African snakes), Natricidae (water snakes), Elapidae (cobras, mambas and allies), Viperidae (vipers), Typhlopidae (blind snakes) and Leptotyphlopidae (worm or thread snakes). Africa's really dangerous snakes are the so-called 'front-fanged snakes', those with tubular fangs at the front of the upper jaw, a system that ensures venom can be injected, under pressure, below the skin on every bite. There are about 120 species of front-fanged snakes in Africa. There are around 55 African vipers, in the family Viperidae; all have long fangs that can fold flat, enabling a large fang to fit in a small mouth. Just over 40 species are placed in the Elapidae (the family of cobras, mambas, African garter snakes and allies); they have relatively short, rigid fangs at the front of the upper jaw. The 20 or so species of burrowing asps (also called stiletto snakes, mole vipers or side-stabbing snakes; genus *Atractaspis*) are in the subfamily Atractaspidinae, within the family Lamprophiidae. Like the vipers, they have long, folding fangs, which led to them being classified as vipers in the past.

In addition, there are at least 10 or 15 so-called 'rear-fanged' snakes that are known to have toxic venom, and several of these have killed people, although the danger they pose to the ordinary rural dweller is almost negligible. These rear-fanged snakes are in two families, Colubridae and Lamprophiidae, and have grooved (not tubular) fangs set back in the upper jaw, often below the eye; when they bite, the venom trickles down the fangs. Africa also has a handful of fangless snakes, often with long but unspecialised teeth, which have been shown to have fairly toxic oral secretions. These toxins were largely identified after the snakes bit and chewed snake handlers, who were reluctant to quickly remove the snake in case they damaged its mouth. We have described those known to have toxic saliva. There will probably turn out be more of these. However, they are not medically significant to the average person in Africa, who does not allow snakes to chew on them. There are also two large pythons in Africa, which are large enough to be able to cause physical injury or even death by constriction and have killed a few people. The remaining 450-odd species of African snake are essentially harmless to humans.

The taxonomy (classification) of Africa's snakes has changed a lot over the years, and it continues to change. It is an exciting and often controversial field, and the uncertain status of some taxa – whether they are separate species, subspecies, or just varieties within a species – is why our numbers are sometimes imprecise. For example, it was long believed there were only four species of true cobra, *Naja*, in Africa: the Egyptian (*Naja haje*), Cape (*N. nivea*), Forest (*N. melanoleuca*) and Spitting (*N. nigricollis*) Cobras. In 1995, in the original edition of this book, we described seven species of African cobra. In this book, the number of cobra species has increased to 21. The additions are not totally new forms, freshly discovered, but are nearly all due to the taxonomic splitting of the original species, using advanced tools to elucidate diverging lineages. In addition, molecular studies show that the water cobras originally placed in the genus *Boulengerina* are nested with the true cobras, *Naja*, and more than two species of water cobras may exist (see Figure 1). The small vipers of the many-horned adder (*Bitis cornuta*) complex are now split into five species. The 'original' Forest Cobra, *Naja melanoleuca*, has been split into five separate species, some of which are hard to tell apart. Further splits may be on the cards.

Fig 1: Three water cobras; the status of the bottom one is unknown (Václav Gvoždik).

In the past, snakes were largely classified by a mixture of general morphology or shape, scale details, colour and teeth, characters that have proved useful for over 250 years of herpetological taxonomy. Recently, a new set of tools has become available to laboratory-based taxonomists: the techniques that enable the isolation of standardised bits of DNA from both the cell nucleus and the mitochondria, and the computing power that allows the statistical analysis of that DNA. With these tools, new forms can be defined. New evolutionary lineages and relationships can be unearthed and old ones modified, through the identification and alignment of, and observed differences in, matching sections of DNA. Although traditional reptile taxonomy is still useful, especially with new forms that clearly differ from other members of the genus, the analysis of DNA has enabled researchers to fine-tune their identification and often split what seemed in the past to be a single species into several species that have evolved separately. Even the approximate date at which the two forms began to diverge and to accumulate differences can be ascertained. This is a very useful development, although it has occasionally led to new species being identifiable only in terms of their DNA and to controversy over what constitutes a separate species; the veteran molecular taxonomist David Hillis has suggested that many proposed species in reality simply represent arbitrary slices of continuous geographic clines.

WHERE ARE THE DANGEROUS SNAKES IN AFRICA?

Dangerous snakes occur almost throughout mainland Africa, apart from areas above the snowline and in some of the higher-altitude forests. However, the number of species and individual numbers of dangerous snakes vary considerably throughout the continent. In general, the number of snake species (including dangerous forms) at any place increases with (a) nearness to the equator, (b) decreasing altitude and (c) increasing rainfall. Thus, a low-altitude town with high rainfall in Central Africa will have many snake species; a high-altitude town with low rainfall in North Africa will have few. In most places in tropical Africa, there will be between 25 and 40 species of snake, and five to eight of these will be dangerous.

Although our knowledge is imperfect, at present Watamu on the Kenyan coast seems to be the snake capital of Africa, with some 48 species of snake, 10 of which are dangerous. However, on the moorlands of the Aberdare Mountains in Kenya, at altitudes of over 2,800m, only two species of snake occur, one of which is dangerous. There are a few places in Africa where there are no dangerous species. Addis Ababa, in Ethiopia, is the only African capital with no dangerous snakes; the six species known from that city are all harmless. The capital of Lesotho, Maseru, has only three species of dangerous snake. But at Wajir, in dry north-east Kenya, one dangerous species, the North-east African Carpet Viper (*Echis pyramidum*), is the most common and abundant species.

The number of species is not the same as the actual number of dangerous snakes, however. This varies a lot, not only from place to place, but also with the seasons, the weather, the time of day and the habitat. Small habitat changes may make a considerable difference. At a town in northern Ghana, not a single snake in a collection of over 400 was a West African Brown Spitting Cobra (*Naja katiensis*), yet at a town 160km north, two out of five snakes collected in a short expedition were of that species. In a survey of snakes near Lake Turkana in northern Kenya, out of 70 snakes collected, 51 were carpet vipers, four were

Red Spitting Cobras (*N. pallida*), three were Puff Adders and only 12 were innocuous species. At Moille Hill, in northern Kenya, a commercial collecting team found 4,000 North-east African Carpet Vipers (*Echis pyramidum*) in four months, and yet only two examples of this species were found in 25 years by a zoologist collecting around Cairo.

Certain dangerous species can be very abundant, even if there are few species of dangerous snake in the area. On the other hand, some areas may have several species of dangerous snake but the common species all turn out to be harmless. And, even if a dangerous species turns out to be common, it may not represent any danger to humans. The Tanzania-based herpetologist C. J. P. Ionides collected over 4,500 Eastern Green Mambas (*Dendroaspis angusticeps*) in a four-year period in south-eastern Tanzania. This is a front-fanged snake with a deadly venom, yet the only person Ionides knew of who had been bitten by an Eastern Green Mamba at that time was himself; he picked up a juvenile thinking it was a harmless species. There are few quantitative studies of dangerous snakes, but certain species, in particular the carpet vipers, Horned Adder (*Bitis caudalis*), Eastern Green Mamba, Eastern Gaboon Viper (*Bitis gabonica*) and Forest Vine Snake (*Thelotornis kirtlandii*) have been found to be abundant in parts of their ranges.

To understand the distribution of dangerous snakes in Africa, it is helpful to have some understanding of the vegetation, as many species are associated with one (or more) vegetation types, which themselves are associated with climate and altitude. Africa can be split simply into three main vegetation types: forest, savanna and desert. Two other 'transition' zones may be usefully defined: the zone where the savanna becomes heavily wooded but the tree canopy is mostly at a single level, for which we have used the term 'woodland'; and the zone where the savanna becomes very dry and sparsely vegetated, with few trees, for which we have used the term 'semi-desert'. We have also used the term 'forest–savanna mosaic' to describe areas where forest is interspersed with savanna. These are broad terms in respect of Africa's complex botanical regions, but they are useful in describing where snakes live. Snakes tend to occur in one of the major vegetation types (forest, savanna or desert) and often in one or both of the transition zones, but relatively few species occur in two of the zones, and no snake occurs in all three. The West African Night Adder (*Causus maculatus*), a versatile snake, is found in forest, woodland, savanna and semi-desert, but this is very unusual. A number of widespread dangerous snakes occur in woodland, savanna and semi-desert, including the Black Mamba, Boomslang (*Dispholidus typus*) and Puff Adder.

Some other vegetation zones that should be mentioned are the Mediterranean regions of Africa north of the Sahara, the ancient forested hills and mountains of eastern and south-eastern Africa (the Eastern Arc), and the temperate regions, hills and small deserts of southern Africa. In these regions there occur some dangerous snakes with very limited distributions. Most of these snakes are small; some occur in one country alone or in one very limited habitat and are called endemics. There is also a small group of dangerous snakes of the Middle East that just reach eastern Egypt but penetrate no further into Africa. Interestingly, the limited information that we have at present indicates that human–snake encounters take place most commonly in savanna or semi-desert and not, as one might suspect, in the great forests. We return to this point in the final section, Snakebite in Africa (pp. 279–308). Figures 2–10 (p. 14) show some of these vegetation types.

Fig 2: African desert habitat (Stephen Spawls).

Fig 3: East African savanna (Stephen Spawls).

Fig 4: African forest habitat (Stephen Spawls).

Fig 5: Grassland habitat (Stephen Spawls).

Fig 6: Montane moorland habitat (Stephen Spawls).

Fig 7: Rocky desert habitat (Stephen Spawls).

Fig 8: Semi-desert habitat (Stephen Spawls).

Fig 9: West African savanna habitat (Stephen Spawls).

Fig 10: Woodland habitat (Stephen Spawls).

USING THE MAPS IN THIS BOOK

Where a species is widespread, there are many locality records and/or there is a fairly clear pattern of distribution, we have shown the overall distribution in solid colour. For species with a poorly understood distribution, or species with relatively few records, coloured dots show the actual locality records. For such species, pay particular attention to the notes on distribution and habitat. In some cases, both solid areas and dots appear on the map, indicating that the distribution pattern is clear in parts of the snake's range but not in others; the dots represent records beyond the present known continuous range of the species. A question mark indicates possible but unproven occurrence. We have also provided three maps showing African countries, vegetation and altitude (Figures 11–13, pp. 15–17), to assist with orientation.

Generally, the distribution of African snakes is poorly documented, and much work remains to be done. Species may well occur outside their known ranges. If you think you have positively identified a snake but find that the map shows it is outside its known range, take careful note of the habitat – a suspected Gaboon viper (*Bitis gabonica* or *B. rhinoceros*) will not be in desert. If you have positively identified the snake and think your specimen represents a range extension,

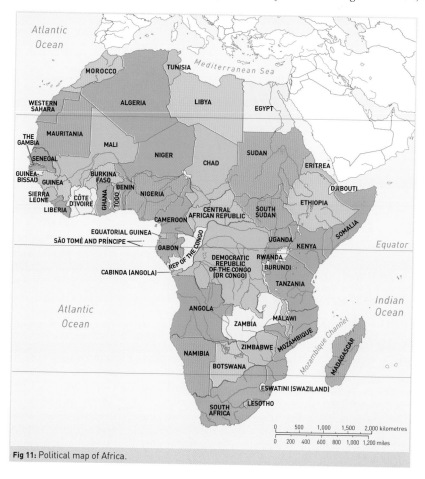

Fig 11: Political map of Africa.

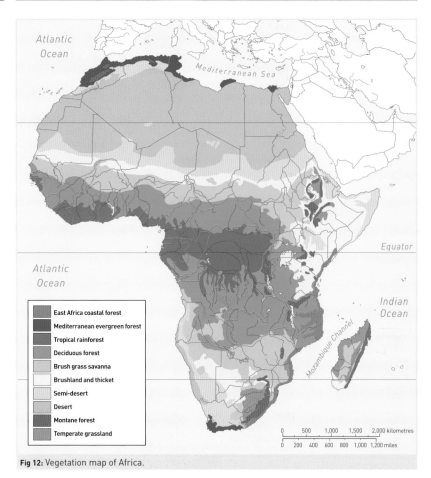

Fig 12: Vegetation map of Africa.

Legend:
- East Africa coastal forest
- Mediterranean evergreen forest
- Tropical rainforest
- Deciduous forest
- Brush grass savanna
- Brushland and thicket
- Semi-desert
- Desert
- Montane forest
- Temperate grassland

and it is dead, then preserve it and send it to the nearest museum or institute dealing with reptiles, noting in particular: (a) the geographical locality where it was found, including nearest town, and latitude and longitude if possible; (b) the date, time and microlocality (under a rock, in a tree, etc.) if possible; and (c) the collector's name and address. Alternatively, send a good photograph. You may well have a medically and biologically important discovery. Several forums exist for the publication of range extensions for African snakes (see the Appendix, pp. 309–24).

A NOTE ON CONSERVATION

Snakes have benefits. Their venoms contain a range of compounds that are being increasingly utilised by the pharmaceutical industry, to make useful drugs. And snakes, even dangerous ones, are important members of Africa's fauna and have a significant place in food webs, helping to maintain the balance of nature. At the same time, the species described here can be dangerous to humans and may have to be removed from places where they pose a threat. If no competent snake handler is available to relocate the snake safely, it may need to be driven away or killed. A large cobra within a school, or sliding around a village, must obviously

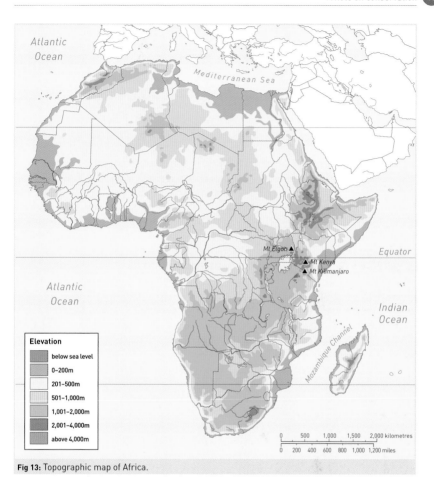

Fig 13: Topographic map of Africa.

be made safe, one way or another. However, a cobra crossing a road in a national park can be watched and appreciated without being molested. Our request, where snakes are concerned, is to use your judgement. Try to find out if the animal you are dealing with is dangerous or not. Use this book. And remember that killing, or trying to kill or catch, a dangerous snake (particularly a large one like a mamba or a cobra) is a risky undertaking. Take great care. If the snake is not near habitation and poses no threat to life, leave it alone.

A brief word on ethics may also be important here. Humanity's attitudes towards snakes, especially dangerous ones, vary considerably, depending on education, culture and to some extent how affluent we are. However, much of it is driven by fear of a deadly hazard. It has been said that we are instinctively afraid of snakes (a point explored in some elegant works by the anthropologist Lynne A. Isbell). To the reasonably well-off, viewing a wild snake in a wild place, or even in the garden, can be an exciting and possibly pleasant experience. To the poor rural dweller, however, snakes are an ever-present deadly menace that can kill, or cripple and potentially financially ruin them.

Most people are unable to distinguish between a harmless snake that may be beneficial, for example by eating rats, and a deadly snake. Some species are notoriously hard to tell apart (see Figure 14). The instincts of many people (and not just in Africa, but worldwide, even in prosperous countries like Australia and the United States) is to immediately kill the snake and neutralise the potential danger. Researchers in California who placed a realistic rubber snake on a road observed a disturbingly large number of drivers try to deliberately run the snake over, even at considerable risk to themselves and other road users. In Africa, if you spot a snake in a residential area and are seen to leave it alone, you are asking for trouble. If it bites someone, you may get the blame or, worse, be accused of being involved with witchcraft and loving snakes. In parts of Africa, being suspected of being a witch or wizard can be a death sentence. And yet there is often a surprising amount of intolerance towards those who kill snakes, directed largely, it must be said, by the affluent towards the poor, and this shows a worrying inability to put oneself in the shoes of others. People who post pictures of dead snakes on the internet, even innocently seeking identification, may be subject to a torrent of abuse. A recent social media post exemplified this: the writer had seen two Kenya Wildlife Service (KWS) rangers kill a small spitting cobra that was approaching a ticket office. The post attracted a storm of abuse towards the rangers and, by association, KWS. Nobody seemed to notice that (a) if the snake had got into the office and spat at or bitten a visitor, there would have been considerable bad publicity, and (b) shepherding away or catching a dangerous snake, especially one that spits venom, is a very difficult thing to do and requires expertise. It is important to see things from both sides.

In conservation terms, however, Africa's snakes are not really threatened by direct killing for safety's sake, or by commercial collection, and relatively few people eat snakes in Africa (although there are a few that do; see Figure 15). Snakes tend to avoid humans, often making strenuous efforts to escape detection. The most serious threat at present to Africa's herpetofauna is from large-scale habitat destruction (although the potential effects of climate change may also prove to be significant). Many of the snakes described here live in small forest patches, often on hills and mountains, and these refuges are threatened by large-scale farming and logging. Others are large, and/or sluggish, and are simply incompatible with any sort of human presence. If Africa's unique fauna is to survive, it needs not just protection in reasonable-sized reserves (which by itself can be counterproductive), but also for the people of the region to feel that the survival of the wilderness and its inhabitants is of benefit to them, either in a practical or aesthetic way. The best way for this to be achieved is for the wilderness to provide those who live nearby with a living, and with material benefits. The more direct those benefits are, the more the people will respect our wild places.

IDENTIFYING A SNAKE

This book should help you identify Africa's dangerous snakes, primarily by using the photographs, maps and text together. We have also provided dichotomous keys (see our comments on these on page 28). But note well: while it is often useful to know what you are dealing with, it should be remembered that successful treatment for snakebite is not dependent upon the actual snake being identified. Although such identification may be useful, snakebite can be treated without positive identification. There are also some useful techniques and rules to assist with snake identification in Africa, and we detail them here. Use them with care.

Fig 14: Rhombic Night Adder, *Causus rhombeatus* (dangerous, on left), and Common Egg-eater, *Dasypeltis scabra* (harmless, on right) (Stephen Spawls).

Fig 15: Central African Rock Python, *Python sebae*, on the menu, Ethiopia (Dietmar Emmrich).

You don't want to make a mistake. Read also our notes at the beginning of the section on look-alikes and common species (p. 252).

There are nearly 600 types of snake in Africa; this book covers about a third of them. If you can't find the snake you want, it may not have been recorded as dangerous. Have a look at the pictures of non-dangerous snakes in the Look-alikes and common species section (pp. 252–78). It is worth browsing through this book, possibly in connection with a regional guide to all the snakes of your area (a list of these is included in the Appendix, pp. 309–24), and possibly combining this with a visit to your local reptile or snake park, before you can make a certain identification. Practice makes perfect. Get used to what the animals look like. Also make use of the internet; this is an excellent source of data (but always bear in mind that it is unregulated and misidentifications occur). There are some useful websites and online forums, also mentioned in the Appendix (pp. 309–24).

Figures 16–22 show details of the head, body and scales of snakes that may be used in identification. In this book we have given details of the midbody scale count, and the ventral and subcaudal scale counts. It goes without saying that scale counts should never be attempted on a living snake. The dorsal scale rows are counted at midbody, roughly halfway between the head and the tail. Starting at one side of a belly scale, count diagonally forwards to the middle of the back, and then diagonally backwards to the other side of the belly scale. Figure 21 shows how this is done, and shows a snake with a midbody scale count of 19. To count the ventral

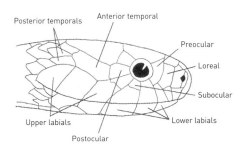

Fig 16: Head scales of snake, side view.

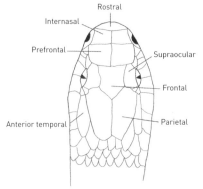

Fig 17: Head scales of a snake, from above.

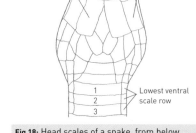

Fig 18: Head scales of a snake, from below.

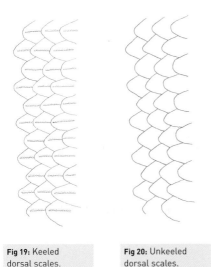

Fig 19: Keeled dorsal scales.

Fig 20: Unkeeled dorsal scales.

scales (the broad belly scales; see Figure 18); start your count with the first ventral scale (marked '1') that touches the lowest dorsal scale row, and count towards the tail; the count finishes on the scale in front of the cloacal scale.

Careful observation of a potentially dangerous snake is a must. But remember, the sighting of a snake is often an emotional situation. It may be difficult to recall exactly what you saw. Try to remain calm, take no risks, and look as carefully and objectively as possible. Photograph the snake if it is safe to do so; use flash if the light is low, take several pictures

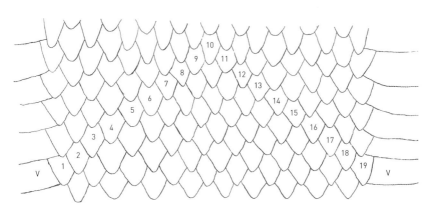

Fig 21: How to count the dorsal scale rows of a snake. V = ventral scale.

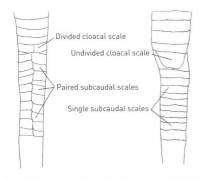

Divided cloacal scale

Undivided cloacal scale

Paired subcaudal scales

Single subcaudal scales

Fig 22: Snake tail cloacal and subcaudal scales.

and keep the camera steady. Make sure the head is visible and in sharp focus. Activate the GPS function if you have one. Compare the pictures to the ones in this book. There are several internet forums where snakes can be identified, and you can send your pictures to them (details in the Appendix, pp. 309–24); apps are also available that aid identification. If you can't photograph the snake, write down what you saw as soon as possible.

If you are looking at a living snake that may disappear, do the following:

1. Make a careful estimate of its size, shape and appearance, without getting too close. Take a photo or write your observations down immediately, if possible while still watching the snake. Be objective and careful. People who have had a quick look at a snake often remember it as being much larger than it really was. If the snake is (or was) lying on a branch, or on the ground, try to note where its head and tail are, and measure this distance later.
2. Take careful note of its colour (this may be difficult to judge at dusk, or if the snake is seen by torchlight or artificial light). Compare the colour, if possible, to nearby vegetation or soil/rock colours. Look carefully to see if the snake has any markings or patterns.
3. Look at its thickness. Is it as thick as, for example, a pencil, a broomstick, your forearm, or a car inner tube? Would you say it was fat, medium or thin? Try to see its head. Are its eyes obvious? Are they large or small, or any obvious colour? Do they have round or vertical pupils? Does it have large or small scales on top of the head? Does the snake have an obvious neck? Is the head broader than the neck? Are its scales dull, glossy or keeled?
4. Note its position and behaviour; these are often useful aids to identification. Was it in a tree, or on the ground? Did it move slowly, quickly or not at all? Did it hiss or strike, and did it move with its head raised? Was it moving by day (hence probably a diurnal snake), or was it under cover during the day or moving by night (probably a nocturnal snake)? Did it try to climb or burrow? What did it do when approached – did it flee, move off slowly, spread a hood, freeze, open its mouth, rub its coils together, strike, lunge forward or spit? All these forms of behaviour may give vital clues towards identification.

If the snake is dead, or captive, identification may be easier and you may be able to get a good photograph, but you must be totally sure that the snake is dead before you try to approach or handle it. Some snakes pretend to be dead (feigning or shamming death), and a fatally injured and thoroughly bashed snake, on the point of death, can still bite; even if it is essentially dead, its jaws may still open and close. Severed snake heads have been known to bite more than an hour after they were cut off. To find out if a snake is dead, one or more of the following tests may be tried. In all cases, use a stick or a broom to move the snake, not your bare hands. Be particularly careful if the snake is large, or might be a cobra or mamba.

1. Turn the snake totally onto its back and watch it. If it tries to turn over, it is not dead. If rhythmic waves or 'shivers' are seen along the body and the tail coils and uncoils, the snake is probably fatally injured, but it is not safe to handle yet.
2. Drop the body into a bucket of water and push the head under. If the head makes no move to come up for 20 minutes or so, it is dead.
3. Lay the body out in sunlight, on a hot surface, and watch for movement.
4. If you're still not certain that the snake is dead, use a spade or a long-bladed knife to chop the head off. Don't touch the head for at least an hour afterwards.

Note well: it is no good poking the eye of a snake to see if it is dead. Snakes have no eyelids and their eyes never shut.

If you are totally satisfied that the snake is dead, then do the following:

1. Lay the snake out and measure its length. If you have no measuring tool, the following dimensions may be useful: the long stride of an adult man is just under a metre, and that of an adult female is about 80cm; a soft drink bottle is about 25cm long; the average table is about 75cm high.

2. Compare the appearance, colour and pattern against the pictures in this book. Remember to look at and photograph the underside, which may provide valuable clues, especially for identifying cobras and vipers. Remember that after death, cobras' hoods may disappear, and snakes that have inflated necks when threatened (such as the Boomslang and vine snakes, *Thelotornis*) will have thin necks after death.

3. Look at the head; take close-up pictures of the head and neck, above and below if feasible. Check (a) width, (b) position and size of the eye, (c) type of pupil, (d) type of scales on top (many small scales usually means the snake is a viper) and (e) neck – is it thinner than the head, or the same thickness? Look at the scales. Are they dull or shiny, and do they have keels or are they smooth (see Figures 19 and 20)?

4. An expert may be able to check if the snake has obvious fangs. To do this (with great care), grip the snake by the neck and use a strong, thin twig, piece of wire or thin-bladed screwdriver to open the mouth. Push the wire to the back of the mouth, as shown in Figure 23. Taking great care not to catch a hand or a finger on the teeth, push the wire up against the upper jaw. Slide it forward and see if it catches on any enlarged teeth (Figure 24). This method is not an infallible way of detecting a dangerous snake. The snake's fangs may be broken or folded well back, so that they don't show, and some dangerous snakes (for example, the night adders, *Causus*) have small fangs that may be hard to see. The fangs may also be hidden in a fleshy sheath. In addition, some harmless snakes have very long teeth, which can be confusing. However, if your wire catches up on some long teeth roughly below the eye, the snake is probably rear-fanged. If there are enlarged immoveable teeth at the front of the upper jaw, the snake is probably an elapid (Figure 25, p. 24). If there are some very long, curved teeth at the front of the upper jaw, the snake is probably a viper (Figure 26, p. 24). Note also the colour of the inside of the mouth. The

Fig 23: Jameson's Mamba, *Dendroaspis jamesoni* (Gerald Dunger).

Fig 24: Boomslang, *Dispholidus typus* (Harald Hinkel).

Black Mamba has a black mouth lining, as does the Western Green Snake (*Philothamnus irregularis*), the Common Egg-eater (*Dasypeltis scabra*) and the Hook-nosed Snake (*Scaphiophis albopunctatus*). Most other snakes have a white mouth.

5. If you have a tentative identification, check the distribution and habitat, to see if they match your specimen.

6. If there is a possibility of taking the snake to an expert or an institution for identification but it may take a while, you will need to preserve the snake – just the head will do if the snake is large. A small snake or a severed head can be simply dropped into preservative. Suitable liquids are 10% formaldehyde solution (formalin), methylated spirit, very strong brine (salt solution), petrol or alcohol (spirits such as gin or whisky will do). If you have no preservative or suitable container, then make some long cuts along the body, ensuring you slit the gut, and hang it in a tree in the sun to dry. Watch out for predators trying to take it away. Alternatively, send your photographs. Don't forget to include details of exactly where you found the snake.

Fig 25: Black Mamba, *Dendroaspis polylepis* (Stephen Spawls).

Fig 26: Gaboon Viper, *Bitis gabonica* (Gerald Dunger).

It is important to know that, in Africa, there is no single way in which a dangerous snake can be told from a harmless one – you have to know all the species. However, there are certain general rules that may be used to identify African snakes, and we list them below. You should use them **with caution**. Be aware that there are also some extremely dangerous myths about how to identify venomous and non-venomous snakes. These frequently circulate on the internet, often with misleading endorsements of their benefits. A common myth – totally incorrect – is that dangerous snakes have vertical pupils and harmless snakes have round ones. Look at Figures 27–30 (pp. 25–26), showing an innocuous Speckled Green Snake (*Philothamnus punctatus*) and dangerous Red Spitting Cobra, both with round pupils, and an innocuous Large-eyed Snake (*Telescopus dhara*) and dangerous Puff Adder, both with vertical pupils. Other incorrect myths are that harmless snakes have smooth scales and dangerous ones keeled scales; that dangerous snakes have broad heads and harmless ones narrow heads; and that dangerous snakes have single subcaudal scales and harmless ones have double ones. Bear in mind the apposite saying: the internet exposes us to the stupidity of others even more efficiently.

Fig 27: Speckled Green Snake, *Philothamnus punctatus* (Bill Branch).

Fig 28: Red Spitting Cobra, *Naja pallida* (Tim Spawls).

Fig 29: Large-eyed Snake, *Telescopus dhara* (Bill Branch).

Fig 30: Puff Adder, *Bitis arietans* (Florian Finke).

Thus some functional rules for African snakes (only); cross-check these against our pictures:

- Most slim snakes with conspicuous stripes running along the body are probably not dangerous (see the sand snakes and relatives in the look-alikes section, pp. 252–78).
- Any snake over 2m long is probably dangerous.
- Any grey, green or greenish tree snake over 1.2m long is almost certainly dangerous; it will be an elapid or rear-fanged snake. If it inflates the front half of the body when threatened, it will probably be a Boomslang or a vine snake (although the harmless tree snakes of the genus *Thrasops* also inflate themselves, as do the smaller harmless green snakes, *Philothamnus*).
- Any snake that spreads a hood, flattens the neck or raises the forepart of the body off the ground when threatened is almost certainly dangerous; it will probably be an elapid (but it might be a night adder or a beaked snake, *Rhamphiophis*).
- Any snake with conspicuous bars, cross-bands, rings or chevrons (V-shapes), especially on the neck, the front half of the body, on the back or on the belly, is probably dangerous.
- Any snake with dark cross-bars or blotches on the underside of the body, especially the front half of the body or the neck, is almost certainly dangerous; it is probably a cobra.
- Any fat-bodied snake, especially one that lies quietly when approached, or any snake with a triangular or subtriangular head, is almost certainly dangerous; it is probably a viper.
- Small black or grey snakes with very tiny eyes, no obvious neck and a short, fat tail that ends in a spike are probably burrowing asps, and are dangerous.
- Small, mostly green and/or yellow and black snakes with broad heads and thin necks, in trees or bushes, are probably bush vipers and are dangerous.
- Any small snake with conspicuous pale bands all along a dark body will be a juvenile garter snake (*Elapsoidea*), and is dangerous.
- Any snake with rectangular, subrectangular or triangular markings on the back or sides, or rows of semicircular or circular markings along the flanks, is probably a viper (dangerous).
- A snake that forms C-shaped coils and rubs them together, making a noise like water falling on a hot plate, will either be a carpet viper (*Echis*, dangerous), North African desert viper (*Cerastes*; northern half of Africa, dangerous), bush viper (*Atheris*; forests of tropical Africa, dangerous), egg-eater (*Dasypeltis*; anywhere in Africa, harmless) or Floodplain Viper (*Proatheris superciliaris*; south-eastern Africa, dangerous).
- Any small snake with a blunt, rounded head, a blunt, rounded tail, eyes that are either invisible or visible as small black dots beneath the skin, and tiny scales that are the same size all the way around the body will be either a blind snake (*Afrotyphlops* species, fairly stout) or a worm snake (*Leptotyphlops* species, thin like a bootlace); both are harmless.

On a slightly more technical note:
- All African vipers (apart from the night adders) have many small scales on top of the head – this is a good positive identification of a viper, as the only other African snakes with tiny scales of the same size all over the top of the head are the sand boas (*Eryx* species). Sand boas are harmless, are found in

the northern half of Africa, in dry areas, and their heads are more or less the same thickness as their necks (vipers have broad heads and thin necks). Vipers (apart from the night adders) also usually have keeled scales, as do Boomslangs, but note that harmless egg-eaters and a few other innocuous species also have keeled scales.

- Most elapids (cobras, mambas, garter snakes, and tree and water cobras) do not have a loreal scale. The loreal is a scale between the eye and the nostril, but does not touch either (see Figure 16), nor does it touch the lip scales or extend onto the top of the head. If a snake has a loreal, it will not be an elapid (but remember, some dangerous snakes do have loreals, including the Boomslang, vine snakes and night adders).
- All the vipers (except the night adders) have fairly long fangs, as do burrowing asps; all the elapids and the night adders have fairly short fangs.

And, importantly, there are no venomous lizards in Africa.

Some final thoughts. The first is that it can sometimes be extremely difficult to identify a snake with certainty. Recently we sent a clear picture of a cobra, photographed in Kenya, to a cobra expert. He made a provisional identification. We then sent two more pictures, and he changed his mind. You may need to be flexible. And, as previously mentioned, successful medical treatment of a snakebite is not necessarily dependent upon the snake being identified.

Secondly, on our photographs. In a field guide, a secondary consideration beside the illustrative merits of a picture is usually its aesthetic quality; for example, is the animal alive, is it posed naturally, does it have any blemishes, is the background natural and uncluttered, can you see all of it? However, in this book we intend that users can identify, with as great a certainty as possible, Africa's dangerous snakes. So in some cases we have used pictures that are not aesthetically pleasing, but which we feel aid identification; for example, dead snakes, damaged snakes, snakes in cages, snakes upside down, and snakes held by hand or in grab sticks. We offer no apology for this; if such photographs aid medical personnel to make an identification with certainty, then our pictures will have served their purpose.

Finally, dichotomous keys are usually used to identify museum specimens. The keys in this book do not cover every snake in Africa, only the ones perceived to be dangerous, and thus some couplets end with 'not a dangerous snake'. We have tried to make the keys user-friendly, and thus we have included some visual characteristics. In the final analysis, however, if you want to identify some animals to species level with certainty (for example, burrowing asps), you will probably need a binocular microscope, you may need to have some idea of where the snake originated, and you may have to find fangs and count scales (never try this on a living snake). We hope these keys may be useful not only for indentifying snakes in the field, but also to those maintaining snakes, particularly for antivenom production, for a very important reason. There have been cases where commercially sourced venom has come from misidentified snakes, meaning of course that it is then ineffective, or even useless, for the treatment of a bite from the correctly identified snake. Antivenom manufacturers must be confident that the venom they use has come from snakes that have been unerringly identified.

Dangerous
front-fanged snakes

Mt Mabu Bush Viper,
Atheris mabuensis,
Mozambique
(Bill Branch).

In this and the next section on dangerous rear-fanged and fangless snakes, we describe the 130-plus species of dangerous snakes known from Africa. Each subsection starts with a brief overview of the family, a key where relevant, the subfamily (where relevant) and a generic introduction, including a synopsis of the venom and its effects. Each species is described under the following headings: Identification, Habitat and Distribution, Natural History and Medical Significance, and Taxonomic Notes where relevant. All species are accompanied by a map and photographs, where available.

Recent taxonomic work, using both fossil evidence and DNA, has created some stimulating debate over the order of appearance of the various groups of reptiles, snakes included. However, in this book we have not followed the taxonomic order of appearance of the various groups, but have started with the most medically significant. Thus our running order is: vipers, elapids, burrowing asps, and finally dangerous rear-fanged and fangless snakes. The following section, the 'look-alikes' (pp. 252–78), is similarly not in taxonomic order.

The two keys on p. 30 should enable the technical identification of any African snake to (a) its superfamily and (b) the smaller taxonomic groups. These are then further broken down at the beginning of the relevant sections. These keys are slightly convoluted, but should enable both the separation of dangerous and harmless snake taxa, and the individual identification of the dangerous species. As we have repeated in later pages, the keys are technical; the snake will need to be dead (do not start trying to key out a living snake), and you will probably need a binocular microscope and some familiarity with snake scalation (see Figures 16–22, pp. 20–21).

Key to the African snake superfamilies

1a Body worm-like; head round and blunt; tail blunt; eyes visible only as minute dark dots under the head skin; body scales all the same size......Typhlopoidea, blind and worm snakes; all harmless; see pictures in the section on look-alikes (pp. 252–78)

1b Body not worm-like; head not round; tail not blunt; eyes well developed; enlarged belly scales present......2

2a Ventral plates as broad or almost as broad as the body; no vestigial limbs; midbody scale rows fewer than 50......Colubroidea, typical snakes (pp. 226–41)

2b Ventral plates broader than the body scales but much narrower than the body; vestiges of hind limbs present as short claws on either side of the vent; midbody scale rows more than 70......3

3a Small, adults less than 1.2m; subcaudal scales single; no obvious neck......Booidea, boas; all harmless (p. 254)

3b Large, adults over 1.2m; subcaudal scales paired; obvious neck......Pythonoidea, pythons; non-venomous, but some may be dangerous due to their size (pp. 222–25)

Key to the African families, relevant subfamilies and relevant genera of the Colubroidea

1a One or more pairs of enlarged caniculate or tubular poison fangs present in the front of the upper jaw......2

1b No enlarged caniculate or tubular poison fangs present in the front of the upper jaw; fangs when present usually grooved and set below the eye......5

2a Poison fangs relatively small and immobile, not folded back when not in use......3

2b Poison fangs large, moveable, folded back in a sheath when not in use......4

3a Confined to South Africa; black with a red or yellow vertebral stripe; adults always less than 65cm......Subfamily Atractaspidinae, genus *Homoroselaps*, harlequin snakes; slightly dangerous (pp. 218–20)

3b Not confined to South Africa; not black with a red or yellow vertebral stripe; adults often larger than 65cm......Family Elapidae, cobras, mambas and allies; all dangerous (pp. 119–93)

4a Eye very small; pupil round; body usually uniformly dark (a few with white head or tail-tip markings); no loreal scale......Subfamily Atractaspidinae, genus *Atractaspis*, burrowing asps; all dangerous (pp. 194–217)

4b Eye relatively large; pupil usually vertically elliptic (but round in *Causus*, night adders); loreal scale present......Family Viperidae, vipers; all dangerous (pp. 31–118)

5a No loreal scale present; grooved rear fangs nearly always present......Subfamily Aparallactinae, African burrowing snakes (pp. 249–50); largely not dangerous (except Kwazulu-Natal Black Snake, *Macrelaps microlepidotus*, which is possibly dangerous; see account, p. 249)

5b Loreal scale present; grooved rear fangs present or absent......Families Colubridae/ Lamprophiidae/Natricidae, colubrids/African snakes/house snakes/water snakes (excluding the Aparallactinae and Atractaspidinae); a few dangerous species pp. 226–41, pp. 242–51 and pp. 272–77)

Family **Viperidae** Vipers

A family of dangerous snakes with long, perfectly tubular, folding venom fangs, and of international medical significance. The vipers or adders (the names are interchangeable, although it is suggested that the name adder be restricted to those that give birth to live young) are split into three subfamilies: the Viperinae (Old World vipers), Crotalinae (pit vipers of America and Asia) and Azemiopinae (two species in one genus, curious slim Asian vipers).

Subfamily **Viperinae** Old World vipers

A subfamily of 13 genera and just over 100 species, found in Asia, Europe and Africa; five genera are confined to Africa and six are in both Africa and Asia. At present, some 56 species occur in Africa. The various genera, and most of the species, can be readily identified in the field using a combination of appearance, size, locality and behaviour, but we have provided technical keys below. Viper venoms tend to cause shock and have local cytotoxic or systemic haemorrhagic effects, although the venoms of some of the small southern African *Bitis* (and in particular the Berg Adder, *Bitis atropos*) are neurotoxic.

Key to the African viper genera

Note one exception to this key: a single African population of Puff Adders (*Bitis arietans*) is known from north of the Sahara, in Morocco and Western Sahara.

1a Pupil round; nine large scales on top of the head......*Causus*, night adders (pp. 101–11)
1b Pupil vertical; many small scales on top of the head......2

2a Subcaudal scales single......3
2b Subcaudal scales paired......4

3a Head pear-shaped; usually on the ground in dry country and semi-desert of the northern half of Africa; usually shades of brown or grey......*Echis*, carpet or saw-scaled vipers (pp. 83–95)
3b Head not pear-shaped; usually in trees and bushes, in forest and woodland of tropical Africa; usually some combination of green, yellow and black......*Atheris*, bush vipers (pp. 32–51)

4a On the high-altitude moorlands of central Kenya (Aberdares/Mount Kenya); slim; brown or grey with black markings......*Montatheris hindii*, Kenya Montane Viper (p. 52)
4b Not on the high-altitude moorlands of central Kenya; not slim; not brown or grey with black markings......5

5a A large single supraocular shield present; south-eastern Africa only......*Proatheris superciliaris*, Floodplain Viper (p. 53)
5b No large single supraocular shield; not only in south-eastern Africa......6

6a In sub-Saharan Africa, and all small species in Kenya or further south......*Bitis*, African vipers (pp. 55–82)
6b In the Sahel or North Africa, and all small species north of 10°N......7

7a Upper dorsals with knobbed (lumpy) keels; usually in or near the Sahara......*Cerastes*, North African desert vipers (pp. 96–100)
7b Upper dorsals with narrow keels; usually north of the Sahara......8

8a In the Sinai; has horns......*Pseudocerastes fieldi*, Western False Horned Viper (p. 112)
8b Not in the Sinai; without horns......9

9a Snout tip turned upwards; adults less than 60cm......*Vipera latastei*, Lataste's Viper (p. 117)
9b Snout tip blunt; adults larger than 60cm......10

10a Midbody scale count 26–27......*Daboia mauritanica*, Moorish Viper (p. 113)
10b Midbody scale count 25......*Macrovipera lebetina transmediterranea*, North African Blunt-nosed Viper (p. 115)

Bush vipers *Atheris*

A tropical African genus of attractive, largely green, broad-headed vipers. Most are arboreal and inhabit forest and dense woodland. They are small, never longer than 80cm. They are usually regarded as medically insignificant and appear to cause very few documented snakebites in Africa. This is for a number of reasons: they are secretive and live largely in trees, often in inhospitable areas (although they may be relatively common), and they are rarely on the ground near habitation at night or moving through farmland (although, being in a tree, if confronted they are more likely to deliver a bite to the upper body, a risk noted with arboreal pit vipers in South America).

No antivenom is produced against their bite, although African carpet viper (*Echis*) antivenoms seem to provide paraspecific neutralisation of their venoms. However, two species have large ranges, namely the Green Bush Viper (*Atheris squamigera*) and Western Bush Viper (*Atheris chlorechis*). The Green Bush Viper has caused some fatalities (although none with any thorough clinical details), and some alarming symptoms have been experienced following bites by the Western Bush Viper and Great Lakes Bush Viper (*Atheris nitschei*). So perhaps they are more dangerous than they are given credit for; a viper that feeds on mammals may well have venom that is very toxic to humans. Several bites to herpetologists and snake keepers are documented. The venom appears to be locally cytotoxic, and bites result in pain (often intense), swelling, shock, incoagulable blood and necrosis. Any doctor dealing with a bush viper bite should consider pain relief, testing for clotting abnormalities, and the possible necessity of blood transfusions. As mentioned, the use of antivenom prepared for carpet vipers can be considered.

At present, at least 15 species of bush viper are known; nine were known when the original edition of this book was published, although there is debate about the status of some forms. Their evolutionary scenario is interesting; probably one or more ancestral forms occurred through the great Central African forest, and the subsequent shrinking of the forest left populations isolated in hilltop and other relict forest patches; these populations then accumulated changes. Several forms can be instantly identified to species by their locality. These small snakes show a remarkable similarity in colour and body form to the arboreal pit vipers of Asia and Central and northern South America – an example of convergent evolution.

Key to the genus *Atheris*, bush vipers

1a Subcaudals fewer than 22......*Atheris barbouri*, Barbour's Short-headed Viper (p. 34)
1b Subcaudals more than 22......2

2a Predominantly brown with very prickly head scales; found west of Ghana......*Atheris hirsuta*, Hairy Bush Viper (p. 40)
2b Does not have the above combination of characters......3

3a Ventral scales fewer than 138; south of 15°S, in Mozambique......*Atheris mabuensis*, Mount Mabu Bush Viper (p. 44)
3b Ventral scales usually more than 138; north of 15°S......4

4a Lateral scales serrated......5
4b Lateral scales not, or only feebly and irregularly, serrated......10

5a Supraocular scales forming elongate 'horns'......6

5b No supraocular 'horns'......7

6a Occurs from Udzungwa Mountains northwards in Tanzania; up to 20 transverse head scales......*Atheris ceratophora*, Horned Bush Viper (p. 36)

6b Occurs south of the Udzungwa Mountains in Tanzania; up to 28 transverse head scales......*Atheris matildae*, Matilda's Bush Viper (p. 45)

7a Four suprarostral scales in first (or only) row; dorsals rounded at the apex; each dorsal scale tipped with yellow; in hills of central Kenya......*Atheris desaixi*, Mount Kenya Bush Viper (p. 39)

7b Three to five suprarostrals in first (or only) row; dorsals pointed at apex; dorsal scales not tipped with yellow; not in central Kenya......8

8a Gular scales strongly keeled; lateral scale rows 4–6 weakly serrated; dorsum yellow-brown to purple-brown with dark-centred, pale yellowish rhombic vertebral markings......*Atheris katangensis*, Shaba Bush Viper (p. 43)

8b Gular scales smooth or feebly keeled; lateral scale rows 2–6 or –8 strongly serrated; dorsum green with irregular black markings or green to blackish with symmetrical yellow markings......9

9a Scales on top of head anteriorly smooth or feebly keeled; 18–20 scales across back of head between posterior supralabials; dorsal body scales with keels extending to the tip; green with variable black markings......*Atheris nitschei*, Great Lakes Bush Viper (p. 46)

9b Scales on top of head anteriorly strongly keeled; 24–26 scales across back of head between posterior supralabials; dorsal body scales with keels not extending to the tip; dorsum dark green to blackish, often with symmetrical yellow markings on back of head and zigzag yellow dorsolateral lines and spots......*Atheris rungweensis*, Mount Rungwe Bush Viper (p. 48)

10a Scales across top of head between posterior supralabials usually more than 23; three or more scales between eye and nasal; midbody scale rows 25–36......*Atheris chlorechis*, Western Bush Viper (p. 37)

10b Scales across top of head between posterior supralabials fewer than 23; one or two scales between eye and nasal; midbody scale rows 14–25......11

11a Scales on neck lanceolate or acuminate; lateral scale rows 2–5 frequently fused; scales across top of head between posterior supralabials 10–12......12

11b Scales on neck not lanceolate or acuminate; lateral scale rows 2–5 frequently duplicated; scales across top of head between posterior supralabials 15–18......13

12a Three large suprarostrals; interorbitals 6–10, strongly keeled; two scales between eye and nasal; supralabials 9–10; midbody scale rows 15–18; lanceolate dorsal scales do not extend beyond midbody......*Atheris hispida*, Rough-scaled Bush Viper (p. 41)

12b Two very large suprarostrals; five interorbitals, median ones feebly keeled; a single scale between eye and nasal; six supralabials; midbody scale rows 14; lanceolate dorsal scales extend beyond midbody......*Atheris acuminata*, Acuminate Bush Viper (p. 43)

13a Underside checked black, blue and white; dark postocular stripe present......*Atheris broadleyi*, Broadley's Bush Viper (p. 35)

13b Underside not checked black, blue and white; dark postocular stripe not always present......14

14a Usually at least one labial scale in contact with the orbit; in Cameroon has 14–16 midbody scale rows......*Atheris subocularis*, Cameroon Bush Viper (p. 51)

14b Usually no labial scale in direct contact with the orbit; in Cameroon has 18–23 midbody scale rows......*Atheris squamigera*, Green Bush Viper (p. 49)

Barbour's Short-headed Viper (Udzungwa Viper)
Atheris barbouri

Identification: A small, stout brown forest-dwelling viper endemic to the Udzungwa and Ukinga Mountains, Tanzania. Unusually for a bush viper, it is terrestrial and thus adapted by being brown and not having a prehensile tail. It has a triangular head with a distinctly short and rounded snout; the tail is short and non-prehensile. The scales of the head and body are strongly keeled; in 19–23 rows at midbody; ventrals 115–127; cloacal scale entire; subcaudals single, 15–23. Total length of largest male 35.2cm, largest female 36.9cm; the smallest snake examined

Atheris barbouri

was 16.4cm long. The body is brown to dark olive above, with a pair of zigzag, pale yellow dorsolateral stripes that extend from the back of the head to the end of the tail. An irregular chain of darker rhombic blotches may be present on the back. Faint black chequering may be present on the tail. The ventrum is greenish white to olive. Females may be more speckled than males.

Habitat and Distribution: An enigmatic Tanzanian endemic, from the hills of southern Tanzania, found in mid-altitude woodland of the Udzungwa and Ukinga Mountains at around 1,700–1,900m, in thick bush and bamboo undergrowth, but also found in fields and gardens of tea farms. Known localities include Bomalang'ombe, Dabaga, Kifulilo, Lugoda, Massisiwe and Mufindi in the Udzungwa Mountains, and Madehani and Tandala in the Ukinga Mountains. Very little is known about the ecology and habitat requirements of this species; it would appear to be forest-dependent, and while it may be found in forest-edge situations, such as tea plantations, or occasionally in fields, it is unlikely to survive in the absence of forest nearby, nor is it likely to survive extensive forest destruction.

Natural History: Very poorly known. The vertical pupil suggests nocturnal activity, but it has been observed while active in sunlight after a rainstorm. Specimens were collected in January–March, June, September and October. All recent specimens have been collected on the forest floor, but earlier specimens

Barbour's Short-headed Viper, *Atheris barbouri*.
Left and centre: Tanzania (Michele Menegon). **Right:** Tanzania (Luke Mahler).

were recorded among agricultural plots and gardens at the edge of forest or formerly forested land. It is not adapted for climbing or burrowing, suggesting it lives in leaf litter on the forest floor. The diet appears to be earthworms, and possibly small frogs. It probably lays eggs. Three females collected in February 1930 each contained 10 eggs, the largest of which measured 10 × 6mm; a fourth was non-gravid. Three females collected at Mufindi in June 1983 held fairly large follicles, as did a female in October.

Medical Significance: Unlikely to be significant, due to its small range, rarity and small size. Nothing is known of the venom, but it is unlikely to be life-threatening. No bite cases are documented.

Broadley's Bush Viper *Atheris broadleyi*

Identification: A large bush viper that is variable in colour, from the forests of western Central Africa; first described in 1999. The head is broad, triangular and covered in small, strongly keeled scales. The small eye, with a vertical pupil, is set far forward. The tail is long and prehensile, 13–18% of total length (shorter in females). The scales are keeled; in 17–23 rows at midbody; ventrals 157–169; subcaudals 45–61 (higher counts in males). Maximum known size 77cm, average probably 45–65cm; hatchling size unknown, but probably 15–20cm. The colour is

Atheris broadleyi

dull green or rufous; greenish animals have yellow cross-bars, and rufous animals dark cross-bars. The underside is sky blue, suffused with white, and darkening towards the tail. Juveniles have a whitish tail tip. Specimens from the eastern side of the range have a dark bar behind the eye, which broadens towards the back of the jaw.

Broadley's Bush Viper, *Atheris broadleyi*.
Left and right top: Cameroon (Dwight Lawson). **Right bottom:** captive (Mike McLaren).

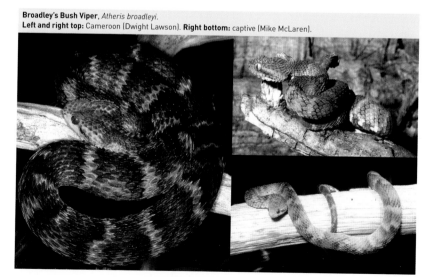

Habitat and Distribution: In forest with palm swamps and scrub and grassy clearings, between sea level and 1,000m. Known from extreme south-west Cameroon (on the Nigerian border) eastwards to the south-western corner of Central African Republic, and south to Equatorial Guinea, north-western Gabon and north-central Republic of the Congo.

Natural History: A largely arboreal, nocturnal, slow-moving snake, it waits in ambush among vegetation, but has been found on the ground after rain. It has been observed to be relatively common in parts of its range (24 specimens were collected in 48 hours in Cameroon; the sample contained more females than males). Mating in captivity was in late June and early July, and 8–13 offspring were observed.

Medical Significance: Might be a significant hazard due to its large size (for a bush viper), relative abundance in parts of its range and willingness to hunt on the ground, but no bites or symptoms are documented. Symptomatic treatment is recommended, with pain relief, ensuring the bitten limb is comfortable, and monitoring for necrosis and clotting abnormalities.

Horned Bush Viper (Usambara Bush Viper)
Atheris ceratophora

Identification: A small horned bush viper of the Eastern Arc Mountains in Tanzania, and thus unmistakable. It has a noticeably triangular head covered with small keeled scales, with a small cluster of 'horns' – two or three elongate scales above the eye. The neck is narrow. The dorsal scales are strongly keeled; in 19–27 rows at midbody; ventrals 134–152; subcaudals 46–58. Maximum size 55cm, average 40–50cm; hatchlings 12–16cm. The tail is prehensile. The colour is very variable; this snake may be uniform olive-green, black (with or without light cross-bars), orangey-yellow with black mottling near the tail, golden yellow with increasing dull green and black mottling towards the tail, barred black and yellow, dull brownish yellow, or bright yellow with black mottling. Hatchlings

Atheris ceratophora

Horned Bush Viper, *Atheris ceratophora.* **Left:** Tanzania (Lorenzo Vinciguerra). **Right:** Tanzania (Michele Menegon).

Horned Bush Viper, *Atheris ceratophora*. **Left:** Tanzania (Michele Menegon). **Right:** Tanzania (Lorenzo Vinciguerra).

are either shiny or dull black, with a vivid yellow tail tip (which may be used as a caudal lure, waved to attract insect-eating prey). Juveniles gain their adult colours after the third moult of the skin, and at this stage their horns become noticeable.

Habitat and Distribution: Lives in forest and woodland at low to medium altitude, below 700m in the Udzungwa National Park, but over 2,000m in the Usambara National Park. This species is endemic to Tanzania, known only from the Usambara, Nguru, Uluguru and Udzungwa Mountains. A forest-dependent species, but much of its habitat is in protected areas.

Natural History: Theoretically nocturnal; certainly both terrestrial and arboreal. Most specimens have been collected during the day when animals are either basking or are detected by their movement as they crawl on the forest floor. In captivity, some individuals remain motionless for weeks on the same branch. It is usually found on the ground, or coiled in a clump of vegetation up to 2m above the ground. Most specimens have been found between January and April, when it is warm; during the colder months, when minimum temperatures may drop to 6°C, the animals may go into a period of inactivity, if not true hibernation. Few details are available on its feeding in the wild; amphibians and lizards are recorded, but captive specimens take mice readily. Limited evidence suggests that breeding takes place in September and October; young have been found in April, and captured females gave birth in March. The scientific name refers to the horn-like scales (*cerato* = horn).

Medical Significance: Unlikely to be medically significant, being uncommon and living in remote montane forests. A known bite case caused minor pain and discolouration at the bite site. Symptomatic treatment is recommended, with pain relief, keeping the bitten limb comfortable and monitoring for necrosis and clotting abnormalities.

Western Bush Viper *Atheris chlorechis*

Identification: A fairly large green bush viper from the forests of West Africa; the only green bush viper found west of Ghana. The head is broad, relatively deep and triangular, and covered in small, strongly keeled scales. The small eye,

Western Bush Viper, *Atheris chlorechis*.
Left top: captive (Paul Freed). **Right top:** Ivory Coast (Mark-Oliver Rödel). **Left bottom:** hatchling (Paul Freed).
Right bottom: captive (Bill Branch).

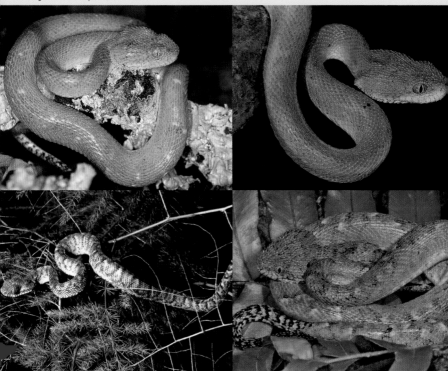

with a vertical pupil, is set far forward. The tail is long and prehensile, 15–19% of total length (shorter in females). The scales are keeled; in 25–37 rows at midbody; ventrals 151–165; subcaudals 48–64 (higher counts in males). Maximum known size about 70cm, average probably 40–65cm; hatchling size unknown, but probably 13–18cm. The colour is light to vivid green, uniform, or spotted or blotched to a greater or lesser extent with yellow; the tail and parts of the body are sometimes speckled black. The underside may be light green, bluish or yellow.

Atheris chlorechis

Juveniles may be blue-grey, tan or dull green, sometimes with dark and light cross-bars and a yellow tail tip. A melanistic (black) specimen and a uniform yellow animal have been reported.

Habitat and Distribution: Lives in forest and thick woodland at low to medium altitude, from sea level to 560m. A forest-dependent species, although it will utilise secondary forest and abandoned farmland in forest. It occurs from Sierra Leone and southern Guinea eastwards to Ghana; there are a couple of records from Togo and extreme south-west Nigeria.

Natural History: Nocturnal and arboreal, although it will descend to the ground. Most specimens have been found in forests and forest clearings, on the ground or coiled in or on vegetation at heights up to 2m above the ground. An individual in an abandoned cassava plantation in Ghana dropped to the ground when approached and hid in the undergrowth. Females give birth to 5–9 live young, in March or April, at the beginning of the rainy season.

Medical Significance: This snake has a large range, but being arboreal and secretive it probably causes few bites. Little is known of the toxicity or composition of the cytotoxic venom, but it can cause severe symptoms, and yields of up to 100mg of wet venom have been reported. In a documented bite case by a captive snake, the victim experienced severe swelling, pain and coagulopathy. Impending compartment syndrome prompted a fasciotomy, but the victim then lost 5 litres of blood in six hours. Six vials of Fav-Afrique antivenom stopped the blood loss (although this antivenom was not raised using *Atheris* venom), but acute kidney injury had occurred and full renal function took three weeks to return. A bite from an Ivory Coast snake caused intense pain, swelling and major necrosis, and required surgical intervention. Symptomatic treatment is recommended, with pain relief, keeping the bitten limb comfortable and monitoring for necrosis and clotting abnormalities.

Mount Kenya Bush Viper *Atheris desaixi*

Identification: A large, thick-bodied black and yellow bush viper, discovered in 1967 and found only in the forests of high central Kenya. The head is broad, triangular and covered in small, strongly keeled scales. The small eye, with a vertical pupil, is set far forward. The tail is long and prehensile, 13–15% of total length in females and 16–18% in males. The scales are keeled; in 24–31 rows at midbody; ventrals 160–174; subcaudals 41–54 (higher counts in males). Maximum size about 70cm, average 40–60cm; hatchlings 17–22cm. The body is greeny-black

Atheris desaixi

to charcoal black, each scale edged with yellow or yellowy-green, creating either a speckled effect or a series of yellow loops. On the hind body and tail the speckles may fuse into yellow zigzags. The belly is yellow on the front half, becoming progressively suffused with purplish black to the rear and under the tail; the tail

Mount Kenya Bush Viper, *Atheris desaixi*. **Left and centre:** Kenya (Stephen Spawls). **Right:** Kenya (Bio-Ken Archive).

tip is blotchy yellow. Occasional yellow-brown or dull green individuals occur; they have faint darker festoons. Hatchlings are predominantly yellow, with a white tail tip, but gradually darken to adult colour at around 30cm in length.

Habitat and Distribution: Endemic to the central highlands of Kenya, in mid-altitude evergreen forest between 1,000m and 1,800m. There are three populations: one in the Ngaia (Ngaya) Forest, at 1,000m on the north-east flank of the Nyambene Range; one around Igembe, at 1,300m in the northern Nyambene Range; and one in forest just west of Chuka, south-eastern Mount Kenya, between 1,500m and 1,800m. It might be more widespread in the relatively unexplored forests around the central highlands. This is a snake under threat from collecting for the pet trade and deforestation. Its distribution needs investigation and, if feasible, an area of its habitat needs rigorous protection.

Natural History: A slow-moving arboreal snake. Its activity patterns are not known; it might be diurnal or nocturnal, or both. It is usually found draped in low vegetation, around 2–3m from the ground, around the edges of small clearings in forest, but has been seen 15m up a tree. The Ngaya specimen was in a lantana bush at the forest edge. Some Nyambene specimens were in yam plantations. This species is perfectly camouflaged and difficult to observe. It is very willing to strike when first caught, and will form C-shaped coils like a carpet viper; the coils are shifted against each other in opposite directions, producing a hissing sound, like water falling on a hot plate. It struggles fiercely in the hand, but soon tames in captivity. Little is known of its biology, but it appears to feed on small mammals. A female from the Nyambene Range gave birth to 13 young in August; the smallest was 17.5cm and the largest 21.1cm.

Medical Significance: Unlikely to be medically significant, being uncommon and living in remote montane forests. A known bite case caused swelling and pain. Symptomatic treatment is recommended, with pain relief, keeping the bitten limb comfortable and monitoring for necrosis and clotting abnormalities.

Hairy Bush Viper *Atheris hirsuta*

Identification: A small, slim bronze-coloured West African bush viper with spiky scales, first described in 2002 and known solely from two males, from Ivory Coast and Liberia. The head is short and ovoid, the eye is huge with a vertical pupil, and the neck is thin. There are many small scales on top of the head; the scales are noticeably raised and elevated anteriorly, particularly on the head and neck. The body is subtriangular in section and the tail is 19–20% of total length. The scales are strongly keeled; in 15–16 rows at midbody; ventrals 159–160;

Atheris hirsuta

subcaudals 58. The dorsal scales are very elongate and taper to an elevated point. The two specimens were 43cm and 48cm long. The colour is bronze, with yellow- and dark-tipped dorsal scales forming a series of light and dark cross-bars; the underside is brownish, sparsely suffused with golden yellow.

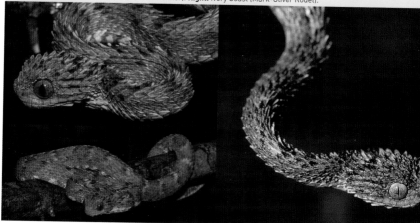

Habitat and Distribution: The type specimen was found in the Taï National Park, extreme south-west Ivory Coast, in secondary forest at 200m; the second animal was on Mount Swa, Nimba County, central Liberia, also in secondary forest at 585m. The first specimen was on a road in the early morning after heavy rain, while the second was climbing a tree, 2m up, at night; both were found in September, which is near the end of the rainy season.

Natural History: In general appearance this snake resembles the Rough-scaled Bush Viper (*Atheris hispida*) and its habits are probably similar; that is, primarily nocturnal, mostly arboreal, gives birth to live young and feeds on small vertebrates. However, nothing is actually known. The type was described as 'more aggressive' than the sympatric Western Bush Viper (*A. chlorechis*).

Medical Significance: Nothing known. Symptomatic treatment is recommended, with pain relief, ensuring the bitten limb is comfortable, and monitoring for necrosis and clotting abnormalities.

Rough-scaled Bush Viper (Prickly Bush Viper)
Atheris hispida

Identification: A large, slender bush viper from western East Africa, with bizarrely long scales, giving it a spiky appearance. The head is subtriangular, the eye fairly large, the pupil vertical and the iris brown, heavily speckled with black (looks uniformly dark under most conditions). There are many small scales on top of the head. The body is cylindrical in section and the tail long, 17–21% of total length. The scales are heavily keeled, prickly and leaf-shaped (this is a good field character); in 15–19 rows at midbody; ventrals 149–166; subcaudals

Atheris hispida

Rough-scaled Bush Viper, *Atheris hispida*. **Left:** Uganda (Jelmer Groen). **Right:** captive (Paul Freed).

49–64 (high counts usually males). Maximum size just over 70cm, average 40–60cm; hatchlings 15–17cm. Males are usually olive-green, with a black mark on the nape (in the shape of an 'H', 'V' or 'W', or just a blotch), and sometimes a dark line behind the eye. The ventrals are greenish, darkening towards the tail. Females are usually yellowy-brown or olive-brown, with a similar dark nape mark, and yellow-brown below.

Habitat and Distribution: Associated with forest, woodland and thicket, sometimes in waterside vegetation at 900–2,400m. It has a bizarre, disjunct distribution. There is one Tanzanian (Minziro Forest, north-west) and one Kenyan (Kakamega Forest) record. In Uganda, it is known from forest patches around Mityana and in the south-west Ruwenzoris, in Kigezi Game Reserve and Bwindi Impenetrable National Park. There are no Rwandan or Burundi records, but there is a cluster of records in the eastern DR Congo, from the Ituri Forest southwards to Lutunguru, Rutshuru and Bunyakiri, west of Lake Kivu. It probably occurs in forest and forest patches in the intervening areas.

Natural History: Poorly known. It is arboreal and an expert climber, moving relatively quickly through the branches. It lives in tall grasses, Papyrus (*Cyperus papyrus*), bushes, creepers and small trees. In Kakamega Forest, these snakes were found slightly higher up than the sympatric Green Bush Viper (*Atheris squamigera*), in drier bushes, but were also found in reeds. It will bask on top of small bushes or on flowers. Although probably nocturnal, it will opportunistically strike from ambush if prey passes by. It is an irascible snake, struggling furiously if held, and forming the body into C-shaped coils like the Mount Kenya Bush Viper (*A. desaixi*). Its mouth is often full of small back mites. It gives birth to live young, with clutches of 2–12 recorded; two Kakamega females had clutches of two and nine young in mid-April, the time of highest rainfall. The diet is not well known; the holotype had a snail in its stomach, while captive specimens take rodents and tree frogs. Population numbers of this species in Kakamega seem to peak and crash, for reasons as yet unexplained; in some years they are abundant, while in others they cannot be found.

Medical Significance: The venom is unstudied and no bite cases are documented.

Acuminate Bush Viper *Atheris acuminata*

Identification: A small, slim bush viper, described in 1998 from a single specimen from Uganda and not seen since. The head is subtriangular, the eye fairly large, the pupil vertical and the iris mottled green, with gold bordering the pupil. There are many small scales on top of the head. The body is subtriangular in section and the tail is 18% of total length. The scales are keeled; in 14 rows at midbody; ventrals 160; subcaudals 54. The dorsal scales are very elongate and taper to a point (the name 'acuminate' means tapering to a point). The specimen was 44cm

Atheris acuminata

long, with the tail 8.1cm, and yellow-green above, with a vague H-shaped black marking on the crown and a short black blotch behind the eye. The tail was black-blotched and the belly pale greenish yellow, black-blotched posteriorly.

Habitat and Distribution: Described from Kyambura Game Reserve, adjacent to Queen Elizabeth National Park, just south of Lake George, Uganda, at an altitude of 950m, in riverine forest. There is little data in conservation terms, but it was found in a protected area.

Natural History: Nothing known. If it is a valid species, it is probably similar to the Rough-scaled Bush Viper (*Atheris hispida*).

Medical Significance: Nothing known.

Taxonomic Notes: This snake is virtually identical to the Rough-scaled Bush Viper, and is probably just an aberrant example of that species.

Acuminate Bush Viper, *Atheris acuminata*. **Below:** preserved type (Don Broadley).

Shaba Bush Viper *Atheris katangensis*

Identification: A small green or tan bush viper from south-eastern DR Congo and adjacent areas. The head is short and quite deep, the eye large, the pupil vertical and the neck thin. There are many small scales on top of the head. The body is subtriangular in section and the tail is 13–16% of total length. The scales are strongly keeled; in 24–31 rows at midbody; ventrals 133–144 (higher counts in males); subcaudals 38–42 in females and 45–49 in males. The six known specimens ranged from 22–40cm. The colour is dull green or yellow-brown,

Atheris katangensis

with faint lighter cross-bars or lozenges on the back; the underside is light yellow or greenish. Juveniles have a whitish tail tip.

Shaba Bush Viper, *Atheris katangensis*. **Left and right:** Zambia (Frank Willems).

Habitat and Distribution: Originally described from gallery forest along rivers in the Upemba National Park, in south-eastern DR Congo, between 1,250m and 1,400m. Two recent examples of this species were observed north of Mwinilunga in north-west Zambia and from north-western Angola.

Natural History: Nothing known, but presumably similar to other bush vipers; that is, primarily nocturnal, mostly arboreal, gives birth to live young and feeds on small vertebrates.

Medical Significance: Nothing known.

Mount Mabu Bush Viper *Atheris mabuensis*

Identification: A very small brown or dull green bush viper from Mounts Mabu and Namuli in north-central Mozambique. The head is triangular, the eye large, the pupil vertical and the neck thin. There are many small scales on top of the head. The body is subtriangular in section and the tail is 14–17% of total length. The scales are strongly keeled; in 22–26 rows at midbody; ventrals 128–137; subcaudals 39–47. The five known specimens ranged from 18cm to 38cm. Some specimens have a black V-shape on the head. The dorsal colour is

Atheris mabuensis

olive, blue-grey or brown, and along the centre of the back is a pair of light lines forming a twisted helix shape, interspersed with black blotches. In some specimens the dorsal scales have a distinctive yellow tip. The adults seem to retain the juvenile yellow tail tip. The underside is blue-grey with fine, heavy black stippling, except on the yellow tail tip.

Habitat and Distribution: All specimens were found on the forest floor in wet forest on Mounts Mabu and Namuli, Zambezia Province, north-central Mozambique, between 1,000m and 1,550m.

Natural History: Specimens were found on the forest floor in leaf litter during the day, and have also been found under ground cover. Captive specimens either hid

Mount Mabu Bush Viper, *Atheris mabuensis*. **Left and right:** Mozambique (Bill Branch).

themselves in leaf litter or coiled in branches or leaf clumps up to 20cm above the ground. They ate small lizards and frogs, and used the yellow tail as a lure, waving it to attract prey. The 38cm female was sexually mature. A 15cm captive grew to 36cm in three years.

Medical Significance: Little known. Local hunters describe bites from these snakes as painful but not lethal.

Matilda's Bush Viper *Atheris matildae*

Identification: A relatively large, heavy-bodied bush viper from south-western Tanzania; adults often larger than 70cm. First described in 2011. The head is broad, triangular and covered in small scales. The small eye, with a vertical pupil, is set far forward. The tail is long and prehensile. The dorsal scales are strongly keeled; in 26–27 rows at midbody; ventrals 142–150; subcaudals 44–50. Maximum size is probably around 80cm. Its colouration, though quite variable, is generally black with zigzag, bright yellow dorsolateral lines running along its length. A   black

Atheris matildae

patch around the nasal, rostral, mental and first infralabials is present in most of the observed males, but also in a few females and immature individuals. Males in

Matilda's Bush Viper, *Atheris matildae*.
Left: Southern Highlands, Tanzania (Michele Menegon). **Centre and right:** Tanzania (Michele Menegon).

general tend to be darker, with the belly suffused with black. Adult females tend to be more yellow, in some cases with an immaculate throat and belly. The side of the head can be completely yellow or have black patches at the tips of the scales. It is distinguished from most other *Atheris* vipers by the presence, in both males and females, of enlarged supraciliary scales that form this species' yellow and black horns. This species closely resembles the only other bush viper with horns, the Horned Bush Viper (*Atheris ceratophora*), but is larger and heavier, and has differences in scalation.

Habitat and Distribution: Known from just two forest fragments in the southern highlands of Tanzania, north of the northern tip of Lake Malawi (Lake Nyasa).

Natural History: Poorly known. Individuals were found both climbing on bushes and trees and moving on the ground; it appears that larger individuals become more terrestrial and hide in rodent burrows. It presumably feeds largely on small vertebrates and gives birth to live young.

Medical Significance: Nothing is known of the toxicity or composition of the venom, but, as with other larger bush viper species, a bite from this species might be serious. Symptomatic treatment is recommended, with pain relief, keeping the bitten limb comfortable and monitoring for necrosis and clotting abnormalities.

Great Lakes Bush Viper *Atheris nitschei*

Identification: A large, stout black and green bush viper of the central Albertine Rift Valley. The head is triangular and the neck thin. The eye is fairly large, set well forward and has a vertical pupil, but this is hard to see; the whole eye just looks black, although the iris is brown. There are many small scales on top of the head. The body is cylindrical and stout, and the tail is fairly long and strongly prehensile, 15–17% of total length in males and 13–16% in females. The scales are keeled; in 23–34 rows at midbody; ventrals 140–163; subcaudals 35–59 (higher counts in males).

Atheris nitschei

Great Lakes Bush Viper, *Atheris nitschei*.
Left: Rwanda (Michele Menegon). **Centre:** Rwanda (Max Dehling). **Right:** striking (Matthijs Kuijpers).

Maximum size about 75cm, average 45–65cm; hatchlings 15–18cm. Colour various shades of green, blue-green, yellowy-green or olive, heavily speckled, blotched or barred black; there is usually a conspicuous black bar behind the eye, and a dark blotch or V-shape on top of the head. The belly is yellow or greeny-yellow. There is an ontogenic colour change: hatchlings are deep green (almost black), brown or grey-brown, with a white or yellow tail tip; after three or four months they become uniform green, and black blotches then appear gradually.

Great Lakes Bush Viper, *Atheris nitschei*. **Below:** hatchling, Uganda (Jelmer Groen).

Habitat and Distribution: A characteristic species of the Albertine Rift. It lives in medium- to high-altitude moist savanna and woodland, montane forest and bamboo, sometimes associated with lakes and swamps, in waterside vegetation, between 1,000m and 2,800m. It occurs from North Kivu in DR Congo and extreme western Uganda, south along the Albertine Rift, to both sides of northern Lake Tanganyika, where it extends slightly further south on the DR Congo shore. On the eastern shore it is replaced by the Mount Rungwe Bush Viper (*Atheris rungweensis*).

Natural History: Arboreal, living in Papyrus (*Cyperus papyrus*) and reedy swamp vegetation, Elephant Grass (*Pennisetum purpureum*), bushes, small trees and bamboo. It will descend to the ground to hunt. It is probably nocturnal, but known to bask, at heights of 3m or more in creepers, Elephant Grass and Papyrus. If disturbed it slides quickly downwards or simply drops off its perch. In Bwindi Impenetrable National Park, it was found waiting in ambush for diurnal lizards such as Jackson's Forest Lizard (*Adolfus jacksoni*), concealed in grass at the base of roadcut walls. Hatchlings used caudal luring, waving their bright tail tips to attract insect-eating prey. Some specimens are placid, but often it is bad-tempered, hissing, striking and forming C-shaped coils when threatened. It gives birth to live young in clutches of 4–15; a female from Mount Karisimbi gave birth to 12 young in February, and gravid females were collected in October and November in south-west Uganda, and in January in the Ruwenzoris. The diet includes small mammals, amphibians and lizards (including chameleons); captive specimens descended to the ground to stalk rodents.

Medical Significance: Few documented bites, but in countries like Rwanda and Burundi, with intensive agriculture, this species is likely to be present in marginal vegetation along rivers running through farmland. Bites by this species, including one in Rwanda, are characterised by extreme pain, swelling (in one case up the entire arm and onto the trunk; the victim also became unconscious), oedema, clotting abnormalities and necrosis, sometimes resulting in the loss of parts of digits. No antivenom is produced. Symptomatic treatment is recommended, with pain relief, keeping the bitten limb comfortable and monitoring for necrosis and clotting abnormalities.

Mount Rungwe Bush Viper *Atheris rungweensis*

Identification: A large, attractive green and yellow bush viper of south-western Tanzania, extreme northern Zambia and Malawi. The large head is triangular in shape and rather flat in profile, with a distinct neck. The dorsals are keeled and pointed, but with keels ending before the tip; in 22–33 rows at midbody; ventrals 150–165; subcaudals 46–58. Maximum size about 65cm, average 35–55cm; hatchlings 15–17cm. The colour is predominantly light or dark green, usually with oblique yellow cross-bars, festoons or zigzags; in some specimens

Atheris rungweensis

the upwards curve of the festoon has a black infilling. The underside is greenish yellow. Newborn animals are dark brown or grey, with a bright yellow tip to the tail. They may use this brightly coloured tail as a 'lure' to attract prey. After several moults, they develop a uniform green and then more patterned adult colouration.

Habitat and Distribution: Most records are from moist savanna, woodland and hill forest of south-western Tanzania, northern Zambia and Malawi, between 800m and 2,000m. It occurs southwards from the Gombe Stream National Park and Kigoma down the eastern shores of Lake Tanganyika to the Mahale Peninsula, and inland to the Ugalla River. In addition, in forest patches in the high country around Sumbawanga, between the north end of Lake Rukwa and the south end of Lake Tanganyika, and on Mount Rungwe, just north of the north end of Lake Tanganyika; it is also found in the Nyika National Park and Plateau, on both sides of the Zambia–Malawi border, around Mbala in northern Zambia and in the Matipa Complex Forest in northern Malawi.

Natural History: Poorly known. It is arboreal, and the vertical pupil suggests it is nocturnal. An expert climber, found in bushes 1–3m above the ground, or on the ground at edges of forest. It might hunt on the ground or ambush from trees. It gives birth to live young. The diet includes small frogs, although captive animals accept mice.

Mount Rungwe Bush Viper, *Atheris rungweensis.*
Left and centre: Tanzania (Michele Menegon). **Right:** Tanzania (David Moyer).

Medical Significance: Nothing is known of the toxicity or composition of the venom, but, as with other large bush vipers, a bite from this species might be serious. Symptomatic treatment is recommended, with pain relief, keeping the bitten limb comfortable and monitoring for necrosis and clotting abnormalities.

Taxonomic Notes: Originally regarded as a subspecies of the Great Lakes Bush Viper (*Atheris nitschei*).

Green Bush Viper *Atheris squamigera*

Identification: A large bush viper, usually green but sometimes other colours, often with fine yellow cross-bars. It has the widest range of any bush viper, occurring right across forested Central Africa. The head is broad, triangular and covered in small, strongly keeled scales. The small eye, with a vertical pupil, is set far forward. The tail is long and prehensile, 15–17% of total length in females and 17–20% in males. The scales are keeled; in 15–25 rows at midbody; ventrals 133–175; subcaudals 45–67 (higher counts in males). Maximum size about

Atheris squamigera

80cm, average 40–65cm; hatchlings 16–22cm. The colour in much of its range is green or yellow-green (very occasionally brown), often becoming turquoise towards the tail. Sometimes the yellow-tipped scales form a series of fine yellow cross-bars. The belly is greenish blue, with yellow blotches. Juveniles are olive, green or yellow-green, with dark-edged, olive V-shapes on the back. Hatchlings are more strongly banded than adults, but have a bright yellow or pink tail tip, which may be used as a lure. However, in Central Africa, especially in DR Congo, several other colour morphs are known, including orange, bright and dull yellow, slate grey and blue-grey, and with or without cross-bars. The colour varies within broods; for example, a Kinshasa female gave birth to orange, yellow and green offspring.

Habitat and Distribution: All types of forest, and in well-wooded savanna, from sea level to 1,700m. In Nigeria it is found in a wide range of habitats, including suburbia, agriculture and mangroves. It seems to be separated by altitude from the Great Lakes Bush Viper (*Atheris nitschei*) in the country around the Albertine Rift, the Green Bush Viper usually occurring below 1,600m and the Great Lakes Bush Viper above that. Sporadically distributed from Kenya (Kakamega Forest), Tanzania (Rumanyika National Park) and central Uganda (Mount Elgon, Lake Victoria forests, and low-altitude localities along the Albertine Rift), westwards across DR Congo, Republic of the Congo and southern Central African Republic to south-eastern Nigeria, and south to northern Angola. There is an isolated record from Watoka in South Sudan. Doubt has been cast on the historical records of this snake from Ghana, Togo and Ivory Coast; there are no unequivocal recent records, so these are not shown on the map.

Natural History: A largely arboreal, slow-moving snake, but it will descend to the ground at night. Largely nocturnal; it is most active between sundown and 2 a.m.

Green Bush Viper, *Atheris squamigera*.
Left top: Eastern DR Congo (Konrad Mebert). **Right top:** captives (Paul Freed). **Left bottom:** Angola (Alan Channing).
Right bottom: Kenya (Stephen Spawls).

It may bask on top of vegetation in clearings during the day, though is sometimes found on the ground. It is usually found in bushes and small trees, but may climb up to 6m or more above ground level. It is willing to strike when first caught and forms C-shaped coils, like carpet vipers (*Echis*); the coils are shifted against each other in opposite directions, producing a sizzling sound, like water falling on a hot plate. At night, it descends to low levels and waits in ambush with the head hanging down. It can also drink in this position, sipping water from condensing mist or rain running down the body. It feeds largely on small mammals and is also known to take birds, lizards, amphibians and small snakes; adults eat mostly warm-blooded prey, while juveniles eat lizards and shrews. Mating has been observed in October in Uganda; 7–9 young are born in March–April.

Medical Significance: At present this is the only species of bush viper known to have caused human fatalities, its wide range bringing it into contact with forest villagers. Chippaux and Jackson, in *Snakes of Central and Western Africa*, report 'many deaths', although without further data. The venom appears to be cytotoxic. Two documented fatal bite cases, from Central African Republic and Kenya, showed massive swelling and incoagulable blood; death occurred after six days in both cases. A Nigerian victim lapsed into a coma, but recovered. The species is said to have caused 'a number' of fatalities in West Africa. Local pain, often intense, swelling and coagulopathy seem to characterise most cases of bite from this viper. Specific antivenom is not available, although carpet viper antivenom has in some cases been shown to assist, so bites may need symptomatic treatment, including elevation, monitoring of clotting times, transfusions and replacement therapy.

Taxonomic Notes: Several taxa within this widespread species have been variously granted full species status, relegated back into the synonymy of *Atheris squamigera*, and in some cases re-elevated. These forms are largely based on slight differences of scalation and/or colour. They include *A. anisolepis* and *A. laeviceps*. However, the characters that supposedly identify these forms in their habitats are often present in typical *A. squamigera* from other parts of Africa. Pending a re-examination of the taxon, we leave them within the synonymy of *A. squamigera*.

Cameroon Bush Viper (Fischer's Bush Viper)
Atheris subocularis

Identification: A small, slim green bush viper from the forests of western Cameroon. The head is short, the eye large, the pupil vertical and the neck thin. Usually there are no subocular scales, though occasionally there may be a single row or a few tiny irregular ones. There are many small scales on top of the head, some of which curve noticeably upwards, like those of the Rough-scaled Bush Viper (*Atheris hispida*). The body is subtriangular in section and the tail is 19–22% of total length. The scales are strongly keeled; in 14–16 rows at midbody; ventrals

Atheris subocularis

152–163; cloacal scale entire; subcaudals 58–65. Maximum size about 55cm, average around 30–45cm; hatchling size unknown. The colour is dull green or yellow-brown, with faint, pale yellow cross-bars, interspersed with small dark blotches, and the underside is dull lime green, sparsely smudged with black. The head is marked with black blotches, forming an incomplete chevron. Juveniles have a yellowish tail tip; hatchlings are dark brown or black.

Habitat and Distribution: Known only from the forests of western Cameroon, in the South-West Region, between 200m and 400m.

Natural History: Nothing known, but presumably similar to other bush vipers; that is, primarily nocturnal, mostly arboreal (although the specimens used in the redescription were found crossing a road), and gives birth to live young. It presumably feeds on small vertebrates.

Medical Significance: Nothing known.

Cameroon Bush Viper, *Atheris subocularis*. **Left:** Cameroon (Dwight Lawnson). **Right:** captive (Matthijs Kuijpers).

Montane viper *Montatheris*

A genus containing a single species that lives at high altitude in central Kenya. Of all Africa's dangerous snakes, it is probably the least significant medically, as it is very small, and its entire range lies on montane moorlands within two high-altitude national parks where virtually no one lives and nobody walks without shoes. No antivenom is available.

Kenya Montane Viper *Montatheris hindii*

Identification: A small, slender viper from the moorlands of the Aberdare Mountains and Mount Kenya. It has an elongate head, covered in small, strongly keeled scales. The small eye, with a brown iris and vertical pupil, is set far forward. The body is cylindrical and the tail short, 10–13% of total length. The scales are keeled; in 24–28 rows at midbody; ventrals 127–144; subcaudals paired, 25–36. Maximum size about 35cm, average 20–30cm; hatchlings 10–14cm. The dull grey or brown body has a paired series of triangular, pale-edged

Montatheris hindii

black blotches along the back. The underside is yellow, cream or grey-white, but heavily mottled black, to the extent of appearing black in places. There is an irregular, dark brown arrow- or V-shaped mark on the crown of the head. A wide, dark stripe passes through the eye to the temporal region. The upper and lower labials are white.

Habitat and Distribution: This unique, beautiful tiny Kenyan endemic is found only at high altitude (2,700–3,800m) in treeless montane moorland. There is an isolated population on the Aberdare Mountains and another on Mount Kenya. It has probably the smallest range of any dangerous African snake.

Natural History: Poorly known. It is terrestrial, and somewhat sluggish unless warmed up. It is active by day, usually between 10 a.m. and 4 p.m, as the night-time temperatures in its mountain habitat are usually below freezing. It shelters in thick grass tufts that provide cover and insulation from the extreme cold, but

Kenya Montane Viper, *Montatheris hindii*. **Left, centre and right:** Kenya (Stephen Spawls).

may also hide under ground cover (rocks, vegetable debris) or in holes. Due to the cold winds and rarefied air, much time is spent inactive in shelter. On days of warm sunshine it emerges to bask on patches of warm soil or a grass tussock, and will then hunt. It is irascible and willing to bite if threatened. A female from the Aberdares gave birth to two live young, of 13.1cm and 13.5cm in length, in late January, a captive female produced three young (length 10.2–10.9cm) in May, and small juveniles (15–17cm) were collected in February. It eats lizards, including chameleons and skinks, and small frogs, and may take small rodents as well; a Mount Kenya specimen ate a shrew while in a pitfall trap. It is quite common in suitable habitat, and females seem to be found more often than males, probably because they need to bask more frequently when gravid. The enemies of this small viper include predatory birds; Augur Buzzards (*Buteo augur*) have been seen to take them. The taxonomic relationships of this small viper and its evolutionary history are mysteries.

Medical Significance: Very low; see the notes on the genus (p. 52).

Floodplain or lowland viper *Proatheris*

A genus containing a single medium-sized terrestrial viper, found on river floodplains in south-eastern Africa. Sometimes common, and fishermen and farmers might be at risk.

Floodplain Viper (Lowland Viper) *Proatheris superciliaris*

Identification: An attractive small terrestrial viper, with black spots and fine yellow stripes, associated with Lake Malawi and the rivers running out of it. It has a distinct, rather elongate triangular head that is covered with small, keeled, overlapping scales. The tail is short and distinct in females, but longer and less distinct in males. The dorsal scales are strongly keeled and overlapping, the outermost row enlarged and feebly keeled to smooth; in 27–29 (rarely 26 or 30) rows at midbody; ventrals smooth, 131–159; cloacal scale entire; subcaudals smooth, in 32–45

Proatheris superciliaris

pairs. Maximum size about 60cm in females, 55cm in males, average 25–45cm; hatchlings 13.5–15.5cm. The dorsum is grey-brown with three rows of dark brown spots separated laterally by a series of elongate yellowish bars that form an interrupted line on either side of the body. Three dark, chevron-shaped marks cover the front of the head. The belly is off-white, with numerous black blotches in irregular rows. The lower surface of the tail is straw yellow to bright orange. The species' colour and pattern make it extremely difficult to detect unless it is in motion.

Habitat and Distribution: Lives on floodplains and in moist savanna, between sea level and about 800m. It has a curious distribution. It occurs around the shores of the southern end of Lake Malawi, and Lake Chilwa, and south along the Shire

Floodplain Viper, *Proatheris superciliaris.*
Left top: Malawi (Gary Brown). **Left bottom:** Mozambique (Mark-Oliver Rödel). **Right:** underside (Gary Brown).

River into Mozambique where it meets the Zambezi, then down the Zambezi to its mouth. From there it extends northwards on the Mozambique coastal plain to Quelimane, south to Beira, and inland to Muda and Gorongosa. It also occurs at the north end of Lake Malawi, just extending into Tanzania at Mwaya and Ipyana. The type, collected by Wilhelm Peters and described in 1854, was from 'Terra Querimba', which is taken to mean the mainland opposite the trading post on the Quirimba Islands, but this is more than 700km from the nearest documented record.

Natural History: Terrestrial. Just before the breeding season, it can be seen basking at the mouths of rodent burrows. As is suggested by its vertical pupil, this snake is most active in the early evening, when its prey, mostly amphibians but also small rodents, is also active. During the cold season (April–July), when night temperatures drop to 6°C, it may bask in front of its retreat during daylight hours. When threatened, it throws its body into C-shaped coils, like egg-eaters (*Dasypeltis*) and carpet vipers (*Echis*), and can strike rapidly. Mating occurs in July, and 3–16 young are born November–December.

Medical Significance: Apparently a rare snake, with a restricted range, so not a significant cause of snakebite. No fatalities are known. No antivenom is available. However, case histories of bites inflicted on snake keepers indicate that this is a very dangerous snake; the venom has some severe effects, the victims suffering intense pain and swelling, necrosis, coagulopathy, thrombocytopenia and acute kidney injury. Recovery was relatively slow.

African vipers *Bitis*

A genus of stout-bodied, broad-headed, nocturnal African vipers, all beautifully marked and fairly easily identified in the field. Eighteen species are known; six are heavy-bodied and large, reaching a metre or more, while 12 are small, rarely larger than 60cm. Most of the small forms have restricted ranges in inhospitable areas; they are not very significant medically, and no definite fatalities are recorded. Of the large species, two have small ranges, both in the Ethiopian highlands, and little is known of their venom. However, the other four (Rhinoceros Viper, *Bitis nasicornis*; Puff Adder, *B. arietans*; and Eastern and Western Gaboon Vipers, *B. gabonica* and *B. rhinoceros*) have large ranges and are among Africa's most dangerous snakes. They are slow-moving, active on the ground at night, well camouflaged and easily trodden on, and they deliver a large dose of a powerful cytotoxic venom when they strike. A bite from one is a medical emergency, and must be treated in a well-equipped hospital. Antivenom is available. Symptoms include intense local pain, swelling, blistering, compartment syndrome, shock, blood coagulation abnormalities and necrosis. The Puff Adder is arguably Africa's most dangerous snake. It has a huge range in the savanna where people live, it is large, common and irascible, and it bites a lot of rural people and their animals.

All members of the genus give birth to live young. They all move slowly but strike quickly, from ambush, the stout body lending stability to the strike. These vipers have a curious small pocket behind the nostril, the supranasal sac. This sac is similar to the pit organ of the rattlesnake and other pit vipers. It may be able to detect radiant heat, and thus serve to help the snake target warm prey in the dark; recent experiments have shown that blinded Puff Adders can still detect and strike at warm-blooded prey.

Key to the genus *Bitis*, African vipers

1a Prominent horns (not just raised scales) on the nose......2
1b No prominent horns on the nose......3

2a Head pale, with a fine dark line down the centre......*Bitis rhinoceros*, Western Gaboon Viper (p. 63)
2b Head usually green or blue, with a broad black arrowhead in the centre......*Bitis nasicornis*, Rhinoceros Viper (p. 59)

3a A long (not minute) horn above the eye, consisting of a single elongate scale......4
3b No horn above the eye, or, if present, the horn consists of a cluster of scales, or is minute......5

4a Two light dorsolateral flank stripes enclose a double row of dark blotches; in highland Kenya......*Bitis worthingtoni*, Kenya Horned Viper (p. 81)
4b No dorsolateral flank stripes, and a single row of central flank blotches present; in southern Africa......*Bitis caudalis*, Horned Adder (p. 67)

5a Black with fine yellow vermiculations; in high montane forest in Ethiopia......*Bitis harenna*, Bale Mountains Viper (p. 66)
5b Not black with yellow vermiculations; not necessarily in high montane forest in Ethiopia......6

6a A pale line between the eyes; dorsal pattern consists of V-shapes pointing towards the tail......*Bitis arietans*, Puff Adder (p. 57)
6b No pale line between the eyes; dorsal pattern not consisting of V-shapes......7

7a Dorsal pattern consists of yellow butterfly shapes and dark hexagons; in highland forest in Ethiopia......*Bitis parviocula*, Ethiopian Mountain Viper (p. 64)

7b Dorsal pattern does not consist of yellow butterfly shapes and dark hexagons; not in highland forest in Ethiopia......8

8a Huge, up to 1.8m; head usually pale and unmarked except for a light central line; midbody scale rows 34 or more......*Bitis gabonica*, Eastern Gaboon Viper (p. 61)

8b Small, less than 80cm; head not pale and unmarked; midbody scale rows fewer than 34......9

9a Three-pronged trident shape between eyes; in central Angola......*Bitis heraldica*, Angolan Adder (p. 80)

9b No three-pronged trident shape between eyes; not in central Angola......10

10a Number of scale rows one head length behind the head equal to or slightly more than the number at midbody......*Bitis xeropaga*, Desert Mountain Adder (p. 79)

10b Number of scale rows one head length behind the head fewer than the number at midbody......11

11a Supraorbital region of the head not raised......12

11b Supraorbital region of the head raised......13

12a Outer row of dorsal scales more or less keeled; head elongate; 11–16 scales between the eyes and 1–2 scales between the eyes and the upper labials; subcaudals smooth......*Bitis atropos*, Berg Adder (p. 72)

12b Outer row of dorsal scales smooth; head not elongate; 6–9 scales between the eyes and 2–4 scales between the eyes and the upper labials; subcaudals usually keeled towards the tail tip......*Bitis peringueyi*, Péringuey's Adder (p. 70)

13a Restricted to southern Namibia and Little Namaqualand, extreme western Northern Cape and Western Cape, South Africa; a single minute horn-like scale above the eye......*Bitis schneideri*, Namaqua Dwarf Adder (p. 69)

13b Not restricted to southern Namibia and Little Namaqualand, extreme western

Northern Cape and Western Cape, South Africa; does not have a single horn-like scale above the eye......14

14a Ventrals always fewer than 131 in females and fewer than 126 in males......15

14b Ventrals usually more than 130 in females and more than 126 in males......16

15a Always more than 20 paired dorsal blotches on the body; horns usually well developed, especially in males; restricted to the southern tip of the Western Cape, South Africa......*Bitis armata*, Southern Adder (p. 76)

15b Never more than 20 paired dorsal blotches on the body; horns usually absent or represented by small conical scales; restricted to the Port Elizabeth area of the Eastern Cape......*Bitis albanica*, Albany Adder (p. 75)

16a Horns always well developed as a tuft of 4–7 elongate scales above the eye; usually 27 midbody scale rows; usually 15–16 scales around the eye; body always with prominent paired dorsal blotches, usually 25 or more......*Bitis cornuta*, Many-horned Adder (p. 74)

16b Horns absent or poorly developed; usually 29 midbody scale rows; usually fewer than 14 scales around the eye; body always brown or reddish, often patternless, or usually fewer than 24 paired dorsal blotches......17

17a Horns always absent; body usually plain yellowish brown, rarely with faint dorsal blotches; usually less than 33cm maximum, in montane grassland of the Sneeuberg Range, Eastern Cape......*Bitis inornata*, Plain Mountain Adder (p. 78)

17b Horns often absent in females and poorly developed in males; body commonly reddish, usually with 21–24 faint dorsal blotches (sometimes up to 30), rarely plain; usually over 35cm maximum; widely distributed in the western and southern part of the Northern Cape, South Africa......*Bitis rubida*, Red Adder (p. 77)

Puff Adder *Bitis arietans*

Identification: Africa's largest viper. It occurs throughout the savanna and is a fat snake, with a broad, flat triangular head, a pale line between the eyes, and V-shapes on the back. The head is covered in small, overlapping, strongly keeled scales. The small eye, with a vertical pupil, is set far forward, and the nostrils are upturned. The neck is thin, the body fat and depressed, and the tail fairly short, 10–15% of total length in males and 6–9% in females. The dorsal scales are keeled; in 27–41 rows at midbody; ventrals 123–147; subcaudals 14–39 (usually more than 25 in

Bitis arietans

males). Maximum size depends on locality; in most of Africa rarely larger than 1.2m, average 0.6–1m. Much larger examples, up to 1.8m or possibly more, occur in the dry savanna of northern Uganda, northern and eastern Kenya, Somalia and eastern Ethiopia. Hatchlings 15–23cm. The colour is variable; it may be grey, brown, rufous or yellow, usually with a series of V-shapes down the back that may morph into cross-bars. The underside is yellow or white, interspersed with short black dashes. There is usually a light line between the eyes. In highland areas, males are often brighter than females. Unusual colour morphs (snakes with vertebral stripes, no head markings, or cross-bars instead of stripes) occasionally occur.

Puff Adder, *Bitis arietans*.
Left top: Tanzania (Stephen Spawls). **Right top:** Ghana (Stephen Spawls).
Left bottom: threat display, Morocco (Konrad Mebert). **Right bottom:** Ethiopia (Stephen Spawls).

Puff Adder, *Bitis arietans*. **Left:** Kenya (Stephen Spawls). **Right:** underside (Stephen Spawls).

Habitat and Distribution: Throughout the African savannas, and also in woodland, semi-desert and near-desert, from sea level to 2,200m, sometimes higher. Absent only from high altitude above 2,400m and closed forest. It occurs from southern Mauritania east to Eritrea and Somalia, and south to South Africa. There is an isolated population in Western Sahara and southern Morocco, and an extralimital one in the south-west Arabian Peninsula.

Natural History: Terrestrial, but may climb low bushes or trees. It is usually nocturnal, though is sometimes active by day in the rainy season. In high-altitude areas it may bask. It hides in thick grass, under ground cover, down holes, in leaf drifts, under bushes, etc. It is often poorly concealed under small bushes or in grass tufts, but well camouflaged and thus liable to be trodden on. It hunts by ambush, waiting for prey, which is ambushed with a rapid strike. It crawls in a straight line if unhurried ('caterpillar crawl'), leaving a single broad track; if stressed it can move in the usual serpentine, side-to-side movement. When threatened, it inflates the body – hence the common name – hisses loudly, and raises and draws back the front third of the body. If a target is within range, a rapid strike follows. The withdrawal is equally rapid and the snake may overbalance, hence the incorrect and highly dangerous belief that Puff Adders only strike backwards and it is safe to stand in front of one. Captive individuals often remain bad-tempered and noisy. Males indulge in combat, neck-wrestling each other. Females produce a pheromone that attracts males. They give birth to live young. The clutch size depends on female size and altitude; usually 10–35 are born, but a Kenyan female had 156 young, a record for any vertebrate. The diet is varied and includes mammals (from rats up to small antelope), birds, amphibians and reptiles, including snakes and tortoises. Large prey animals are struck, released and followed; small prey may be simply seized and swallowed. This snake's enemies include monitor lizards, large cobras, predatory birds and small carnivorous mammals.

Medical Significance: A very dangerous snake, and its relative abundance, good camouflage and disinclination to move if approached mean that rural people are often bitten, although a recent survey indicates that it is not as willing to bite as is sometimes believed. The venom is potently cytotoxic. Symptoms include severe local pain, extensive swelling and blistering, compartment syndrome, necrosis, hypovolaemia, shock and blood coagulation abnormalities. Although rapid death from Puff Adder bites is rare, and there is usually time to get the victim to hospital, treatment there may involve antivenom therapy and be prolonged, complicated and intense. A Puff Adder bite is a medical emergency; rapid hospitalisation is mandatory. Local necrosis is common, often resulting in permanent damage and occasionally death; amputation is often necessary. There is some regional variation; in West Africa, Puff Adder bites are often characterised by spontaneous bleeding, bruising and skin haemorrhages, due to thrombocytopenia. However, there is usually no coagulopathy. In East and South Africa, in addition to bleeding, bruising and swelling, coagulopathy leading to incoagulable blood has been noted, and in rare cases cerebral thrombosis. These regional differences have led to suggestions that there is more than one species of Puff Adder. However, except in unusual circumstances (venom injection directly into a blood vessel, or anaphylactic shock), rapid death is unlikely. There is some debate about how useful antivenom is for Puff Adder bites, but the consensus is that it should be used as early as possible, particularly if local swelling is spreading rapidly or is extensive. Shock should be treated with intravenous fluid replacement and antivenom. Fasciotomy is very rarely indicated.

Rhinoceros Viper (Rhino-horned Viper) *Bitis nasicornis*

Identification: Unmistakable if seen clearly; a large, stout viper, with a narrow, flat triangular head covered in small, strongly keeled scales. On the end of the nose is a cluster of two to three pairs of horn-like scales. The small eye is set well forward, the pupil is vertical, and the iris is green or gold with black flecks. The neck is thin, the body triangular in section and the tail short, 13–18% of total length in males and 7–10% in females. The body scales are rough and heavily keeled, the keels so hard and prominent that they have been known to inflict cuts

Bitis nasicornis

on snake handlers when the snake struggles. The scales are in 30–43 rows at midbody; ventrals 117–140; cloacal scale entire; subcaudals paired, 12–32 (higher counts in males). Maximum size about 1.4m, average 0.8–1.1m; hatchlings 18–25cm. The colour pattern is complex. Down the back runs a series of 15–18 large oblong blue or blue-green markings, each with a yellow line down the centre. Irregular black rhombic blotches enclose these markings. On the flanks are a series of dark crimson triangles, sometimes narrowly bordered with blue or green. Many lateral scales are white-tipped. The top of the head is blue or green, with a vivid black arrow mark. The belly is dirty white to dull green, extensively marbled and blotched in black and grey. Specimens from the centre and west of the range tend to be more blue, and those from the east more greenish. The tongue is black with a purple base.

Habitat and Distribution: Forest, woodland and forest–savanna mosaic of West and Central Africa, from sea level to 2,400m. A true forest snake, not usually in woodland, sensitive to habitat destruction, and usually doesn't cling on in deforested areas. It occurs in West Africa, from southern Sierra Leone, Liberia and Guinea east to Ghana. Known from forests in Togo, and occurs east from Nigeria to western Uganda, and south to northern Angola. Isolated records exist from

Rhinoceros Viper, *Bitis nasicornis.*
Left top: Tanzania (Stephen Spawls). **Right top:** Uganda (Stephen Spawls).
Left bottom: Eastern DR Congo (Konrad Mebert). **Right bottom:** underside (Stephen Spawls).

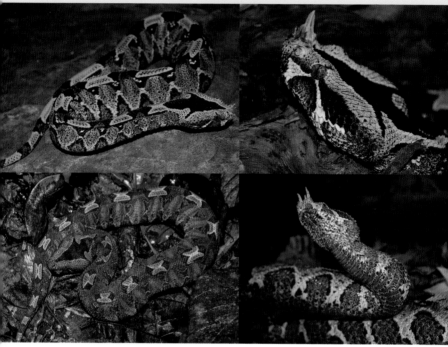

Bioko, the Imatong Mountains in Sudan, the north-western shore of Lake Victoria in Uganda and northern Tanzania (Minziro Forest), and the Kakamega area in Kenya. Dubious records from Lake Bangweulu in Zambia and the Usambara/ Uluguru Mountains in Tanzania are not shown on the map.

Natural History: Terrestrial, but regularly climbs into thickets and trees; in a series of 66 specimens from Kakamega, Kenya, over a third were taken in trees or bushes, sometimes up to 2.5m above ground level. A slow-moving snake, but strikes quickly. It swims, and has been found in shallow pools. It is nocturnal, tending to hide during the day among leaf litter, around fallen trees, in holes or among root tangles of large forest trees. It will also climb into thickets, clumps of leaves or cracks in trees. This species hunts by ambush, and probably spends most of its life motionless, waiting for a suitable prey animal to pass within striking range. It is more bad-tempered than the Gaboon vipers (*Bitis gabonica* and *B. rhinoceros*) but less so than the Puff Adder (*B. arietans*). If threatened, it can produce an astonishingly loud hiss

Rhinoceros Viper, *Bitis nasicornis.*
Below: Ghana (Barry Hughes).

– almost a shriek. The diet consists mostly of mammals (rodents and shrews) and amphibians. The females give birth to live young in March–April (start of the rainy season) in West Africa; there are few details in East Africa, but a Kakamega female had 26 young in March. From six to 38 young have been recorded.

Medical Significance: As this species is a forest snake that often climbs and spends much time motionless, bites are rare; it tends to stay out of humanity's way. But it is a dangerous snake, present through the Central African forest, with cytotoxic venom causing extensive local swelling, necrosis and coagulopathy; the lethal dose is reportedly lower than that for both the Gaboon vipers and the Puff Adder. Bite victims need to be rapidly transported to a major hospital and treated; antivenom therapy may be necessary and clotting times need to be monitored. Many of the reported cases have involved swelling and coagulopathy, but the victims recovered. However, an American hobbyist bitten by a Rhinoceros Viper died within a few hours, as a consequence of anoxic brain damage.

Eastern Gaboon Viper (Gaboon Adder) *Bitis gabonica*

Identification: Unmistakable; a fat viper of the Central and southern African forests, with a broad, flat white triangular head and a remarkable geometric pattern on its body. Between the raised nostrils is a pair of tiny horns. The small eyes, with a vertical pupil and a cream, yellow-white or orange iris, are set far forward. The tongue is black with a red tip. The body is very stout and depressed, and the tail short, 9–12% of total length in males and 5–8% in females. The scales are heavily keeled; in 35–46 rows at midbody; ventrals 124–140; subcaudals 17–33 (higher counts in males). Maximum size about 1.75m, possibly larger. Average 0.9–1.5m (the mean size of a large Nigerian collection was 1.3m), females usually larger than males; hatchlings 25–37cm. The head is white or cream with a fine, dark central line and black spots on the rear corners, and a dark blue-black triangle behind and below each eye. Along the centre of the back is a series of pale, subrectangular blotches, interspaced with dark, yellow-edged hourglass markings. On the flanks is a series of fawn or brown rhomboidal shapes, with light vertical central bars. The belly is pale, with irregular black or brown blotches. Specimens from the south-eastern area of its continuous range (Zambia, south-east DR Congo) often have a pinkish or orange cast.

Bitis gabonica

Habitat and Distribution: Coastal forest and thicket, woodland, forest–savanna mosaic, well-wooded savanna and forest, from sea level to 2,200m. It occurs eastwards from south-west

Eastern Gaboon Viper, *Bitis gabonica.*
Below: Uganda (Jelmer Groen).

Eastern Gaboon Viper, *Bitis gabonica.*
Left top: DR Congo (Colin Tilbury). **Right top:** Tanzania (Michele Menegon). **Left bottom:** Uganda (Stephen Spawls).
Right bottom: hybrid between Eastern Gaboon Viper and Rhinoceros Viper (Matthijs Kuijpers).

Nigeria to eastern DR Congo and western Uganda, south to northern Angola and Zambia (Gaboon vipers from the West African forests are another species, the Western Gaboon Viper, *Bitis rhinoceros*). Apparently isolated populations occur in South Sudan (Imatong Mountains), along the Lake Victoria shore in Uganda, in the Kakamega area in Kenya, in eastern Tanzania and northern Mozambique, in parts of Malawi, from high eastern Zimbabwe across the Mozambique plain, and on the north-eastern South African coast, into southern Mozambique.

Natural History: A slow-moving, usually placid nocturnal viper, spending much of its time motionless, hidden in leaf litter, thick vegetation, under bushes or in thickets, often near a game trail. It sometimes climbs into the understorey. Not usually in dense forest; more often in clearings, sunlit thickets and forest margins, where prey is more common. Adults largely hunt by ambush, waiting, often for a long time (a South African specimen remained motionless for 90 days), until suitable prey passes. However, juveniles actively forage, as indicated by the presence of newborn rats in their stomachs. Large prey animals are struck, released and followed; small prey may be simply seized and swallowed. Males are more active than females, especially after rain, even on cool nights. Individuals may bask in the day. Female give birth to live young; in southern Tanzania broods varied from 11 to 34, but clutches of up to 60 are known. Females do not eat while they are heavily gravid. A 35cm juvenile was taken in Kakamega Forest in September; hatchlings were found in mid-December to February in southern

Tanzania. The diet is mostly small mammals, from shrews and rodents up to African Brush-tailed Porcupines (*Atherurus africanus*) and small antelope. Birds (several Tanzanian specimens had eaten chickens), amphibians and reptiles are sometimes taken. Males fight in the mating season, neck-wrestling, one male trying to force the other's head down while they hiss continuously and strike with closed mouths. This species sometimes hybridises with the Puff Adder (*Bitis arietans*) in areas where the habitat overlaps.

Medical Significance: This is a very dangerous snake, being slow-moving and well concealed, and liable to be trodden on by those walking though vegetation. Anyone moving material on the ground is also at risk, although it is a placid snake, and often reluctant to strike. It has a deadly cytotoxic and haemorrhagic venom, although bites are rare due to its placid nature. Symptoms include spontaneous systemic bleeding, swelling, bruising, blistering, necrosis, and haemostatic and cardiovascular abnormalities (hypotension, shock, arrhythmias, ECG changes). Bites must be treated in hospital with antivenom and clinical support.

Taxonomic Notes: This snake and the Western Gaboon Viper (*Bitis rhinoceros*) were originally regarded as subspecies (*B. gabonica gabonica* and *B. g. rhinoceros*). However, they are reproductively isolated.

Western Gaboon Viper *Bitis rhinoceros*

Identification: A fat viper of the West African forests, from Guinea to Togo, with a broad, flat white triangular head and a remarkable geometric pattern on its body. Between the raised nostrils is a pair of horns. The small eyes, with a vertical pupil and a cream, yellow-white or silvery iris, are set far forward. The tongue is black with a red tip. The body is very stout and depressed, and the tail short, 9–12% of total length in males and 5–8% in females. The scales are heavily keeled; in 28–46 rows at midbody; ventrals 128–147; subcaudals 17–33

Bitis rhinoceros

(higher counts in males). Maximum size about 1.7m, possibly larger. Average 0.8–1.3m, females usually larger than males; hatchlings 25–37cm. The head is white or cream with a fine, dark central line and black spots on the rear corners, and a single dark triangle behind each eye; this is the only way this species differs externally from the Eastern Gaboon Viper (*Bitis gabonica*). Along the centre of the back is a series of pale, subrectangular blotches, interspaced with dark, yellow-edged hourglass markings. On the flanks is a series of fawn or brown rhomboidal shapes, with light vertical central bars. The belly is pale, with irregular black or brown blotches.

Habitat and Distribution: Coastal forest and thicket, woodland, forest–savanna mosaic, well-wooded savanna and forest. It occurs eastwards from western Guinea and Sierra Leone to Ghana; also in forests in south and central Togo. Further east it is replaced by the Eastern Gaboon Viper. A specimen was found on Legon Campus, just east of Accra, Ghana.

Western Gaboon Viper, *Bitis rhinoceros*. **Left:** Ivory Coast (Stephen Spawls). **Right:** Sierra Leone (Bill Branch).

Natural History: Fairly similar to the Eastern Gaboon Viper (i.e. slow-moving, nocturnal, usually motionless, hidden in leaf litter, eats mostly mammals and birds, and gives birth to live young). A fascinating study in Tai National Park in Ivory Coast made use of a habituated troop of Sooty Mangabeys (*Cercocebus atys*) to detect these snakes; the foraging monkey troop leapt aside and gave loud alarm calls when they found a viper. Vipers were located on average every 10 days, and it was estimated there was roughly one snake per 20 hectares (i.e. five per square kilometre).

Medical Significance: This is a very dangerous snake, with a deadly cytotoxic and haemorrhagic venom, although bites are rare due to its placid nature. Risk factors are as for the previous species. Symptoms include spontaneous systemic bleeding, swelling, bruising, blistering, necrosis, and haemostatic and cardiovascular abnormalities (hypotension, shock, arrhythmias, ECG changes). Bites must be treated in hospital with antivenom and clinical support.

Ethiopian Mountain Viper *Bitis parviocula*

Identification: A fat viper endemic to the forests of southern Ethiopia. The small eyes, with a vertical pupil and a yellow iris, are set fairly far forward. The tongue is deep blue and black. The body is very stout and depressed, and the tail short. The scales are heavily keeled; in 37–39 rows at midbody; ventrals 141–146; subcaudals 20–21. Maximum size about 1.3m, possibly larger. Average 0.7–1m; hatchlings 17–24cm. The head is broad and yellow, with a dark triangle between the eyes and a vague hammer shape behind this. The ground colour is yellowish or

Bitis parviocula

greeny-grey, often darkening posteriorly. Along the centre of the back is a series of black hexagons, interspersed with yellow butterfly shapes, with white-spotted black subtriangular shapes below. On the flanks is a series of upward-pointing, pale-edged dark triangles. The underside is greenish or yellow, with irregular black or brown blotches.

Habitat and Distribution: Known from broadleaved forest, coffee plantations and deforested areas of southern Ethiopia. There are few formal records. West of the Ethiopian Rift Valley it is probably widespread in the area between the towns of Bedelle, Metu, Bonga and Mizan Teferi, as specimens are known from all those towns and a fair amount of forest remains; it is also known from Sheka Forest Biosphere Reserve, north of Mizan Teferi. It probably occurs in the extensive forest west of the Metu–Mizan Teferi road. East of the Ethiopian Rift Valley, a single enigmatic museum specimen is recorded from Dodolla, at 2,400m on the northern side of the Bale Mountains; the town itself is in high grassland, but suitable forest exists not far south of there. Local people in both the town of Wondo Genet (west of Dodolla) and the Harenna Forest of the Bale Mountains claimed to recognise this snake from photographs.

Natural History: Poorly known, but presumably similar to the other large forest *Bitis*; that is, slow-moving, nocturnal, usually motionless, hidden in leaf litter, and eats mostly mammals and birds. Captive specimens seem fairly placid, and resemble Gaboon vipers (*Bitis gabonica* and *B. rhinoceros*) in temperament, although a restrained juvenile hissed and struggled furiously. Farmers in the forest east of Metu indicate that this snake is most common in valley bottoms and often shelters in cracks and recesses in large trees. A number of specimens in the pet trade were collected while coffee plantations were being cleared and burnt; in Ethiopia, coffee is grown as an understorey tree within forest. A male and female

Ethiopian Mountain Viper, *Bitis parviocula*.
Left top: Ethiopia (Tomáš Mazuch). **Right top:** juvenile (Stephen Spawls).
Left bottom and right bottom: captive (Stephen Spawls).

were killed near dusk just west of Bonga on a road in September; they may have been mating. Captive specimens gave birth to 11–16 offspring in May; this is near the start of the main rainy season at Bonga. Smaller and larger clutches, between 6 and 21, are informally reported, in August and September. Captive animals feed readily on rodents and chicks.

Medical Significance: Although the danger to humans posed by this snake is slight, due to its geographically very restricted range in medium- to high-altitude forest, it is a large viper, closely related to other large African vipers, and its venom is likely to be highly toxic. The people living in the Metu area, where the type specimen was collected, state that there are several fatalities every year from bites from this snake. The lethal dose is only slightly greater than that for Puff Adder (*Bitis arietans*) venom, and the venom has a haemorrhagic effect. Although no specific antivenom is produced for this snake, the polyvalent antivenom produced by South African Vaccine Producers (SAVP) is reported as effective against its venom. Bites are fairly unlikely, but treatment as for a Puff Adder bite should be effective; a bite by a captive specimen in the USA caused only local envenoming.

Bale Mountains Viper *Bitis harenna*

Identification: A fat black and yellow viper known (so far) only from a small area of south-central Ethiopia, east of the Gregory Rift Valley. Described in 2016; only a small handful of specimens are deposited in museums. The small eyes, with a vertical pupil and a dark iris, are set fairly far forward. The body is stout and depressed, and the tail short. The scales are heavily keeled; in 38–39 rows at midbody; ventrals 145; subcaudals 20. One museum specimen is 66cm long, but uncollected specimens of about 1m have been observed; hatchlings are probably around 15–

Bitis harenna

20cm. The head is black or blue-black, edged with a yellow arrowhead shape, and there is a yellow line across the head behind the eyes. The body is black, with a fine reticulate network of yellow or cream lines; the underside is cream or yellow, marked with dark subrectangular blotches.

Habitat and Distribution: Several living specimens have been observed at the edges of coniferous and broadleaved forest, so presumably it favours clearings in forest. The type specimen is labelled from Dodolla, a town in the high grasslands in the northern foothills of the Bale Mountains, Ethiopia, at 2,400m, but it may have been transported there. The living animals have all been observed in the Bale Mountains National Park, along the road that descends from the Sanetti Plateau southwards towards Dolo Mena. All were within the forest between 2,100m and 2,400m. It probably occurs throughout the forest that extends 10–30km on either side of this road, and may extend further into the forest patches to the west. Local people at Dodolla and also in the town of Wondo Genet (more than 100km to the west, on a forested hill at 2,400m) claimed to recognise this snake from photographs.

Bale Mountains Viper, *Bitis harenna*.
Left and centre: Ethiopia (Jonas Arvidsson). **Right:** Ethiopia (Evan Buechley).

Natural History: Largely unknown, but presumably similar to the other large forest *Bitis*; that is, slow-moving, usually motionless, hidden in leaf litter, eats mostly mammals and birds, and gives birth to live young. Observers have remarked on its physical similarity to Puff Adders (*Bitis arietans*) of the same size. Its habitat is wet and relatively cold, and for several months of the year receives little sunlight, so the species may be cold-adapted, or perhaps is quick to utilise opportunities to bask. It may be selectively diurnal; specimens have been observed to be active by day.

Medical Significance: The danger to humans posed by this snake is very slight, due to its geographically very restricted range in medium- to high-altitude forest where there are relatively few people. However, it is a large viper and probably closely related to other large African vipers, so the venom is likely to be a potent cytotoxin. Although no specific antivenom is produced for this snake, the polyvalent antivenom produced by South African Vaccine Producers (SAVP) may be effective, as it has been shown to be effective against the bite of the related Ethiopian Mountain Viper (*Bitis parviocula*). Treatment should be as for a Puff Adder bite (see p. 58).

Horned Adder *Bitis caudalis*

Identification: A small, attractive adder of the dry country of southern Africa, with a single horn above each eye. Very rarely, hornless specimens may be found. The head is triangular, the body stout and slightly depressed, and the tail short, 7–8% of total length in females and 8–12.5% in males. The head and body scales are strongly keeled; in 21–33 (usually 23–31) rows at midbody; ventrals 120–155; cloacal scale entire; subcaudals 16–40 (higher counts usually males). Maximum length around 60cm, average 25–45cm; hatchlings 10–15cm. The colour

Bitis caudalis

is tremendously variable, usually shades of brown, grey or rufous, but the pattern typically consists of a series of darker subrectangular patches down the spine on a lighter background, and a line of subcircular blotches with lighter centres along the flanks. There is sometimes a vague arrow- or hammer-shaped mark on the back of the head. The underside is white or cream, with dark markings on the chin and throat.

Horned Adder, *Bitis caudalis*.
Left top: Botswana (Stephen Spawls). **Right top and left bottom:** Namibia (Bill Branch).
Right bottom: Zimbabwe (Stephen Spawls).

Habitat and Distribution: In desert, semi-desert and dry or occasionally moist savanna, from sea level to over 1,600m on dry mountains. It has the largest range of any of the small *Bitis* species, extending south from south-western Angola almost throughout Namibia and the southern two-thirds of Botswana, down the western side of South Africa, south to the Little Karoo, and also in north-east South Africa, into Zimbabwe extending north past Bulawayo to Shangani, and from the border north to Masvingo. There is a curious record from Luanda, north-west Angola.

Natural History: A fairly slow-moving terrestrial snake, although it can strike quickly. It can also sidewind, moving rapidly across sand. It is active by night;

Horned Adder, *Bitis caudalis*. **Left:** Botswana (Stephen Spawls). **Right:** Namibia (Bill Branch).

by day it conceals itself in sand, shuffling down until it is buried, either totally or with the top of the head and eyes exposed, or it may hide in shade under bushes, or up against or under ground cover, logs and rocks, etc. If approached it may flatten itself, but if disturbed it hisses loudly and strikes repeatedly. It gives birth to live young, usually 3–18 but up to 27 have been recorded. It feeds on small vertebrates, largely lizards, but occasionally rodents and frogs, which it ambushes during the day or actively hunts by night. It has been seen to lure prey animals by wriggling its tail tip, which resembles a centipede. Its relative abundance varies dramatically; 134 were collected in five months in south-west Botswana, while in Bulawayo only three were found in 25 years.

Medical Significance: This snake is widespread, well able to conceal itself, and active on the ground at night, so it poses some threat to humans, particularly in the stock farming areas of dry southern Africa. Many people in south-western Botswana had either been bitten or knew somebody who had been bitten by this snake, often while going outside to the toilet in the night during the summer months. It has a cytotoxic venom. However, few bites have been formally documented and no human fatalities are known. Bite symptoms include local swelling with intense pain, vomiting and lymphangitis. Local necrosis, almost certainly caused by bad clinical management, has been recorded. No antivenom is produced, and no known antivenom gives protection against the venom of this species. Bites should be treated conservatively by elevation and painkillers; consider antibiotics if infection is present.

Namaqua Dwarf Adder *Bitis schneideri*

Identification: A very small, stocky adder of the dry coastal strip of western South Africa and southern Namibia. The head is short, flat and triangular, sometimes with a raised scale above the eye like a blunt horn. The body is stout and slightly depressed, and the tail short, 7–10% of total length in females and 10–12.5% in males. The head and body scales are strongly keeled; in 21–27 rows at midbody; ventrals 104–129; cloacal scale entire; subcaudals 17–27 (higher counts usually males). Maximum length around 28cm, average 18–25cm; hatchlings 11–13cm. The colour often

Bitis schneideri

matches the sand of its habitat, being greyer in the south and more reddish in the north. The pattern consists of three rows of darker, light-centred spots, but these may be faded so the snake just looks mottled light grey. The underside is grey to dirty cream, speckled and spotted to varying extent. In some specimens the tail tip is black.

Habitat and Distribution: Occurs in semi-stable vegetated sandy regions of the southern and transitional Namib Desert, where it is a close ecological equivalent of Péringuey's Adder (*Bitis peringueyi*), but lives in more stable, vegetated sand dunes, between sea level and 500m. It is found in coastal regions from Papendorp and the mouth of the Olifants River in the Western Cape, South Africa, north to Luderitz in southern Namibia. It may extend up to 60km inland in the Sperrgebiet (diamond concession) of southern Namibia.

Namaqua Dwarf Adder, *Bitis schneideri*.
Left: South Africa (Tyrone Ping). **Centre and right:** South Africa (Bill Branch).

Natural History: A fairly slow-moving terrestrial snake, although it can strike quickly. It sidewinds exceptionally well, moving rapidly across sand. It is crepuscular and nocturnal, active during early evening, at night and sometimes just after dawn. By day it conceals itself in sand, shuffling down until it is buried, either totally or with the top of the head and eyes exposed. Little is known of its reproduction. Broods are small (4–7 young). Lizards, particularly diurnal skinks and lacertids and nocturnal geckoes, form the main part of its diet, although small rodents and amphibians may also be taken. The tail is sometimes exposed and waved as a lure to attract prey.

Medical Significance: Fairly low, being a very small, secretive snake living in a sparsely inhabited sandy area, although bites have been suffered by barefoot beachgoers. No fatalities are known. The venom appears to be cytotoxic. No specific antivenom is produced, and no known antivenom gives protection against the venom of this species. Symptoms of documented bites vary considerably, from intense pain to no pain at all, from no swelling to swelling of the entire bitten area, and both local sensitivity and local numbness. In one case involving swelling there was also discolouration and oozing of fluid from the punctures and, after 24 hours, a 5mm haematoma developed at the bite site. Recovery was complete. Bites should be treated conservatively by elevation and painkillers; consider antibiotics if infection is present.

Péringuey's Adder *Bitis peringueyi*

Identification: A very small, flat adder with eyes on the top of its head, which inhabits the dry coastal strip of Namibia. In general appearance it resembles the Sahara Sand Viper (*Cerastes vipera*), which has also evolved to hunt concealed from sand. The head is flat and triangular, the neck thin, the body stout and depressed, and the tail very short, 6–7% of total length in females and 7–9% in males. The head and body scales are strongly keeled (except the lowest dorsal rows); in 22–31 rows at midbody; ventrals 117–144; cloacal scale entire; subcaudals

Bitis peringueyi

15–30 (higher counts usually males). Maximum length around 33cm, average 22–27cm; hatchlings 10–13cm. The colour often matches the sand of its habitat, being sandy-grey to yellow-grey, brown or orange-brown. The pattern consists of three irregular rows of darker spots, sometimes with light edging. The underside is white, uniform or spotted. In about 25% of specimens the tail tip is black, both above and below.

Habitat and Distribution: Lives in sparsely vegetated, loose, wind-blown dunes of the Namib Desert, and thus doesn't extend inland much beyond 150km or so; found from sea level to around 1,000m. It occurs from the vicinity of Luderitz in southern Namibia north to the border with Angola, and might occur where the Namib Desert extends into south-west Angola in the Parque Nacional do Iona, although there are no definite Angolan records.

Natural History: A terrestrial snake, fairly slow-moving when crawling (although it can strike quickly), but it sidewinds exceptionally well, moving rapidly across sand and leaving a series of parallel tracks; it has been seen to climb 45° slopes by sidewinding. Unusually, it is active at any time of the day or night, depending upon the weather, temperature and season. It largely hunts from ambush, shuffling down into the sand until concealed, with only the eyes, top of the head and tail tip exposed; in this position it is hidden from prey and isolated from surface heat. It gives birth between February and April to 2–10 young. It feeds largely on lizards, in particular the Shovel-snouted Lizard (*Meroles anchietae*) and barking geckoes, sometimes waving the tail as a lure to attract prey. In a region of low, infrequent rainfall, the snake obtains most of its water from the food it eats. However, coastal fogs frequently extend 60km into the Namib and provide additional moisture. The snake flattens its body to increase the surface area, and sips droplets of condensing fog from its scales.

Medical Significance: Very low, being a very small, secretive snake living in an almost uninhabited sandy area. At risk are those who wear open-toed footwear or go barefoot. The venom is cytotoxic. No antivenom is produced, and no known

Péringuey's Adder, *Bitis peringueyi*. **Below:** Namibia (Johan Marais).

antivenom gives protection against the venom of this species. Symptoms of documented bite cases included local pain, swelling and lymphangitis. A bite to a child in Namibia caused destruction of the ciliary ganglion, causing permanently dilated pupils (mydriasis). Bites should be treated conservatively by painkillers; consider antibiotics if infection is present.

Berg Adder *Bitis atropos*

Identification: A small, variable, attractive adder of the rocky hinterland, hills and mountains of the southern and eastern side of South Africa, eastern Zimbabwe and far western montane Mozambique. The head is triangular but elongate, the body stout and slightly depressed, and the tail short, 7–10% of total length in females and 10–12% in males. The head and body scales are strongly keeled; in 27–33 rows at midbody; ventrals 118–144; cloacal scale entire; subcaudals 15–31 (higher counts usually males). Maximum length around 60cm, average 35–50cm; hatchlings 10–15cm.

Bitis atropos

The colour and pattern vary regionally. The ground colour is usually grey, brown or rufous. In several forms there is a dorsolateral series of triangular black markings down either side. These are edged below by a white or yellowish line, below which is another series of smaller dark spots. Between these spots are dark, Y-shaped blotches. The head bears a dark arrow-shaped mark on the crown, and two pale streaks on either side. The chin and throat are usually flesh pink or yellowish, while the ventrum is off-white with dusky infusions, or occasionally slate grey to black. The latter is particularly evident from high-altitude populations, where snowfalls are common (e.g. Swartberg in the Western Cape, and the Drakensberg in Kwazulu-Natal). Some specimens, particularly in the Mpumalanga Drakensberg, are a drab brown, while some Zimbabwe animals have vivid rufous markings.

Habitat and Distribution: Occupies a variety of habitats, but prefers cool climates with high rainfall. It is thus restricted to higher elevations in the northern parts of its range, but may occur on small rock outcrops at sea level in the Western Cape; so it ranges from sea level to over 3,000m. In the Western Cape it lives in coastal and mountain fynbos (heathland), where it experiences a Mediterranean climate (cold, wet winters and warm, dry summers). Further north it is found in montane and coarse grassland with cold but dry winters and warm, rainy summers. Its distribution is startlingly disjunct. In Zimbabwe it is found in both the Nyanga and Chimanimani Mountains, being more common in the latter, and also extends onto the Mozambique side of the mountains. It reappears close to 600km further south, as separate populations in the Mpumalanga and Limpopo Escarpment, and the Maloti-Drakensberg of Lesotho and parts of Kwazulu-Natal, Free State and Eastern Cape adjoining the Lesotho border. Further south this snake occurs in the Cape Fold Mountains of the Eastern Cape and Western Cape.

Natural History: A terrestrial snake, but it climbs rocks. It might be nocturnal, but in the montane regions of its range this small adder can often be found basking on rock ledges and in open patches, presenting a risk to mountaineers and hill

Berg Adder, *Bitis atropos*.
Left top: South Africa (Bill Branch). **Right top:** Zimbabwe (Chris Kelly).
Left bottom: captive (Stephen Spawls). **Right bottom:** South Africa (Bill Branch).

walkers. It is an irascible snake when disturbed, hissing loudly and striking actively, a temperament not suited to, or attenuated by, captivity. Between 8 and 16 young are born to large females in the Western Cape, while litters range from 5 to 11 further north; birth takes place in late summer. The diet is varied, including small birds, rodents, shrews, lizards and even other snakes. Juveniles and adults are also fond of amphibians, particularly rain frogs and small toads.

Medical Significance: This snake is quite common in parts of its range, so it presents a threat to those whose work or pleasure takes them into the hills. The venom is unusual among vipers in containing a major neurotoxin, which is not neutralised by antivenom prepared against other snakes, including elapids. Its toxicity is similar to that of Horned Adder (*Bitis caudalis*) venom, and it is produced in similar amounts. No specific antivenom is available. Many bites are followed by a progressive ophthalmoplegia, including ocular paralysis, ptosis (drooping eyelids) and fixed dilated pupils. Associated neurotoxic symptoms of nausea, vomiting, vertigo, difficulty in breathing and the loss of taste and smell may be experienced. Respiratory paralysis has been recorded. Local symptoms include severe pain and rapidly spreading swelling. In some cases these alone occur. Tissue necrosis is rare, and the few examples have been localised to the bite site and may have been exacerbated by poor first aid. Neurological effects are usually transient, but may take several weeks to resolve; one victim spent several days on a ventilator, while another two had dilated pupils for over a year following the bite. No deaths are reliably recorded, but bites are serious and should be treated in hospital. In a study of 14 bite cases in South Africa, all 14 patients developed local cytotoxic effects and 13 of them developed systemic effects to some degree. These included prominent vomiting, disturbances in cranial nerve function that included loss of accommodation, a global decrease in motor power needing mechanical ventilation for respiratory failure, and hyponatraemia, sometimes with associated convulsions. This is an underrated dangerous snake.

Many-horned Adder (Hornsman) *Bitis cornuta*

Identification: A small, spiky-horned adder of the western coast of South Africa and extreme southern Namibia (populations further east have now been moved into other taxa). The head is short, flat and triangular, and characterised by spiky tufts of 2–7 elongate horns over the eyes. The body is stout and slightly depressed, and the tail short, 7–9% of total length in females and 10–12.5% in males. The head and body scales are strongly keeled; in 23–29 rows at midbody; ventrals 120–152; cloacal scale entire; subcaudals 18–37 (higher counts usually

Bitis cornuta

males). Maximum length around 55cm (although a captive female reached 75cm), average 30–50cm; hatchlings 13–16cm. The ground colour is grey, brown or rufous, with a double series of darker subquadrangular blotches down the centre of the back and a single series of similar blotches on each flank. There is sometimes a poorly defined grey dorsolateral stripe. There is a vague dark arrow-shape on the head, and often two lighter bars on the side of the head, behind and in front of the eye.

Habitat and Distribution: Lives in rocky desert, hills, rocky outcrops and gravel flatlands of the coastal plain, in the Namib Desert, into semi-desert (Karoo), and extending south into montane fynbos heathland. It occurs from Meob Bay and Mount Uri-Hauchab in Namibia south to Lambert's Bay, South Africa, and inland to the Cederberg, western South Africa, from sea level to 800m.

Natural History: A fairly slow-moving terrestrial snake, although it can strike quickly. It appears to be crepuscular. It is usually found in rocky situations, although it can sidewind. On rocky hills it is usually restricted to scattered boulders on the upper slopes. It often shelters in rodent burrows or under stones. Nervous by disposition, it will hiss loudly and strike so energetically that most of the body is lifted off the ground. Mating in the Western Cape occurs in October–November, but it is earlier (May–June) in southern Namibia. From five

Many-horned Adder, *Bitis cornuta*.
Left: South Africa (Tyrone Ping). **Right:** South Africa (Bill Branch).

to 12 young are born in late summer or early autumn (January–April). It eats small mammals and lizards (skinks, agamas and lacertids); birds are occasionally ambushed from cover.

Medical Significance: Fairly low, being a very small, secretive snake living in a relatively small and sparsely populated area. The venom is probably cytotoxic. No antivenom is produced, and no known antivenom gives protection against the venom of this species. No documented bite cases are known, but anecdotal evidence from pet keepers indicates that bites are characterised by pain and local swelling. Bites should be treated conservatively by painkillers; consider antibiotics if infection is present.

Taxonomic Notes: The following four species, all associated with the Cape Fold Mountains in southern South Africa, have at some time in the past each been regarded as subspecies or synonyms of this snake; a 1999 reassessment indicated that they are all evolving separately and are thus good evolutionary species, and are largely allopatric (living in separate habitats), although odd problematic intermediate specimens are known.

Albany Adder *Bitis albanica*

Identification: A small, very rare adder with raised 'eyebrows' and sometimes with horns, and a minute range near Port Elizabeth in South Africa. The head is short, flat and triangular, and characterised by raised supraocular scales or short, spiky tufts of elongate horns over the eyes. The body is stout and slightly depressed, and the tail short, 8–9% of total length in females and 9.5–12% in males. The head and body scales are strongly keeled; in 27–29 rows at midbody; ventrals 120–138; cloacal scale entire; subcaudals 21–27 (higher counts usually males). Maximum length

Bitis albanica

around 30cm, average 20–26cm; hatchling size unknown. The ground colour is grey, with a double series of darker subquadrangular blotches down the centre of the back, a series of lighter flank blotches and sometimes a poorly defined grey dorsolateral stripe. There is a dark arrow shape on the head. The underside is cream-grey with darker blotching.

Habitat and Distribution: Lives in the succulent thickets and grassland of the Mediterranean-type vegetation of South Africa's southern coast, locally known as 'bontveld vegetation', on limestone and calcareous ancient dunes. It is endemic to the Eastern Cape,

Albany Adder, *Bitis albanica*.
Below: South Africa (Luke Kemp).

South Africa, and found from Port Elizabeth north-east past Grahamstown to Committees Drift, between sea level and 500m.

Natural History: A rare, fairly slow-moving terrestrial snake. Little is known of its habits. One specimen was found crossing a road in the daytime; others were concealed in thickets. It is presumably similar to other species previously believed to be *Bitis cornuta* (i.e. hunts from ambush and gives birth to live young). Lizard scales and rodent fur were found in the gut of two museum specimens.

Medical Significance: Insignificant, being a small snake with a tiny range. The venom is probably cytotoxic. No antivenom is produced; no documented bite cases are known. Bites should be treated conservatively by elevation and painkillers; consider antibiotics if infection is present.

Southern Adder *Bitis armata*

Identification: A small adder with small horns above the eye, and with a small range around Cape Town, South Africa. The head is short, flat and triangular, and characterised by short, spiky horns over the eyes. The body is stout and slightly depressed, and the tail is short, 8–9% of total length in females and 9.5–11% in males. The head and body scales are strongly keeled; in 25–29 rows at midbody (usually 27); ventrals 115–128; cloacal scale entire; subcaudals 19–31 (higher counts usually males). Maximum length around 42cm, average 25–35cm;

Bitis armata

hatchlings 12–15cm. The ground colour is grey or brown, with a double series of darker, often rufous subquadrangular blotches down the centre of the back, a series of similarly coloured but narrower flank blotches, and sometimes a poorly defined grey dorsolateral stripe or series of lozenges. There is a dark, sometimes arrow-shaped mark on the head. The underside is grey-white with darker edging.

Habitat and Distribution: Lives in thickets and grassland of the Mediterranean-type vegetation, on underlying limestone, of South Africa's southern coast.

Southern Adder, *Bitis armata*, **Left:** South Africa (Johan Marais). **Right:** South Africa (Tyrone Ping).

It is endemic to the Western Cape, South Africa, and found in isolated and suitable habitats from Langebaan Lagoon eastwards to De Hoop Nature Reserve and the Breede River mouth, usually at low altitude, but one record was from around 300m.

Natural History: A rare, fairly slow-moving terrestrial snake. Little is known of its habits. It is said to be crepuscular, and shelters under limestone rock slabs and within thickets. Females give birth to live young; up to seven are born in the late summer. It eats lizards and small rodents.

Medical Significance: Insignificant, being a small snake with a tiny range. The venom is probably cytotoxic. No antivenom is produced; no documented bite cases are known. Bites should be treated conservatively by elevation and painkillers; consider antibiotics if infection is present.

Red Adder *Bitis rubida*

Identification: A small, often rufous adder with horns or raised 'brows' above the eye, and with a small range in the Western Cape and environs, South Africa. The head is short, flat and triangular, sometimes with very short, spiky horns over the eyes. The body is stout and slightly depressed, and the tail short, 8–9% of total length in females and 9.5–11% in males. The head and body scales are strongly keeled; in 25–29 rows at midbody (usually 29); ventrals 126–143; cloacal scale entire; subcaudals 22–35 (higher counts usually males).

Bitis rubida

Maximum length around 48cm, average 25–40cm; hatchlings 12–16cm. The ground colour is rufous, grey or reddish brown, with a double series of darker brown subquadrangular blotches down the centre of the back. The flanks are mottled with irregular darker blotches. Some animals (especially rufous specimens) often have very subdued or near-invisible dorsal markings. There is a darker arrow-shaped mark on the head in some individuals. The underside is cream to grey with darker grey-white infusions.

Red Adder, *Bitis rubida*. **Left and right:** South Africa (Bill Branch).

Red Adder, *Bitis rubida*. **Below:** South Africa (Bill Branch).

Habitat and Distribution: Most records are from rocky hill and mountain slopes, but also on sparsely vegetated gravel plains, between 300m and 1,400m, in the mountain heathland (fynbos) and the dry habitat region known as Succulent Karoo. It is endemic to the Western Cape and adjacent southern areas of the Northern Cape; it occurs from the Cederberg Mountains southwards through the Cape Fold Mountains around Ceres, then eastwards through the Little Karoo to Willowmore.

Natural History: A very rare, fairly slow-moving terrestrial snake. Little is known of its habits. It is said to be crepuscular, or possibly nocturnal in the summer. It shelters under rocks and in the lee of bushes. Females give birth to live young; up to 11 are born in the late summer. It eats lizards and small rodents.

Medical Significance: Insignificant, being a small snake with a tiny range. The venom is probably cytotoxic. No antivenom is produced; no documented bite cases are known. Bites should be treated conservatively by elevation and painkillers; consider antibiotics if infection is present.

Plain Mountain Adder *Bitis inornata*

Identification: A small, dull brown adder with very subdued markings, and with a tiny range in the Sneeuberg Range and vicinity, South Africa. The head is short, flat and triangular, sometimes with very small horns over the eyes. The body is stout and slightly depressed, and the tail short, 9–10% of total length. The head and body scales are strongly keeled; in 27–30 rows at midbody; ventrals 126–138; cloacal scale entire; subcaudals 21–33 (higher counts usually males). Maximum length around 33cm, average 25–30cm; hatchlings 12–15cm. The ground

Bitis inornata

colour is dull brown to yellow or rufous-brown, without markings or with 19–24 very subdued or near-invisible dorsal blotches. The underside is cream to light brown, speckled to a greater or lesser extent.

Habitat and Distribution: High-altitude grassland, between 1,600m and 1,800m. It is endemic to the Eastern Cape, in the Sneeuberg Range and adjacent mountains, from north of Graaff-Reinet eastwards to the Mountain Zebra National Park, near Cradock.

Natural History: A very rare terrestrial snake; there was a lapse of 140 years between Andrew Smith describing the first specimen and the second one being found in

Plain Mountain Adder, *Bitis inornata*. **Left and right:** South Africa (Bill Branch).

the Mountain Zebra National Park. It is fairly slow-moving, but hisses and strikes if disturbed. It is crepuscular, being active in the evenings and mornings during the summer months; it shelters in grass tussocks, in thickets and under rocks. It hibernates during the winter when its high-altitude habitat is affected by frost and snow. It gives birth to 5–8 live young. The diet is known to include small lizards, and it probably also takes rodents.

Medical Significance: Insignificant, being a small snake with a tiny range. The venom is probably cytotoxic. No antivenom is produced; no documented bite cases are known. Bites should be treated conservatively by elevation and painkillers; consider antibiotics if infection is present.

Desert Mountain Adder *Bitis xeropaga*

Identification: A small, slender adder with geometric markings, found in the dry hills of the Orange River basin along the southern Namibian–South African border. The head is pear-shaped, the body relatively slim (for an adder) and the tail short, 8–10% of total length. The dorsal scales are moderately keeled; in 25–27 rows at midbody; ventrals 147–155; subcaudals 22–33. Whether sexual dimorphism occurs is uncertain as very few females have been collected. Maximum length 61cm, average 35–55cm; captive hatchlings reported to be 10–12cm. The

Bitis xeropaga

body is ash to dark grey, occasionally light tan, with a series of 16–34 transverse, often very square-edged, dark bars, flanked on either side by whitish spots. A pale triangular mark completes the bar adjacent to the ventrals. The top of the head is vaguely speckled. The chin and throat are white, and the underside light grey to dusky, with darker speckling.

Habitat and Distribution: Inhabits sparsely vegetated rocky mountain slopes and dry plains, in semi-desert and desert, between 100m and 800m. It is found in southern Namibia and western South Africa, where it is mainly restricted to the arid mountains bordering the lower reaches of the Orange River basin, from

Desert Mountain Adder, *Bitis xeropaga*.
Left: South Africa (Tyrone Ping). **Centre:** South Africa (Wolfgang Wuster). **Right:** South Africa (Luke Kemp).

Dreikammberg in south-western Namibia and the Richtersveld in South Africa, east to Augrabies Falls in the north-western Northern Cape. It also extends along the lower reaches of the Fish River Canyon in Namibia. There is an apparently isolated population on the mountains around Aus and Kuibis.

Natural History: A rare terrestrial snake, and fairly slow-moving, but strikes quickly. Little is known of its habits. It is likely to be nocturnal or crepuscular. It presumably shelters under rocks, although captive specimens often lie out on their cage rocks. It gives birth to 4–5 live young in the late summer (captive animals have produced broods of up to nine). Lizards and small rodents are taken readily in captivity and probably form the main part of the diet in the wild. Water is readily taken from foliage or pools.

Medical Significance: Insignificant, being a small snake with a tiny range. The venom is probably cytotoxic. No antivenom is produced; no documented bite cases are known. Bites should be treated conservatively by elevation and painkillers; consider antibiotics if infection is present.

Angolan Adder *Bitis heraldica*

Identification: A small, attractive, spotted adder, endemic to the high central plateau of Angola. The head is somewhat lumpy and triangular, with fairly prominent eyes. The top of the head is covered with small overlapping, keeled scales. The body is stout and slightly depressed, and the tail short, 9–12% of total length. The head and body scales are strongly keeled; in 27–31 rows at midbody; ventrals 124–131; cloacal scale entire; subcaudals 18–25 (higher counts usually males). Maximum length around 41cm; average size unknown, but probably 25–35cm. The

Bitis heraldica

ground colour is a rich brown or grey, with a series of 25–40 roundish to rhombic chestnut-brown spots along the centre of the back, with rufous between them (giving the appearance of a vertebral stripe), bordered on each flank by a series of irregular dark spots, and below them a series of upward-pointing orangey chevrons. The rear of the head is darkly blotched, with what looks like a three-pronged trident between the eyes. The belly is creamy- or orangey-white, heavily marked with dark grey spots that predominate on the anterior borders of the

Angolan Adder, *Bitis heraldica*.
Left top, right top and right bottom: Angola (Dayne Brayne).

scales. The underside of the tail is orange or yellowish white, with or without grey spots.

Habitat and Distribution: Lives on the high plateau of south-central Angola, at altitudes over 1,000m, reportedly from rocky montane and grassland areas. It is known from only a handful of localities in Huila, Bíe and Huambo Provinces; records from the north-west are now considered doubtful.

Natural History: A rare terrestrial snake, known from just a handful of specimens. Little is known of its habits. It is likely to be nocturnal or crepuscular, and its habits are presumably like those of other small vipers (i.e. it shelters under ground cover, gives birth to live young, hunts from ambush and eats small vertebrates).

Medical Significance: Insignificant, being a small snake with a tiny range. The venom is probably cytotoxic. A 45-year-old male who was bitten on the ankle while walking in the early morning experienced localised pain, which extended across the leg, as well as considerable swelling of the afflicted area and fever; he was unable to walk for four days. Symptoms were intense for a week but resolved after a month without complications. No necrosis was observed. No antivenom is produced. Bites should be treated conservatively by elevation and painkillers; consider antibiotics if infection is present.

Kenya Horned Viper *Bitis worthingtoni*

Identification: A small, stout horned viper, with a broad, flat, triangular head, covered in small overlapping, strongly keeled scales. The small eye, with a vertical pupil, is set far forward; the iris is silver, flecked with black. There is a single horn on a raised 'eyebrow'. The neck is thin, the body stout, and the tail thin and short. The scales are rough and heavily keeled; in 27–29 (occasionally 31) rows at midbody; ventrals 135–146; 19–33 undivided subcaudals. Maximum size about 50cm (possibly larger), average 20–35cm; hatchlings 10–12cm. The

Bitis worthingtoni

Kenya Horned Viper, *Bitis worthingtoni*. **Left:** Kenya (Stephen Spawls). **Centre and right:** Kenya (Daniel Hollands).

ground colour is grey, along each flank is a dirty white or cream dorsolateral stripe, and above and below this, on each side, is a series of semicircular, triangular or square black markings. There is a dark arrow on top of the head. The belly is dirty white, heavily stippled with grey.

Habitat and Distribution: A Kenyan endemic, found in the high grassland and scrub of the Gregory Rift Valley in Kenya. It favours broken rocky country and scrub-covered hill slopes along the edge of the escarpment, right up to the forest edge, but has also been found on the valley floor, and at the edges of acacia woodland. It is restricted to high altitudes (usually over 1,500m, but up to 2,400m) along the high central Gregory Rift Valley. The southernmost record is from the north-west Kedong Valley, from where it extends north along the floor and eastern wall of the rift valley through Naivasha and Elmenteita to Njoro. It then extends up the western wall and out of the rift to Kipkabus and Eldoret (most northerly record). It occurs on the Kinangop Plateau, around Kijabe, on the hills west of Lake Naivasha, and probably on the eastern Mau Escarpment. It might be more widespread, such as westwards towards Narok.

Natural History: Poorly known. It is terrestrial and slow-moving, but can strike quickly. It is mostly nocturnal, prowling after dusk, but will strike from ambush at any time. This snake is often found sheltering in leaf litter among the stems of the Leleshwa shrub (*Tarchonanthus camphoratus*), which grows mainly along the lower rocky slopes of the escarpment edge, but it may also hide under rocks and logs, etc. Some specimens are bad-tempered when disturbed, constantly hissing and puffing, and struggling wildly when restrained, but others seem more placid; the Maasai people who know the snake say it is reluctant to strike. Captive specimens feed readily on rodents and lizards, which are struck from ambush or stalked. It gives birth to 7–12 live young in March or April, at the start of the rainy season.

Medical Significance: Low, being a very small, secretive snake living largely in ranching areas, although local herdsmen often know of people who had been bitten, and describe the bite as very painful and causing swelling, which fits with it being cytotoxic. No antivenom is produced, and no known antivenom gives protection against the venom of this species. Symptoms of a documented bite case included pain and swelling; a dog bitten in Kenya had a very swollen face. Bites should be treated conservatively by elevation and painkillers; antibiotics should be given if there are signs of infection.

Carpet or saw-scaled vipers *Echis*

These small but highly dangerous snakes are found across huge areas of the Old World, from northern Sri Lanka through India, Pakistan and the Middle East, across northern Africa from Egypt to Mauritania, and south to parts of the West African coast and to the Tana River in Kenya. They are arguably the most medically significant snakes in the world, on account of their huge range, their relative abundance in some areas, their willingness to bite and their dangerous venom (for such a small species). They are active on the ground at night and easily trodden on, and they hide in holes and under ground cover in agricultural land. They are implicated in hundreds of thousands of snakebite cases; farmers and rural people are particularly at risk. In the words of Professor David Warrell, African snakebite expert and a man precise with his words, 'this genus of vipers is of enormous medical importance'.

They are all small snakes (none larger than 80cm or so; most specimens are between 25cm and 50cm) with pear-shaped heads covered with small scales, small eyes with vertical pupils set well forward, thin necks and fairly stout, cylindrical bodies. Most of the body scales are keeled. Most carpet vipers are grey, brown or rufous, with various patterns, often including triangles or eyespots on the flanks. All have a distinctive threat display, forming C-shaped coils with the body; they then rub their scales together, the interscale friction creating a hissing sound, like water falling on a hot plate. The purpose of this stridulation is to enable the snake to make a loud warning noise without hissing with the mouth, and thus reduce water loss in a dry area. They are confident in defence, very quick to bite and will move towards a perceived threat while striking vigorously. The African species all lay eggs, but elsewhere some species give birth to live young.

Carpet vipers feed on a variety of prey, which may be the secret of their success; their venom (both within species and from one species to another) varies according to the prey they favour and the locality. This is medically significant; in parts of their range carpet vipers have venoms that are rarely fatal (mostly on the eastern side of Africa), while in other areas they are killers. In general, their venoms cause both local and systemic envenoming. There is pain and severe local swelling, blistering and necrosis, with severe haemostatic disorders leading to spontaneous systemic bleeding, from the membranes in the mouth and the internal organs. As David Warrell says, for carpet vipers 'coagulopathy is universal'. When death occurs it is often three to five days after the bite. Treatment for all carpet viper bites must be carried out in a well-equipped hospital, and will almost certainly involve antivenom (which is available), replacement therapy and treatment for incoagulable blood. Start with what is probably the single most effective, and one of the simplest, clinical procedures known: the 20-minute whole blood clotting test (20WBCT).

The taxonomy of the carpet vipers has changed a lot and continues to do so, as might be expected with snakes with huge ranges and often fragmented populations. Originally, all carpet vipers were considered to be of two species. One of these is Burton's Carpet Viper (*Echis coloratus*), occurring in Egypt and the Arabian Peninsula. The other is the 'typical' carpet or saw-scaled viper, *Echis carinatus*. As originally defined, this snake occupied a gigantic but fragmented range, from Sri Lanka up through India, westwards to the Arabian Peninsula and right across the northern half of Africa to Mauritania. Taxonomic work has now split *Echis coloratus* into two species and the *Echis carinatus* 'superspecies' into 10 species to date, six of which occur in the northern half of Africa.

There is debate on the status of the African species of the 'carinatus' group. Some researchers take the reductionist view that there are only two species: the West African Carpet Viper (*Echis ocellatus*) in the savannas of West and Central Africa, and the North-east African Carpet Viper (*Echis pyramidum*), occurring sporadically right around the Sahara and extending into the Horn of Africa. But this is over-simplistic. However, whether all six African taxa are separate species or represent gradual variation of a widespread form over a huge range remains to be seen.

Carpet vipers are snakes of semi-desert and dry savanna, with a few exceptions (Burton's Carpet Viper favours rocky desert, the West African Carpet Viper is in moist savanna and even forest margins, and in Kenya the North-east African Carpet Viper enters barren lava fields and rocky desert). Almost certainly a single ancestral African carpet viper had a much more continuous range during parts of the last 20 million years. The vegetation of the area that is now covered by the Sahara fluctuated, and at times would have provided suitable savanna habitat for these snakes. Subsequent climatic fluctuations between more and less arid conditions has in some cases fragmented the range of these snakes, or in other cases opened up corridors that allow separate populations to come together. Consequently, there are many apparently isolated populations (relicts) clinging on in patches of suitable habitat all over the northern half of Africa, and there is also evidence that formerly separate populations are now joined (particularly in the Sahel). Work continues; several times in the last few years populations have been moved from one taxon to another by researchers. This is a complex evolutionary scenario, not fully understood and still changing as the Earth's climate changes.

Key to the genus *Echis*, carpet or saw-scaled vipers

Note: there are species here that are difficult to distinguish with a key, as intermediates and other problematic specimens exist, but the range often gives clues as to the species. Use the pictures and the maps.

1a Ventrals 174–205; subcaudals 42–52; only in eastern Egypt......*Echis coloratus*, Burton's Carpet Viper (p. 85)
1b Ventrals usually fewer than 189; subcaudals 43 or less......2

2a Usually has distinctive white blotches ('eyespots') on flanks; largely on the western side of the northern half of Africa......3
2b Usually doesn't have white eyespots, but often has triangular flank shapes; largely either in north-eastern Africa or north of 12°N in central and western Africa......5

3a Midbody scale rows 27; ventrals 123–143; often rufous or pinkish; in dry country from Senegal east to central Mali......*Echis jogeri*, Mali Carpet Viper (p. 93)
3b Midbody scale rows 25–33; ventrals 133–168; usually shades of brown or grey; in dry country from western Mali east to south-western Chad and possibly the south of Sudan......4

4a Midbody scale rows 25–33; ventrals 133–157; in dry country from western Mali east to north-west Nigeria......*Echis ocellatus*, West African Carpet Viper (p. 90)
4b Midbody scale rows 29–33; ventrals 146–168; in dry country from central Nigeria east to south-western Chad and possibly the south of Sudan......*Echis romani*, Roman's Carpet Viper (p. 92)

5a Midbody scale rows 24–25; ventrals 144–149; only in extreme north-eastern Somalia......*Echis hughesi*, Somali Carpet Viper (p. 95)
5b Midbody scale rows 25–33; ventrals 155–198; not in extreme north-eastern Somalia......6

6a Underside white but usually without spots; occurs westwards from the western border of Sudan......*Echis leucogaster*, White-bellied Carpet Viper (p. 89)
6b Underside white, almost always with many small red-brown spots; occurs in North and north-east Africa......*Echis pyramidum*, North-east African Carpet Viper (p. 86)

Burton's Carpet Viper (Painted Carpet Viper) *Echis coloratus*

Identification: A small, robust viper with a pear-shaped head and a thin neck; in Africa, known only in eastern Egypt. The body is cylindrical and the tail short, 8–11% of total length. The scales are small on the head, rough and heavily keeled; in 31–35 rows at midbody; ventrals 174–205; subcaudals 42–52. Maximum size about 80cm (possibly slightly larger), average 30–60cm; hatchlings 12–14cm. This species is quite variable in colour and pattern, especially in Arabia, but eastern Egyptian and Sinai specimens are pale grey, brownish or biscuit-coloured, with

Echis coloratus

a series of pale grey or whitish saddles along the back, rufous-brown patches between the saddles, and dark spots on the flanks. There is usually a diffuse dark stripe extending backwards from the eye.

Habitat and Distribution: Usually in rocky desert. It is fond of dry stone hills, rock pavement, dry wadis and ravines, often in areas where there is some vegetation. It seems to avoid areas of loose sand, but is sometimes found on steep slopes and often near water sources. This species is found from sea level to quite high altitudes (2,500m) in parts of its range. In Africa, it is known only from eastern Egypt. A few specimens have been collected from the hills and rocky wadis just east and south-east of Cairo; from there it extends eastwards across the Sinai Peninsula, north into Israel, and southwards along the west bank of the Gulf of Suez, just entering extreme north-east Sudan. Also on the western and southern coasts of the Arabian Peninsula.

Natural History: Terrestrial. It is nocturnal, active from sunset onwards, particularly during the early hours of the night. During the day it hides in rock fissures and under large rocks, rock piles and ground cover; snakes studied in Israel chose ambush positions that were often near water, both under cover and on elevated objects (rocks, logs, etc.), and occasionally climbed into vegetation. This snake appears to be more good-natured than other carpet vipers, but if severely provoked

Burton's Carpet Viper, *Echis coloratus*. **Left:** Egypt (Stephen Spawls). **Right:** captive (Stephen Spawls).

will form a series of C-shaped coils like other carpet vipers, and shift them against each other in opposite directions; this friction between the scales produces a sound like water falling on a very hot plate. At the same time the snake may be moving backwards or forwards. If further agitated, it will strike. It lays 4–10 eggs, which are glued to hard surfaces; this is an adaptation to living in a rocky environment. The diet includes rodents, birds, lizards and frogs, indicating that it hunts by day and night, both from ambush and by prowling. One popular name of this snake (Burton's Carpet Viper) honours the great explorer and linguist Sir Richard Burton, who collected a specimen and presented it to the British Museum.

Medical Significance: Its small range in Africa and tendency to live in rocky desert lessen the dangers posed by this snake. However, in parts of the Middle East it has been found to be common in agricultural areas, and it is active on the ground at night, increasing the risk. The venom is cytotoxic and haemorrhagic. Antivenom is available. Experimentally in mice, the venom has an IV toxicity of 0.575mg/kg. Local pain and swelling are common and, in severe cases (about 10%), swelling may involve the whole limb and is associated with bruising, haemorrhagic blisters and necrosis. Systemic symptoms appear 15–120 minutes after the bite and include nausea, vomiting, headache, and spontaneous bleeding from the gums, nose, gastrointestinal and urinary tracts and recent wounds. Most patients have incoagulable blood by the time they are admitted to hospital. Do the 20WBCT. About 20% have thrombocytopenia; a few patients develop acute renal failure. Haemostatic dysfunction persists for up to nine days unless antivenom is given. Other abnormalities include hypotension, shock, loss of consciousness, ECG changes, neutrophil leucocytosis, proteinuria and microscopic haematuria with casts. A few fatalities are reported. Compared to envenoming by other *Echis* species, Burton's Carpet Viper seems to be more likely to cause thrombocytopenia and acute kidney injury, but the case fatality is lower. Administration of antivenom as late as two days after the bite may have dramatic and immediate effect. Blood clotting times should be carefully monitored in victims.

Taxonomic Notes: Snakes of this species from the extreme eastern side of its range in eastern Oman and the United Arab Emirates, were elevated in 2004 into a new full species, the Oman Saw-scaled Viper (*Echis omanensis*).

North-east African Carpet Viper (Saw-scaled Viper)
Echis pyramidum

Identification: A small, fairly stout snake of north-east Africa, with a pear-shaped head and a thin neck. The top of the head is covered with small scales, the prominent, pale yellowish eyes with vertical pupils are set near the front of the head, and the tongue is reddish. The body is cylindrical or subtriangular in section and the tail short, 10–11% of total length. The scales are rough and heavily keeled; in 25–31 rows at midbody; ventrals 155–186; subcaudals single, 27–43 (higher counts usually males). Maximum size about 65cm (possibly slightly larger),

Echis pyramidum

North-east African Carpet Viper, *Echis pyramidum*.
Left top: Egypt (Tomáš Mazuch). **Right top:** Ethiopia (Stephen Spawls).
Left bottom: Kenya (Stephen Spawls). **Right bottom:** Kenya (Tomáš Mazuch).

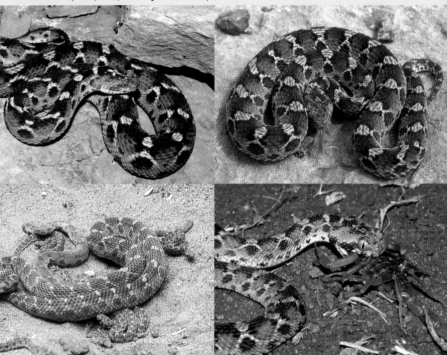

average 30–50cm; hatchlings 10–12cm. The ground colour may be yellowish, brown, grey or rufous. There is usually a series of pale, oblique cross-bars along the back, with dark spaces between, and along each side there is usually a row of dark triangular, subtriangular or circular markings with a pale or white edging. Specimens with very faded or almost invisible markings are known. The belly is pale, finely spotted brown or red.

Habitat and Distribution: Near-desert, semi-desert and dry savanna, between sea level and 1,250m. As presently defined, this snake has a disjunct distribution; this may be as a result of climatic change causing isolation, or undercollecting, although the variability of the genus also causes problems. Apparently isolated populations are known from south-west and coastal Libya, Egypt, south-west Sudan, northern Central African Republic and parts of northern Kenya. An apparently coherent population occurs from south-east Egypt into most of eastern Sudan, east to Eritrea, low north-east Ethiopia and northern Somalia. Another coherent population also occurs from extreme south-east South Sudan into south-western Ethiopia and northern Kenya, south past Lake Turkana to Lake Baringo, and east along the Uaso Nyiro River.

Natural History: Terrestrial, although it occasionally climbs into low bushes to avoid hot or wet surfaces. It moves relatively quickly. It is nocturnal, active from twilight onwards. During the day it hides in holes, under or in logs, or under rocks

North-east African Carpet Viper, *Echis pyramidum*. **Left:** Egypt (Tomáš Mazuch). **Right:** Kenya (Eduardo Razzetti).

or brush piles, or it may partially bury itself in sand or coil up in or around grass tufts. It occasionally climbs trees; in northern Kenya a snake 2m up an acacia tree bit someone. This is a spirited snake; when threatened it forms a series of C-shaped coils, which are shifted against each other in opposite directions, and this friction between the scales produces a sound like water falling on a very hot plate; at the same time the snake may be moving backwards or forwards. If further agitated, it will strike continuously and vigorously, to the point of overbalancing. It can strike a long distance, and may move towards a perceived threat. It can also sidewind at considerable speed. It lays 4–20 eggs; hatchlings have been found in July and August in Kenya. It eats a huge variety of prey; some populations eat mostly invertebrates, but most small vertebrates are also taken, including other snakes, even its own kind. This ability to feed on a wide range of prey may explain why it is abundant in places; in the Moille Hill area of northern Kenya, in an area of 6,500 square kilometres, nearly 7,000 of these snakes were collected in four months. However, in some areas it is uncommon; around Cairo, two specimens were found in 25 years.

Medical Significance: This is a dangerous snake; it is often abundant, very willing to bite if approached and active on the ground at night, and it lives in farming areas. Many people are bitten every year in its range; in some areas fatalities approach 20%. The venom is cytotoxic and haemorrhagic. Antivenom is available. Clinical features include local pain, swelling, blistering, abdominal pain, vomiting, bleeding from injection sites and recent injuries, haematuria, bleeding from the gums, haematemesis, melaena, menorrhagia, generalised bruising, periorbital bleeding, shock, fever, anaemia and leucocytosis, local pain, swelling, blistering and a generalised bleeding tendency. Bites will need to be treated in a well-equipped hospital; start with the 20WBCT. Antivenom therapy may be necessary.

Taxonomic Notes: Several subspecies or full species of this snake have been described, based on varying scale counts, colour differences, unusual localities, etc.; names used include *Echis megalocephalus* and *Echis varius*. The snakes of north-east Africa, north and eastwards from the Awash Valley, show variation that indicates there may be several evolutionary species present, each evolving separately. Our treatment reflects the situation at the time of writing.

White-bellied Carpet Viper *Echis leucogaster*

Identification: A large, distinctly marked carpet viper of the West and Central African Sahel and further north. The body is cylindrical or subtriangular in cross section and the tail short, 10–14% of total length. The scales are small on the head, rough and heavily keeled; in 27–33 rows at midbody; ventrals 158–189; subcaudals 25–40. Maximum size about 80cm, average 30–70cm; hatchlings 12–16cm. This species is quite variable in colour and pattern; the ground colour may be brown, grey or rufous, or shades in between. There is usually a series of pale,

Echis leucogaster

oblique cross-bars or saddles along the back, with dark spaces between. It is paler on the side. Along each side there is usually a row of triangular, subtriangular or circular dark markings with pale or white edging. The belly is pale cream, white or ivory and is usually unspotted (but see taxonomic notes on p. 83).

Habitat and Distribution: Arid savanna, Sahel, semi-desert and well-vegetated dry riverbeds. It is not known from true desert, but occurs on the desert edge, in oases, elevated vegetated areas within deserts, etc. A snake of North, West and north-west Africa, with what appears to be a very disjunct distribution. There is a string of isolated records from Western Sahara, Morocco, Algeria, Tunisia, Libya and northern Chad; this disjunct distribution may be due to

White-bellied Carpet Viper, *Echis leucogaster*. **Left top:** captive (Stephen Spawls). **Right top and left bottom:** Niger (Darrell Raw). **Right bottom:** underside (Darrel Raw).

undercollecting or isolation. An apparently coherent population occurs west from east-central Chad through most of Niger to northern Burkina Faso, central Mali, Senegambia and the southern half of Mauritania, south to northern Guinea.

Natural History: Terrestrial, although it occasionally climbs into low bushes to avoid hot or wet surfaces. It moves relatively quickly for a viper. It is nocturnal, being active from twilight onwards, and most active during the first few hours of the night. During the day it hides in holes, under rocks, or under or in logs, brush piles, etc. A spirited snake; when threatened, it forms a series of C-shaped coils, which are shifted against each other in opposite directions, and this friction between the scales produces a sound like water falling on a very hot plate. At the same time the snake may be moving backwards or forwards. If further agitated, it will strike continuously and vigorously, to a relatively large distance; it may strike so enthusiastically that it overbalances. It may even move towards an aggressor, a most unusual behaviour in a snake. Females lay clutches of 8–22 eggs, between August and October. Captive specimens hatched after two months. It eats a wide range of prey, including invertebrates (especially scorpions and centipedes), small mammals and reptiles.

Medical Significance: This is a dangerous snake; it is often abundant across a huge area of Africa, and it is the most common snake in northern Mali. It is very willing to bite if approached, is active on the ground at night and lives in farming areas. Many people must be bitten every year in its range, yet there seem to be few documented cases. The venom is cytotoxic and haemorrhagic. Antivenom is available. The clinical picture is probably similar to that of the previous species, with swelling, bleeding and incoagulable blood; bites will need to be treated in a well-equipped hospital, and antivenom therapy may be necessary. Start with the 20WBCT.

Taxonomic Notes: One method of distinguishing this species from the very similar North-east African Carpet Viper (*Echis pyramidum*) is by the absence of spotting on the underside. However, recent work indicates that *E. leucogaster* in Mali often do have spots, which suggests that there is a degree of overlap between the two species.

West African Carpet Viper *Echis ocellatus*

Identification: A distinctly marked carpet viper, usually with pale eyespots on the flanks, found in the West African savanna. The body is cylindrical or subtriangular in cross section, and the tail short, 10–14% of total length. The scales are small on the head, rough and heavily keeled; in 25–33 rows at midbody; ventrals 133–157; subcaudals 17–30. Maximum size about 60cm, average 30–50cm; hatchlings 10–13cm. The ground colour is usually brown or grey or shades in between. Individuals usually show one of two main dorsal patterns: one

Echis ocellatus

West African Carpet Viper, *Echis ocellatus*.
Left top: captive (Tomáš Mazuch). **Centre top:** Ghana (Stephen Spawls). **Right top:** underside (Tomáš Mazuch).
Left bottom: Togo (Tomáš Mazuch). **Right bottom:** hatchlings, Ghana (Stephen Spawls).

is a series of dark irregular cross-bars on a lighter background; and the other is a series of pale saddles with darker interspaces. The lower flanks are lighter. Most specimens (but not all) have a diagnostic line of small white eyespots along the flanks. The belly is pale, and usually covered with brown or reddish spots.

Habitat and Distribution: Semi-desert and dry and moist savanna, even into forest margins; between sea level and 1,100m. A snake of the central West African savanna; it occurs from western Nigeria west to north-east Guinea, and southwards in the Dahomey Gap to the coast. An isolated record exists from Trarza Region, extreme south-west Mauritania. As originally defined, this snake had a far larger range, but snakes from central Nigeria eastwards are now assigned to Roman's Carpet Viper (*Echis romani*), and those west of Mali are largely the Mali Carpet Viper (*E. jogeri*).

Natural History: As with other carpet vipers, it is terrestrial, but may climb low bushes to avoid hot or wet surfaces. It moves relatively quickly for a viper. It is nocturnal, being active from twilight onwards, and is most active during the first few hours of the night. During the day it hides in holes, under rocks, or under or in logs, brush piles, etc. When disturbed, it readily forms C-shaped coils, stridulates and strikes vigorously to a long distance, moving backwards or forwards; it may strike so enthusiastically that it overbalances. It may even move towards an aggressor, a most unusual behaviour in a snake. Females lay clutches of 6–28 eggs, usually in March or April; these hatch after two months, to coincide with the rainy season. It eats a wide range of prey, including invertebrates (especially scorpions and centipedes) and vertebrates (amphibians, lizards, other snakes and mammals).

Medical Significance: This is a very dangerous snake. It is often abundant across a huge area of Africa. It was found to be the most common snake in a study in north-west Ghana. It is very willing to bite if approached, is active on the ground at night and lives in farming areas. Many people are bitten every year. Antivenom is available; a district hospital in northern Ghana without antivenom had regular fatalities, while a hospital with antivenom had virtually none. The venom is cytotoxic and haemorrhagic. The clinical picture is pain and severe local swelling, blistering and necrosis, with severe haemostatic disorders leading to systemic bleeding. Bites will need to be treated in a well-equipped hospital; antivenom therapy will be necessary, and replacement therapy and blood transfusions may also be needed. Start with the 20WBCT.

Taxonomic Notes: As mentioned on p. 91, this species, as originally defined, has now been split into three West and Central African forms (*Echis ocellatus*, *E. jogeri* and *E. romani*) with largely mutually exclusive ranges.

Roman's Carpet Viper *Echis romani*

Identification: A relatively large, distinctly marked carpet viper, usually with pale eyespots on the flanks and with high ventral counts, found in the West African savanna, east of Nigeria. The body is cylindrical or subtriangular in cross section, and the tail short, 7–11% of total length. The scales are small on the head, rough and heavily keeled; in 29–33 rows at midbody; ventrals 146–168; subcaudals 18–29 (higher counts in males). Maximum size about 72cm, average 30–50cm; hatchling size unknown, but a 13cm specimen was collected in May. The ground colour is usually rufous,

Echis romani

brown or grey, with a series of grey or brown blotches along the spine, and pale eyespots on the flanks. Nigerian specimens are often distinctly rufous on the flanks. The belly is pale, and usually covered with brown or reddish spots.

Habitat and Distribution: Dry and moist savanna, even into forest margins; between 200m and 1,200m. It occurs from west-central Nigeria and extreme southern Niger eastwards to northern Cameroon, south-western Chad and north-west Central African Republic; two isolated populations are also known from southern Sudan.

Natural History: As with other carpet vipers, it is terrestrial, but it may climb low bushes to avoid hot or wet surfaces. It moves relatively quickly for a viper. It is nocturnal, being active from twilight onwards, and is most active during the first few hours of the night. During the day it hides in holes, under rocks, or under or in logs, brush piles, etc. When disturbed, it readily forms C-shaped coils, stridulates and strikes vigorously to a long distance, moving backwards or forwards; it may strike so enthusiastically that it overbalances, and may even move towards an aggressor, a most unusual behaviour in a snake. This species lays eggs; Nigerian females laid 10–22 eggs, usually in March or April. Eggs hatch after two months, to coincide with the rainy season. It eats a wide range of prey, including invertebrates (especially scorpions and centipedes) and vertebrates (amphibians, lizards, other snakes and mammals).

Roman's Carpet Viper, *Echis romani.*
Left top: Nigeria (Stephen Spawls). **Centre top:** Nigeria (Gerald Dunger).
Right top and left bottom: Nigeria (Stephen Spawls). **Right bottom:** underside, Nigeria (Gerald Dunger).

Medical Significance: This is a very dangerous snake. It is often abundant across a huge area of Africa, and it causes huge numbers of snakebites in Nigeria and Cameroon. It is very willing to bite if approached, is active on the ground at night and lives in farming areas. Many people are bitten every year. Antivenom is available. The venom is cytotoxic and haemorrhagic. The clinical picture is pain and severe local swelling, blistering and necrosis, with severe haemostatic disorders leading to systemic bleeding. Bites will need to be treated in a well-equipped hospital; antivenom therapy may be necessary, and replacement therapy and blood transfusions may also be needed. Start with the 20WBCT.

Taxonomic Notes: As mentioned on p. 91, this species was recently elevated out of the synonymy of *Echis ocellatus*, and was originally regarded as the eastern population of that species. It is named after Benigno Roman, who carried out pioneering work on the snakes of Burkina Faso.

Mali Carpet Viper *Echis jogeri*

Identification: A small, distinctly marked carpet viper, usually with pale eyespots on the flanks and with low ventral counts, found in the West African savanna, west of Burkina Faso. The body is cylindrical or subtriangular in cross section, and the tail short, 7–12% of total length. The scales are small on the head, rough and heavily keeled; in 27 rows at midbody; ventrals 123–143; subcaudals 15–30 (higher counts in males). The maximum size is about 45cm, average 20–35cm;

hatchling size unknown, but probably around 9–10cm. The ground colour is usually rufous, brown or grey, with a series of vague darker cross-bars, and usually pale eyespots on the flanks. The underside is pale, and usually covered with brown or reddish spots.

Echis jogeri

Habitat and Distribution: Dry and moist savanna, between sea level and 800m. It occurs from western Senegal and the Gambia into northern Guinea and western Mali.

Natural History: As with other carpet vipers, it is terrestrial, but it may climb low bushes to avoid hot or wet surfaces. It moves relatively quickly for a viper. It is nocturnal, being active from twilight onwards, and is most active during the first few hours of the night. During the day it hides in holes, under rocks, or under or in logs, brush piles, etc. When disturbed, it readily forms C-shaped coils, stridulates and strikes vigorously to a long distance, moving backwards or forwards; it may strike so enthusiastically that it overbalances, and may even move towards an aggressor, a most unusual behaviour in a snake. Females lay eggs that probably hatch after two months, to coincide with the rainy season. It eats a wide range of prey, including invertebrates (especially scorpions and centipedes) and vertebrates (amphibians, lizards, other snakes and mammals).

Medical Significance: This is a very dangerous snake. It is often abundant across a huge area of Africa. It causes huge numbers of snakebites in Senegal and Mali. It is very willing to bite if approached, is active on the ground at night and lives in farming areas. Many people are bitten every year. Antivenom is available. The venom is cytotoxic and haemorrhagic. The clinical picture is pain and severe local swelling, blistering and necrosis, with severe haemostatic disorders leading to systemic bleeding. Bites will need to be treated in a well-equipped hospital; antivenom therapy may be necessary, and replacement therapy and blood transfusions may also be needed. Start with the 20WBCT.

Taxonomic Notes: As mentioned on p. 91, this species was elevated out of the synonymy of *Echis ocellatus*, and was originally regarded as being part of the western population of that species.

Mali Carpet Viper, *Echis jogeri*.
Left and right: Senegal (Wolfgang Wuster).

Somali Carpet Viper *Echis hughesi*

Identification: A small, pale carpet viper, endemic to north-east Somalia. It has a pear-shaped head and a thin neck. The top of the head is covered with small scales, and the prominent, pale yellowish eyes have vertical pupils and are set near the front of the head. The body is cylindrical or subtriangular in section and the tail short, around 10–12% of total length. The scales are rough and heavily keeled; in 24–25 rows at midbody; ventrals 144–149; subcaudals single, 28–30. Maximum known size about 35cm. The ground colour has never been described in

Echis hughesi

living animals, but is probably grey or brown; there is usually a series of pale, oblique cross-bars along the back, with darker blotches, often with even darker edging, in between. The flanks are marked with irregular darker blotches of varying size. The underside is white, with occasional reddish spots near the ventral scale margins.

Habitat and Distribution: Dry savanna and semi-desert, at low altitude. It is known only from Bari and Nugal Regions of north-east Somalia.

Natural History: Unknown, but presumably similar to other members of the genus: nocturnal and terrestrial, shelters in holes and under ground cover, forms C-shaped coils when disturbed and strikes readily, lays eggs, and eats a range of small vertebrates and possibly invertebrates.

Medical Significance: Unknown, but the venom is probably similar to that of the North-east African Carpet Viper (*Echis pyramidum*).

Taxonomic Notes: Described in 1990, it differs from the North-east African Carpet Viper, which occurs further west, only in having fewer ventral scales. It might just be an ecotype of that species, but it does seem to be an isolated population, and a number of other distinctive endemic species occur in the area. This genus is absent from most of Somalia, except the north, which is curious given the wide range of suitable habitat for these dry-country specialists.

Somali Carpet Viper, *Echis hughesi*.
Left: preserved specimen (Stephen Spawls). **Right:** preserved specimen (Tomáš Mazuch).

North African desert vipers *Cerastes*

A genus of broad-headed brown or biscuit-coloured desert vipers, rarely larger than 70cm. Four species are known; two are widespread over North Africa, one is confined to Tunisia, and the fourth is from the Middle East (almost reaching far eastern Egypt). Two of the three African species have huge ranges, and are implicated in many snakebite cases in North Africa. They are dangerous because they are on the ground, often well hidden in sand, and are active at dusk and in the early hours of the night, so are liable to be trodden on by people without adequate footwear. These snakes appear to have a haemorrhagic venom; bites usually result in local pain and swelling, complicated by necrosis in some cases. Nausea, vomiting, coagulopathy and spontaneous bleeding, including cerebral haemorrhage, have been observed in some cases. Although there are no credible reports of deaths in the last 100 years or so, some documented bites to snake keepers have been life-threatening. Antivenom is available.

Key to the genus *Cerastes*, North African desert vipers

1a Horns consist of a cluster of elongate blunt-topped scales above the eye......*Cerastes boehmei*, Böhme's Sand Viper (p. 99)

1b Either hornless or, when present, horn consists of a single elongate spike......2

2a Small, less than 50cm; never has horns; eyes on the junction of side and top of head; ventrals always fewer than 126......*Cerastes vipera*, Sahara Sand Viper (p. 98)

2b Large, up to 75cm or more; often has horns; eyes on the side of the head; ventrals always more than 129......*Cerastes cerastes*, Sahara Horned Viper (p. 96)

Sahara Horned Viper *Cerastes cerastes*

Identification: A short, stout viper, usually with prominent horns, found in the Sahara and environs. The head is broad, flat and triangular, and covered with small scales. The eye is prominent, on the side of the head, and has a vertical pupil in bright light. The neck is thin. There is often a long horn above each eye; this horn consists of a single scale, not a group of scales. Individuals without horns often have prominent brow ridges. The body is broad and flattened, and the tail short, 7–12% of total length. The scales are heavily keeled; in 23–39 rows at midbody; ventrals

Cerastes cerastes

130–165; cloacal scale entire; subcaudals 23–45. Maximum size in the wild about 75cm, although captive specimens have reached 90cm; most specimens 35–60cm; hatchlings 12–15cm. The ground colour is yellowish, buff, reddish or greyish brown, with a series of almost rectangular brown blotches along the back. A dark line extends backwards from each eye. The tail tip may be black in some individuals.

Habitat and Distribution: Occurs in desert and semi-desert. A versatile snake, usually found in wadi systems with vegetation and sandy soil, but it also occurs on gravel plains, in mountainous country, and even in very rocky areas so long as some sand or soil patches exist. Not usually within dunes, although found on

Sahara Horned Viper, *Cerastes cerastes.*
Left top: captive (Stephen Spawls). **Centre top:** Niger (Jean-Francois Trape). **Right top:** Morocco (Konrad Mebert).
Left bottom: hornless captive (Stephen Spawls). **Right bottom:** captive (Stephen Spawls).

the margins. It is found from sea level to over 1,200m, possibly higher. It occurs throughout the Sahara, from western Mauritania eastwards to Sudan and Egypt, and north to the coast (excluded from the north Moroccan and Algerian coast by the Atlas Mountains). In Egypt it extends across the Sinai; just east of there it is replaced by the Arabian Horned Viper (*Cerastes gasperettii*).

Natural History: Terrestrial and slow-moving, although it can strike very quickly and can also sidewind. It is nocturnal in the summer months, active from dusk onwards, but may bask during the day (especially during winter and early spring). It shelters in rodent burrows or other holes, under ground cover (wood, rocks, etc.) or hidden in a bush or among grass tufts. It often buries (or partially buries) itself in soft sand, sometimes in the lee of a bush or rock, but may simply coil up in a sheltered place out of the sun and wind, where it may be surprisingly visible. This snake is fairly placid, but can strike quickly, to some distance, and when threatened can shift its coils into C-shapes and rub them together (as do carpet vipers), producing a hissing, crackling sound; from this display it will also strike. Females lay 10–23 eggs, which hatch after 45–80 days' incubation. This species is known to move long distances at night in search of prey. The diet includes rodents, lizards and birds (the fact that it eats birds indicates how quickly it strikes). A captive individual in Cairo Zoo took six to eight birds a year, and lived 14 years in captivity. One of the more common hieroglyphic characters in Egypt is a Sahara Horned Viper. In some areas it is greatly feared; it is believed to be able to spring a huge distance into the air to bite or even to be able to fly.

Medical Significance: This is a nocturnal terrestrial snake with a huge range. It is often near villages and encampments, and remains concealed and motionless when approached. It therefore causes a lot of bites; in countries such as Egypt the majority of bites are due to this species, although in some areas it is regarded by the local people as not being particularly dangerous, and it may be tolerated. Reports of deaths from its bite exist in older literature, but appear to be largely mismanaged cases, or possibly caused by another snake. No recent fatal cases are known and antivenom is available. However, the venom is highly toxic. Swelling, nausea, vomiting, coagulopathy and spontaneous bleeding, including cerebral haemorrhage, have been observed; in some cases necrosis has been reported. Recently, disseminated intravascular coagulation, microangiopathic haemolysis and acute kidney injury were described in two bite cases on snake keepers. It is not a fast-acting venom, but bite cases need to be treated at a well-equipped hospital and may involve antivenom therapy.

Sahara Sand Viper *Cerastes vipera*

Cerastes vipera

Identification: A small, flattened, stout viper with strangely angled eyes on top of its head, in sand dunes in the Sahara and environs. The head is broad, flat and triangular, and covered with small scales. The eye is prominent, on the side of the head, and has a vertical pupil in bright light. The neck is thin. The body is broad and flattened, and the tail short, 7–14% of total length. The scales are heavily keeled; in 23–29 rows at midbody; ventrals 102–125; cloacal scale entire; subcaudals 18–26. Maximum size about 50cm, but this is large; most specimens 20–30cm; hatchlings 12–15cm. The ground colour is yellowish, buff, reddish or greyish brown, with a series of almost rectangular brown blotches along the back. A dark line extends backwards from each eye. The tail tip is black in females.

Habitat and Distribution: A desert snake, usually found in dunes or other areas of soft sand; from sea level to over 1,000m, possibly higher. It occurs throughout the Sahara, from western Mauritania eastwards to northern Sudan and Egypt, north to the coast (excluded from the north Moroccan and Algerian coast by the Atlas Mountains). In Egypt it extends across the Sinai, and north into Israel.

Sahara Sand Viper, *Cerastes vipera*. **Left:** Egypt (Darrell Raw). **Right:** Israel (Yannick Francioli).

Sahara Sand Viper, *Cerastes vipera*.
Left: Egypt (Stephen Spawls). **Right:** embedding its fangs into a glove (Konrad Mebert).

Natural History: Terrestrial and slow-moving, although it can strike very quickly and can also sidewind. It is nocturnal in the summer months, active from dusk onwards, but may bask during the day (especially during winter and early spring), and in northern parts of its range may hibernate between November and February. It usually buries itself in soft sand, sometimes with the eyes exposed, but may hide in holes or under ground cover. This snake is fairly placid, but can strike quickly, to some distance, and when threatened can shift its coils into C-shapes and rub them together (as do carpet vipers), producing a hissing, crackling sound; from this display it will also strike. Mating occurs in April and May in Israel, and 3–6 young are born live in the latter half of August. The diet in the wild is mostly lizards, but captive snakes take rodents.

Medical Significance: This is a nocturnal terrestrial snake with a huge range, which hides by partially or totally buying itself. A number of bites probably occur every year as walkers tread on the snake. No fatalities are known; a study in Israel found that most bites resulted in only minor local swelling and small haematomas, and one victim showed slight abnormalities in his coagulation profile. However, bite victims should visit a hospital.

Böhme's Sand Viper *Cerastes boehmei*

Identification: A small, flattened, stout viper, similar to the Sahara Sand Viper (*Cerastes vipera*) but with curious flat-topped horns. Described in 2010, from south-central Tunisia. The head is broad, flat and triangular, and covered with small scales; above the eye is a curious cluster of flat-topped scales, not like the pointed horns of the Sahara Horned Viper. The eye is prominent, on the side of the head, and has a vertical pupil in bright light. The neck is thin, and the body broad and flattened. The scales are heavily keeled; in either 21 or 26 rows at midbody (due to

Cerastes boehmei

Böhme's Sand Viper, *Cerastes boehmei*. **Below:** captive (Library Book Collection/Alamy Stock Photo).

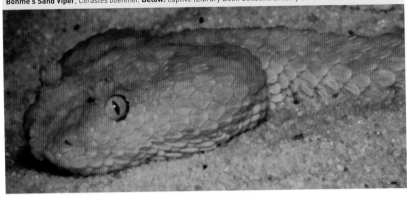

an irregularity of the lower scale rows in the type); ventrals 110; cloacal scale entire; subcaudals 25. The type was just under 22cm in total, and the tail 2.5cm. The ground colour is uniform sandy yellowish, with vague darker dorsal blotches. The tail tip is black and the belly white.

Habitat and Distribution: A desert snake, known from Bani Kheddache and Remada, between 300m and 500m, in south-central Tunisia.

Natural History: Unknown, but its resemblance to the Sahara Sand Viper suggests its habits might be similar; that is, terrestrial and nocturnal, buries itself in sand or hides in holes, forms C-shaped coils when disturbed, gives birth to live young (a captive female gave birth to five young), and eats small desert lizards.

Medical Significance: The venom and its effects are probably similar to those of the Sahara Sand Viper.

Taxonomic Notes: An interesting snake; the presence of horns suggests a relationship to the Sahara Horned Viper (*Cerastes cerastes*), but the general habitus and low ventral count suggests it may be related to the Sahara Sand Viper; it has been suggested that it is either a hybrid or a sport. Further research and specimens may clarify the situation.

Night adders *Causus*

A genus of small, stout, frog-eating vipers, rarely larger than 75cm or so. They are found in sub-Saharan Africa in savanna and forest, and occasionally in semi-desert along watercourses. Seven species are known. Unusually for vipers, they have a round pupil and nine large scales on top of the head (in most other vipers these are small), and they lay eggs. Despite their common name, they are active by day and night. They have a somewhat primitive fang rotation movement and their fangs are relatively short. Several species have exceptionally long venom glands that extend well down the neck; the large quantities of venom thus produced assist in envenoming large frogs.

Night adders can be common, they are active on the ground by day and by night, they are willing to bite and in some areas they are responsible for a high proportion of snakebite cases. The venom is cytotoxic, and bites usually result in local pain, sometimes rapid and intense, as well as swelling, blistering and lymphadenopathy. Necrosis has very rarely been observed, even after blister removal, no systemic effects have been noted in humans, no fatalities are recorded and no antivenom is available. However, research indicates the venom does inhibit blood coagulation; some haemorrhaging may occur and South African polyvalent antivenom, although not raised against night adder venom, does have a neutralising effect on the venom of the Rhombic Night Adder (*Causus rhombeatus*). Treatment for a bite consists of elevation of bitten limbs and pain relief. A study in West Africa found a surprisingly high number (30%) of bites from night adders were 'dry' (i.e. no venom was injected). This is probably because the night adders have very short, broad fangs and do not bite with any great force; they usually hold their struggling prey (amphibians) and wait for the venom to take effect. Thus, the venom is not injected to any great depth. It has also been suggested that the intense, immediate local pain sometimes experienced from night adder bites may serve a defensive purpose, deterring a predator (although, as mentioned in the What happens when a snake bites and how bad will it be? section, pp. 295–96, snake venom is largely not for defence).

Key to the genus *Causus*, night adders

1a Usually vivid green when adult......2
1b Brown, grey or dull olive-green when adult......3

2a Subcaudals single; eyes large; in forest; rostral not upturned......*Causus lichtensteinii*, Forest Night Adder (p. 104)
2b Subcaudals paired; eyes moderate in size; not in forest; rostral slightly upturned......*Causus resimus*, Velvety-green Night Adder (p. 107)

3a Snout clearly upturned; subcaudals fewer than 19......*Causus defilippii*, Snouted Night Adder (p. 103)
3b Snout not upturned; subcaudals more than 14......4

4a Head narrow; usually a pair of narrow, pale dorsolateral stripes present; only in northern Rwanda and south-west Tanzania......*Causus bilineatus*, Two-striped Night Adder (p. 102)
4b Head broad; no dorsolateral stripes......5

5a Upper labials not dark-edged; ventrals 118–150; no dark bar between the eyes; dark dorsal markings very rarely edged with paler scales......*Causus maculatus*, West African Night Adder (p. 106)
5b Upper labials dark-edged; ventrals 130–166; usually a dark bar between the eyes; dark dorsal markings usually edged with paler scales......6

6a Ventrals 134–166......*Causus rhombeatus*, Rhombic Night Adder (p. 109)
6b Ventrals 130–132......*Causus rasmusseni*, Rasmussen's Night Adder (p. 111)

Two-striped Night Adder *Causus bilineatus*

Identification: A small adder, with a head that is narrow and long for a night adder, tapering to a narrow but rounded, not upturned, snout. The head is slightly distinct from the neck, the eye is medium-sized with a round pupil, and the top of the head is covered with nine large scales. The body is cylindrical or slightly depressed and the tail is short, 9–12% of total length. The scales are soft, velvety and feebly keeled; in 17 (occasionally 15, 18 or 19) rows at midbody; ventrals 119–149; subcaudals 18–35 (higher counts usually males). Maximum size

Causus bilineatus

about 65cm, average 30–50cm; hatchlings about 13–14cm. The back is brown, or it may be pinkish or greyish brown, and there are a number of irregular or vaguely rectangular black patches all along the back, interspersed with rufous or brown blotches. There are usually two distinct narrow, pale dorsolateral stripes that run the length of the body. There is a sprinkling of black scales and oblique black bars on the flanks. On the head is a characteristic V-shaped mark, which may be black, or black-edged with a brown infilling. The belly is dark-coloured to dark cream.

Habitat and Distribution: In moist savanna and woodland, between 800m and 1,800m. It occurs in extreme south-west Tanzania, at Tatanda, then south-west across north-eastern and northern Zambia and south-eastern DR Congo, into most of central Angola, with an isolated population in eastern Rwanda.

Natural History: Little known, but probably rather similar to other night adders (terrestrial, slow-moving, active by day and night, inflates the body when threatened and hisses and puffs, and will strike if further molested). This species may be locally abundant. It is known to feed on clawed frogs, which might mean it is more aquatic than other night adders. It lays eggs. It was regarded at one time as simply a colour variant of the Rhombic Night Adder (*Causus rhombeatus*), but the ranges of the two species overlap throughout the Two-striped Night Adder's range, and this species also has slightly lower ventral counts than the Rhombic Night Adder.

Two-striped Night Adder, *Causus bilineatus*.
Left: Zambia (Andre Zoersel). **Right:** Zambia (Philipp Wagner).

Medical Significance: A mildly venomous species, quite common in parts of its range (e.g. northern Zambia) and liable to be trodden on. Antivenom is not available and should not be necessary.

Snouted Night Adder *Causus defilippii*

Identification: A small, stout night adder of south-eastern Africa, with an upturned nose. The short head is slightly distinct from the neck, it has a medium-sized eye with a round pupil, and the top of the head is covered with nine large scales. The tip of the snout is quite distinctly upturned. The body is cylindrical or slightly depressed and the tail is short, 7–9% of total length in males and 5–7% in females. The scales are soft, velvety and feebly keeled; in 13–17 (usually 17) rows at midbody; ventrals 109–130 (higher counts usually females); subcaudals 10–19.

Causus defilippii

Maximum size about 42cm, average 20–35cm; hatchlings about 10cm. The back is brown, or it may be pinkish or greyish brown, and there are 20–30 clear, dark, pale-edged rhombic blotches all along the back, and a sprinkling of black scales and oblique black bars on the flanks. On the head is a characteristic V-shaped mark, usually solid black, a black bar is often present behind the eye, and the upper lip scales are usually black-edged. The belly is cream, pearly white or pinkish grey; it may be glossy black or grey in juveniles.

Habitat and Distribution: Moist and dry savanna and coastal thicket, from sea level to about 1,800m. It occurs south from Malindi on the Kenyan coast, extending across most of south-east Tanzania, Malawi, eastern Zambia, Mozambique and parts of Zimbabwe, into northern South Africa and reaching Kwazulu-Natal on the South African coast; an unequivocal inhabitant of the eastern and southern Africa coastal mosaic.

Natural History: Terrestrial, but may climb into low bushes in pursuit of frogs. It is fairly slow-moving, but strikes quickly. Despite its name, this species seems to be active during the day, as well as at twilight and at night. It is known to

Snouted Night Adder, *Causus defilippii*.
Left: Mozambique (Mark-Oliver Rödel). **Right:** Tanzania (Stephen Spawls).

Snouted Night Adder, *Causus defilippii*.
Left: Mozambique (Mark-Oliver Rödel). **Right:** South Africa (Bill Branch).

bask. When inactive, it hides in holes, brush piles, under ground cover, etc. If threatened, it hisses loudly, inflates the body, elevates the head and may make a determined (but often clumsy) strike. It may also lift the front part of the body off the ground, flatten the neck and move forward with the tongue extended, looking like a small cobra. It eats frogs and toads. Females lay 3–9 eggs, roughly 23 × 15mm. Hatchlings have been collected in the Usambara Mountains in May and June. In the breeding season, males are known to engage in combat, rearing up and wrestling until the weaker male is forced to the ground.

Medical Significance: A mildly venomous species, quite common in parts of its range and liable to be trodden on. Antivenom is not available and should not be necessary. Documented cases are characterised by swelling, sometimes extensive, local pain, which may be intense, often fever and sometimes lymphadenopathy. Recovery usually takes a few days. The victim in one South African case experienced difficulty in breathing, but other factors of the case suggest this was an anaphylactic reaction to the venom; the victim had three previous night adder bites.

Forest Night Adder *Causus lichtensteinii*

Identification: A small, relatively slim green night adder of the great forests of Central and Western Africa. The head is slightly more rounded than the usual night adder head, and is distinct from the neck. It has a prominent eye with a round pupil, and the top of the head is covered with nine large scales. The body is cylindrical or slightly depressed, and (particularly in juveniles) slimmer than in other night adders. The tail is short and blunt, 6–8% of total length in females and 8–11% in males. The scales are soft, velvety and feebly keeled; in

Causus lichtensteinii

15 rows at midbody; ventrals 134–156 (higher counts in females). Unusually, the subcaudal scales are single; they number 14–23 (higher counts in males). Maximum size about 70cm, but this is exceptional; average 30–55cm; hatchlings about 15cm. The back is usually green, of various shades, from vivid pale green to dark or olive-green, or sometimes brown. Some individuals have two or three

orange bars across the tail. There is sometimes a series of vague, pale-centred rhombic back markings, which may be obscured and appear as chevrons, facing forwards or backwards; sometimes there is a white V-shape on the neck. The belly scales are yellowish, cream or pearly. The lips are yellow, the throat is yellowish or white, usually with two or three distinct black cross-bars, and the tongue is blue at the base, then black with an orange band. Juveniles are very variable; deep green, brown or turquoise, uniform or with dark cross-bars; the lips are white. Hatchlings have a white V-shape on the neck, which fades in adults.

Habitat and Distribution: In forest and woodland, swampy areas associated with forest and recently deforested areas, from sea level to about 2,100m. It occurs from the forests of far western Kenya and the northern shore of Lake Victoria in Uganda, west and south-west from the Albertine Rift right across the Central African forests to Nigeria, and south to northern Zambia and north-east Angola; it reappears west of the Dahomey Gap in Ghana, west to southern Sierra Leone. This species is also known from the Imatong Mountains in South Sudan.

Natural History: Terrestrial and secretive. Fairly slow-moving, but strikes quite quickly. Despite its name, it seems to be most active during the day among the shade and leaf litter of the forest floor. It swims well and has colonised islands in Lake Victoria. When inactive, it hides in holes, brush piles, tree root clusters, under ground cover, etc. If threatened, it responds by inflating the body with air, which makes the markings stand out, and hissing and puffing; it may then raise the forepart of the body off the ground, into a coil, with the head back, and strike. It eats frogs and toads, and lays 4–8 eggs.

Forest Night Adder, *Causus lichtensteinii.*
Left top: Eastern DR Congo (Konrad Mebert). **Right top:** Gabon (Bill Branch).
Left bottom: Eastern DR Congo (Konrad Mebert). **Right bottom:** DR Congo (Harald Hinkel).

Medical Significance: A mildly venomous species, quite common in parts of its range and liable to be trodden on. Antivenom is not available and should not be necessary. Documented cases are characterised by swelling that is sometimes extensive, local pain, often fever and sometimes lymphadenopathy. Recovery usually takes a few days.

West African Night Adder *Causus maculatus*

Identification: A small, stout brown night adder, found in a wide range of habitats in Central and West Africa. The fairly short head is slightly distinct from the neck, and it has a rounded snout and a medium-sized eye with a round pupil. The top of the head is covered with nine large scales. The body is cylindrical or slightly depressed and the tail is short, 7–9% of total length in females and 9–11% in males. The scales are soft and feebly keeled; in 17–22 rows at midbody; ventrals 118–154; subcaudals 15–26 (higher counts usually males). Maximum size about

Causus maculatus

70cm (possibly slightly larger), average 30–60cm; hatchlings 13–16cm. The body is usually some shade of brown, pinkish brown or grey above, with dark cross-bars or rhombic markings down the centre of the back and dark bars or spotting on the flanks. Occasional individuals have virtually no markings and the degree of contact between the rhombic markings and the ground colour varies greatly, even in the same area. There is usually a V-shape on the neck, extending onto the head, that is uniformly dark or with dark edges and a light centre. The underside may be white, cream or pinkish grey, uniform or with each scale black-edged so the belly looks finely barred.

Habitat and Distribution: Occurs in a huge range of habitats, from deep forest to woodland, moist and dry savanna and semi-desert; from sea level to 1,800m. It is found westwards from south-west Ethiopia, western Uganda and South Sudan, right across Central and West Africa to Senegal and south-west Mauritania, and south to north-west Angola and south-eastern DR Congo.

West African Night Adder, *Causus maculatus*.
Left: Eastern DR Congo (Konrad Membert). **Right:** Ghana (Stephen Spawls).

West African Night Adder, *Causus maculatus*.
Left: Uganda (Steve Russell). **Right:** Ghana (Stephen Spawls).

Natural History: Terrestrial, but will climb into bushes in pursuit of frogs. It is active by day, by night and at twilight. In West Africa it is most active in the rainy season, and tends to disappear in the dry season. If molested it inflates the body, hisses loudly and makes swiping strikes; it may flatten the body, looking like a small cobra, and demonstrate with an open mouth. It eats frogs and toads. Females lay 6–20 eggs, roughly 26 × 16mm; in West Africa they are laid in February–April. Hatchlings appear there in May–July. This is quite a common snake in parts of its range.

Medical Significance: A mildly venomous species, quite common in parts of its range and liable to be trodden on. Antivenom is not available and should not be necessary. Documented cases are characterised by swelling (sometimes extensive), local pain (often intense), local discolouration, often fever and sometimes lymphadenopathy. There is rarely necrosis. Recovery usually takes a few days. In a remarkable study on a Liberian rubber plantation, over 90% of all snakebites were caused by this species, and about 30% of these were 'dry bites', where no symptoms resulted. There were no deaths.

Taxonomic Notes: In places this snake's range overlaps that of the Rhombic Night Adder (*Causus rhombeatus*), and intermediate specimens are known.

Velvety-green Night Adder *Causus resimus*

Identification: A small, stout viper with a fairly short head that is slightly distinct from the neck, and a medium-sized eye with a round yellow pupil. The top of the head is covered with nine large scales, and the tip of the snout is slightly upturned. The body is cylindrical or slightly depressed and the tail is short, 7–9% of the total length in females and 8–10% in males. The scales are soft, velvety and feebly keeled; in 19–21 rows at midbody; ventrals 131–152; subcaudals 15–25 (males usually more than 20, females usually fewer than 20). Maximum size about 75cm, average 30–60cm; hatchlings 12–15cm. A beautiful snake; the back is usually vivid green (sometimes brown), of various shades. The chin and throat are yellow and the belly scales yellowish, cream or pearly. Black scales often form a V-shaped outline on

Causus resimus

Velvety-green Night Adder, *Causus resimus*.
Left top: Kenya (Vincenzo Ferri). **Right top:** Kenya (Stephen Spawls).
Left bottom: spreading hood, Sudan (Abubakr Mohammed). **Right bottom:** Ethiopia (Stephen Spawls).

the head (especially in juveniles), and scattered black scales may form indistinct rhomboids or V-shapes on the back and oblique dark bars on the flanks. The tongue is pale blue and black. The hidden margin of the scales is often a vivid blue, this colour appearing when the snake inflates its body when threatened.

Habitat and Distribution: Coastal savanna and thicket, dry and moist savanna and woodland, from sea level to 1,800m. It has the most curiously disjunct distribution of any African snake, with what appears to be at least five large, isolated populations. It is known from the Somali and Kenyan coast; from the Lake Victoria basin to the far slopes of the Albertine Rift; from the west coast of Angola; from central Sudan south up the Nile into South Sudan, extreme north-west Kenya and south-west Ethiopia; and in south-western Chad and eastern Cameroon. Isolated records are known from northern DR Congo, northern Central African Republic and a handful of disjunct localities in Nigeria. It seems likely that some of these may prove to be discrete species.

Natural History: Terrestrial, but will climb low vegetation in pursuit of frogs. It is active by day and night, and will bask. This snake is fairly slow-moving, but strikes quite quickly. It swims well. When inactive, it hides in holes, brush piles, under ground cover, etc. If threatened, it responds like other night adders, hissing and puffing and striking. The diet comprises amphibians. It lays 4–12 eggs; captive specimens have been recorded producing clutches at two-month intervals, without any noticeable breeding season, but Somali females laid 4–11 eggs in July.

Medical Significance: Little is known of its venom, but it is presumably similar to that of other night adders. No antivenom is available.

Rhombic Night Adder *Causus rhombeatus*

Identification: A small, stout viper with a fairly short head that is slightly distinct from the neck, a rounded snout and a medium-sized eye with a round pupil. The top of the head is covered with nine large scales. The body is cylindrical or slightly depressed and the tail is short, 9–12% of total length. The scales are soft and feebly keeled; in 15–23 rows at midbody; ventrals 134–166; subcaudals 21–35 (higher counts in males). Maximum size about 95cm (possibly slightly larger), average 30–60cm; hatchlings 13–16cm. The body is usually some shade of brown on

Causus rhombeatus

the back (may be pinkish or greyish brown), occasionally olive-green, and some snakes show a limited ability to change colour. There are usually 20–30 dark, pale-edged rhombic blotches all along the back, and a sprinkling of black scales and oblique black bars on the flanks. The rhombic patches may lack white edging; some snakes are quite patternless. On the head is a characteristic V-shaped mark, which may be solid black, or simply a black outline with brown inside. The belly is cream to pinkish grey; the ventral scales may be uniform, but sometimes each scale grades from light to dark and the belly thus looks finely barred.

Habitat and Distribution: In moist savanna, grassland and woodland, from sea level to 2,200m, nearly always in the vicinity of water sources. It occurs north from the Western Cape in South Africa up the eastern side of Africa, spreading west to Angola and southern DR Congo, and from East Africa westwards across the top of the forest belt to eastern Nigeria. Isolated populations occur in central Ethiopia, southern Sudan and South Sudan. It is largely absent from the south-east, where the Snouted Night Adder (*Causus defilippii*) occurs, and in West Africa is replaced by the West African Night Adder (*Causus maculatus*).

Natural History: Like other night adders it is terrestrial, but will climb low vegetation in pursuit of frogs. It is active by day and night, and will bask. It is fairly slow-moving, but strikes quite quickly, and it swims well. When inactive, it hides in holes, brush piles, under ground cover, etc. If threatened, it responds like other night adders, hissing and puffing and striking. It may also lift the front part

Rhombic Night Adder, *Causus rhombeatus*.
Left: Nigeria (Gerald Dunger). **Right:** Ethiopia (Tim Spawls).

Rhombic Night Adder, *Causus rhombeatus*.
Left top: South Africa (Bill Branch). **Right top:** South Sudan (Alan Channing).
Left bottom: Malawi (Gary Brown). **Right bottom:** spreading hood (Colin Tilbury).

of the body off the ground, flatten the neck and move forward with the tongue extended, looking like a small cobra. It eats frogs and toads, often swallowing them alive, the prey having suffered little obvious initial effect from the venom. One specimen, oddly, had eaten a bird. Females lay 7–26 eggs, roughly 26–37 × 16–20mm; in southern Africa these take about two and a half months to hatch.

Medical Significance: A mildly venomous species, quite common in parts of its range and liable to be trodden on. Antivenom is not available and should not be necessary. Documented cases are characterised by swelling, sometimes extensive, local pain, sometimes local haemorrhaging (more obvious in light-skinned victims), often fever and sometimes lymphadenopathy. There is rarely necrosis. Recovery usually takes a few days. In the early twentieth century, in southern and eastern Africa, this snake had a fearsome reputation (two of its early names were 'death adder' and 'demon adder'), but no human fatalities have been reliably reported. Along with the West African Night Adder, this species has enormously elongate venom glands, and although the venom is no more toxic than that of other night adders, it is produced in larger quantities, to enable the snake to immobilise large toads. Research indicates that its venom does affect blood-clotting times (although genuinely incoagulable blood has rarely been noted following a bite from this snake) and South African polyvalent antivenom does neutralise it. Dogs, even large ones, have been killed by this snake; all showed swelling and necrosis.

Rasmussen's Night Adder *Causus rasmusseni*

Identification: A small adder, either unpatterned or with narrow dorsal bars, described in 2014 from south-central Africa. The head is short, broad and slightly distinct from the neck, the snout is not upturned, and the eye is medium-sized with a round pupil. The top of the head is covered with nine large scales. The body is cylindrical or slightly depressed and the tail is short, 9–14% of total length (longer in males). The scales are soft, velvety and feebly keeled; in 16–18 rows at midbody; ventrals 130–139; subcaudals 24–33. The largest known specimen was

Causus rasmusseni

68.5cm, the others 35–55cm. Hatchling size is unknown. The back is brown or pinkish, either without markings, with very subdued markings or with narrow black cross-bars, and, like other species, with a V-shape on the head. The belly is grey to greyish white.

Habitat and Distribution: In moist savanna and woodland, between 1,200m and 1,600m. It is known from only a handful of records; the range extends from Ugalla Game Reserve in Tanzania to Mbala and Ikelenge in Zambia and Cuando Cubango in south-west Angola.

Natural History: Little known, but probably rather similar to other night adders (terrestrial, slow-moving, active by day and night, inflates the body when threatened and hisses and puffs, will strike if further molested, lays eggs and eats amphibians). This species is named after the Danish herpetologist Jens Rasmussen, who did major revisionary work on African snakes.

Medical Significance: Unknown, but presumably similar to other night adders. Antivenom is not available and should not be necessary.

Rasmussen's Night Adder, *Causus rasmusseni*. **Below:** Angola (Bill Branch).

False horned and spider-tailed vipers *Pseudocerastes*

A genus of broad-headed vipers, found in eastern Egypt and the Middle East, and reaching 80–90cm. Three species are known. Originally there was only one, the False Horned Viper (*Pseudocerastes persicus*), the two subspecies of which have now been elevated to full species level. A third, with a bizarre spider-like tail, was described in 2006 from western Iran. The two species of false horned viper have a huge range, from Egypt east to Afghanistan. They are terrestrial, secretive and largely nocturnal, and must cause a fair number of bites to those who dwell in these dry areas, but little seems to have been documented. In the laboratory, the venom of the species that reaches Egypt, the Western False Horned Viper (*P. fieldi*), surprisingly not only seems to lack the coagulopathic toxins that the eastern form has, but also has neurotoxic components, which is unviperlike. The eastern form, the Persian Horned Viper (*P. persicus*), has a more typical viper venom. This has implications for medical treatment, as the only antivenom available for treating the bites of both *P. persicus* and *P. fieldi*, produced in Iran, is raised from venom from Iranian *P. persicus*.

Western False Horned Viper *Pseudocerastes fieldi*

Identification: A short, stout viper with prominent horns, found in the Sinai Peninsula. The head is broad, flat and triangular, and covered with small scales. The eye is prominent, on the side of the head, and has a vertical pupil in bright light. The neck is thin. There is a stout, backward-curving horn above each eye, consisting of a cluster of scales. The body is broad and slightly flattened, and the tail is fairly short, 11–16% of total length. Oddly for a viper, the flank scales show no or only very slight keeling; the middorsal scales are keeled. The scales are in 21–23 rows at midbody;

Pseudocerastes fieldi

ventrals 127–142; cloacal scale entire; subcaudals 35–38. Maximum size about 90cm, most specimens 40–70cm; hatchlings 15–19cm. The colour is light brown, yellowish, pinkish or mottled grey, with darker (usually brown or reddish) blotches or cross-bands, which are often very straight-edged, and a row of blotches or dots along the sides. These may be very faded, and in some individuals the markings are so faded the snake just looks a uniform colour. There is a dark line through and behind the eye. The tail tip is often black, and the belly surface is white.

Habitat and Distribution: Occurs in semi-desert and desert country, in sandy, gravelly and rocky regions, but prefers areas with some vegetation. It is rarely found in dunes or other areas of soft sand, or on very steep slopes, although it occurs on scree rubble. It tends to avoid human habitation. This species is found from sea level to over 2,200m. In Africa, it occurs in the Sinai (although not in the dune fields in the north), then eastwards to western Iran.

Natural History: Terrestrial and slow-moving, although it can sidewind fairly rapidly. It is usually nocturnal, but sometimes active by day or at dusk, especially

Western False Horned Viper, *Pseudocerastes fieldi*. **Left:** captive (Paul Freed). **Right:** Syria (Tomáš Mazuch).

in the spring. It shelters in rodent burrows or other holes, in rock cracks or under bushes. This snake is fairly placid, but can strike quickly, to some distance. It mates in May or June, and lays 10–21 eggs in August; these contain large embryos and hatch after about a month of incubation. The diet includes mammals and birds; juveniles eat lizards.

Medical Significance: This is a nocturnal terrestrial snake, with a relatively large range. However, it lives mostly in inhospitable, sparsely inhabited country, and in Africa probably causes very few bites, although it is implicated in bite cases in Iraq and Jordan. It is a relatively large snake, so victims of suspected bites by this species should be transported to hospital. Antivenom is available, but see our comments in the generic introduction (p. 112).

Giant Old World vipers (Afro-Asian vipers) *Daboia*

A genus of four large vipers, reaching 1.5m or more, that occur in North Africa, the Middle East and south Asia; one species (Russell's Viper, *Daboia russelii*) is one of India's 'big four' deadly snakes, and causes many bites and fatalities. A single species, the Moorish Viper (*D. mauritanica*), occurs in Africa. It is a major cause of snakebite in the non-desert areas of the Maghreb as it is relatively common, often in agricultural areas, remains motionless if approached, and is active on the ground at dusk or in the dark. The venom appears to be cytotoxic and haemorrhagic, and deaths from the snake's bite have occurred. Antivenom is available, and the venom is relatively slow-acting, but this is a large snake and capable of delivering a lot of venom; a bite must be treated as an emergency and the patient taken to a well-equipped hospital.

Moorish Viper *Daboia mauritanica*

Identification: A large, stocky viper, uniform brown or with a zigzag pattern, in non-desert country of North Africa. The head is broad, flat and triangular, with a narrow snout, and is and covered with small scales. The eye is prominent, on the side of the head, and has a vertical pupil in bright light. The neck is thin, the body is broad and slightly flattened, and the tail is 10–14% of total length. The scales are in 26–27 rows at midbody; ventrals 157–176; cloacal scale entire; subcaudals 40–51. Maximum size about 1.8m, average 1–1.6m; hatchlings 19–28cm. The colour is very variable, but the ground colour is usually some shade of brown, grey-brown or pinkish brown, with a broad wavy or zigzag central stripe of darker brown or black; in the lighter-coloured '*deserti*' form (this was previously regarded as a separate species) the stripe is subdued or even faded to near invisibility, and the snake looks monocoloured. The underside is white or grey, with some darker speckling.

Daboia mauritanica

Habitat and Distribution: Occurs in light woodland (sclerophyllous forest), steppe meadow and semi-desert, often on or near rocky hills; it prefers areas with some vegetation and preferably near open water; in Morocco often in hill valleys. It is found from sea level to over 2,000m. It occurs from northern Western Sahara through Morocco to northern Algeria, Tunisia and north-west Libya, with a credible record from north-east Libya.

Natural History: Terrestrial and fairly slow-moving, but can strike rapidly, to a large distance. However, it is a fairly placid snake, although it can hiss very loudly. It is usually nocturnal, but sometimes active by day or at dusk, especially in the spring. It shelters in thickets (including of introduced prickly pear, as well as natural succulents), under bushes and in rock crevices, and is known to enter caves and abandoned mine galleries in the summer. In the higher parts of its range it hibernates in the winter. It mates in April or May, and lays 13–40 eggs in June or July; in captivity the eggs hatch after 40–60 days. The diet includes mammals and birds; juveniles eat lizards. It both hunts from ambush and actively prowls, investigating recesses for prey.

Medical Significance: This is a nocturnal terrestrial snake with a relatively large range, often in farming country in the mountains and high plateaux of north-west Africa. Studies in Morocco found the majority of snakebites reported were by this snake (although sometimes wrongly attributed to the North African Blunt-nosed Viper, *Macrovipera lebetina transmediterranea*, which as currently defined does not occur in Morocco). The bites were characterised by local swelling, bruising and pain; some victims had thrombocytopenia and other haemorrhagic disorders;

Moorish Viper, *Daboia mauritanica*. **Left and centre:** Morocco (Konrad Mebert). **Right:** Tunisia (Tomáš Mazuch).

one death involved a brain haemorrhage. Necrosis was present in some cases. This snake is relatively large, so victims of suspected bites should be transported to hospital. Antivenom is available.

Taxonomic Notes: The two colour forms of this snake (with obvious and faded/ indistinct markings) were until recently regarded as separate species, *Daboia mauritanica* and *Daboia deserti*. However, recent work indicates that the two are conspecific, although isolating events as a result of climatic fluctuations during the Pleistocene have created at least seven lineages, one from central Morocco eastwards to Libya, and the rest all in western and central Morocco.

Large Palearctic vipers *Macrovipera*

A genus of three large vipers of the eastern Mediterranean and Middle East. One subspecies of the most widespread form, *Macrovipera lebetina*, has been described from Algeria and Tunisia, under the name *M. lebetina transmediterranea*. Few specimens are known. The nominate subspecies is a large, dangerous snake, and its venom causes swelling, pain, necrosis and coagulopathy; death has been caused by acute renal failure. A bite from one is a medical emergency and must be treated in a well-equipped hospital.

North African Blunt-nosed Viper
Macrovipera lebetina transmediterranea

Identification: A very rare, pale, medium-sized adder, from northern Algeria and Tunisia. The head is broad and has a rounded snout, with flat sides and a pronounced angle between the top and sides of the head. The head shields, including the supraoculars, are fragmented and keeled. The neck is thin and the body stout and slightly depressed. The tail is 10–13% of total length. The dorsal scales are strongly keeled; in 25 rows at midbody; ventrals 150–164; subcaudals 37–51 (higher counts in males). The largest adults were just under 1m; the juveniles are probably around 15–20cm. This species is usually

Macrovipera lebetina transmediterranea

grey, with darker, usually brown cross-bars along the centre of the back; these are often offset along the actual spine. There is also a series of 34–41 vertical, tapering brown or rufous bars on the flanks.

Habitat and Distribution: Known solely from a handful of specimens from the vicinity of Oran and Algiers (Algeria) and Tunis (Tunisia), in what were oak woodlands on hill slopes near the coast. Other subspecies in Eurasia inhabit broken country, rocky hills and valleys where there is a reasonable amount of vegetation and also water sources.

Natural History: Described in 1988 from museum specimens, but not documented in the wild since then, so nothing is known of its biology. It is probably similar

North African Blunt-nosed Viper, *Macrovipera lebetina transmediterranea*. **Below:** captive, Tunisia (Juan Calvete).

to other subspecies: largely terrestrial but may climb bushes, active by night in summer, crepuscular in cooler seasons, hunts from ambush, and eats a variety or vertebrate prey. Some subspecies lay eggs that hatch in late summer; others give birth to live young.

Medical Significance: A very rare snake, and thus unlikely to cause many snakebites, although it is highly dangerous. Bites by other subspecies of this snake are known to be life-threatening, and cause bruising, acute pain, swelling, incoagulable blood and necrosis. Antivenom is available. The venom of the typical race (*Macrovipera lebetina lebetina*) has an experimental IV toxicity in mice of 0.80mg/kg, and bites are potentially lethal. However, an analysis of its venom indicates it inhibits the growth of tumours and thus may have a medical use.

Taxonomic Notes: Some authorities doubt the existence of this form in North Africa, as the localities of the types are old and vague and all scale counts lie within the range of other subspecies; however, others accept it. Captive specimens are present in medical research facilities in Tunisia, although their provenance is unknown. It may have been present but now extinct, as its original habitats are now greatly altered, or it might have been introduced from elsewhere in the Mediterranean. Further research may clarify the situation.

Palearctic vipers *Vipera*

A genus of over 20 species of small vipers, some with huge ranges and others with very restricted ranges, across Europe and northern Asia. Most look quite similar, either grey or brown with a wavy darker stripe down the centre of the back. They sometimes hybridise, and the status of some taxa is debated. Their venom largely causes swelling, bruising and pain, and occasionally necrosis; deaths are very unusual but a few are reported. Anaphylactic reactions sometimes occur. Originally, two species were described from North Africa, but recent research indicates that just a single, fairly variable species occurs, which includes a dwarf form that is a high-altitude ecotype.

Lataste's Viper *Vipera latastei*

Identification: A small viper with an upturned snout, from the high-rainfall areas north of the desert in the Maghreb. The head is distinct from the neck, and the head scales are small (apart from the supraoculars). The body is fairly stout and the tail is short, 10–14% of total length (longer tails usually males). The dorsal scales are strongly keeled; in 19–23 rows at midbody (the high-altitude populations usually 19–21); ventrals 124–138; subcaudals 29–47 (higher counts in males). The largest adults at lower altitude reach 60cm (average 25–40cm); at high altitude they

Vipera latastei

rarely exceed 35cm. Juveniles are 12–16cm. The colour is grey or brown, usually with a wavy darker brown line down the centre of the back; this may break up into bars, and the dorsal markings may be so faded as to be almost invisible. The flanks are marked by large, dark, amorphous blotches. Usually a dark bar extends backwards from the eye, broadening on the neck, and this may extend down the flanks. The underside is grey, sometimes very dark, and speckled to a greater or lesser extent.

Habitat and Distribution: Occupies (or originally occupied) quite a wide range of habitats, including coastal dunes, grasslands, agricultural land, open deciduous forests, montane meadows, rocky hills and scree slopes, from sea level to over 2,900m. It occurs from the Tichka Plateau in the western High Atlas of Morocco, eastwards through the High Atlas and Middle Atlas, in the Rif Mountains and along the northern Moroccan coastline, into the Tlemcen Mountains of western Algeria. It reappears in the central Tell Atlas of northern Algeria and in the Medjerda Hills on the north-west Tunisian border. It is also found throughout most of Spain, except the extreme north.

Lataste's Viper, *Vipera latastei*. **Left and right:** Morocco (Konrad Mebert).

Natural History: Largely terrestrial, but it will climb rocks and trees; people in Spain have been bitten while picking fruit by vipers that had climbed trees. It is nocturnal in summer and crepuscular or diurnal in colder months, although the high-altitude animals tend to be almost wholly diurnal. It hibernates in the winter. This is a fairly slow-moving snake, although it can strike quickly and is quick to disappear if disturbed. Mating occurs in spring, and females give birth to up to 10 live young in late summer. The diet includes a wide range of vertebrates, mainly small mammals and lizards.

Medical Significance: In its North African range this is now a very rare snake, and thus unlikely to cause many snakebites. Bites by most snakes of this genus are hardly ever fatal. Symptoms include bruising, local pain and swelling, and sometimes coagulopathy. A few cases involving anaphylaxis have been noted. Although an antivenom raised directly from venom of this snake does not seem to be available, the monospecific European Adder (*Vipera berus*) antivenom (ViperaTAb) has been shown to be extremely effective against its bites. However, antivenom for bites by *Vipera* species is needed in only a small minority of very severe cases.

Taxonomic Notes: For a long time there were assumed to be two species of *Vipera* in North Africa, *V. latastei* (subspecies *V. l. gaditana*) and the so-called Atlas Mountain Viper (*V. monticola*), a dwarf species of the High Atlas Mountains. However, recent genetic work shows that, in fact, there is only a single species, and what was *V. monticola* is merely a dwarf, high-altitude ecotype of *V. latastei*, confined to the High Atlas. Unfortunately, the lower-altitude populations of this snake are virtually, if not totally, extinct and the high-altitude populations are under severe threat, due to long-term changes in land usage to agriculture and urbanisation.

Family Elapidae Elapids (cobras, mambas and relatives)

A family of dangerous snakes with short, immoveable poison fangs at the front of the upper jaw. Many are of international medical significance, and some have powerful neurotoxic venoms that can cause death rapidly if untreated. Around 375 species and over 50 genera are known worldwide in tropical regions. There are none in Europe, but there has been a remarkable radiation of elapids in Australia, where they are the largest group of snakes. Members of the family include the mambas (*Dendroaspis*), cobras, King Cobra (*Ophiophagus hannah*), kraits (*Bungarus*), sea snakes (Hydrophiinae), coral snakes, African garter snakes (*Elapsoidea*) and taipans (*Oxyuranus*). All African land elapids belong to the subfamily Elapinae. The subfamily Hydrophiinae consists of the Australian terrestrial elapids and the sea snakes, one species of which, the Yellow-bellied Sea Snake (*Hydrophis platurus*), occurs off Africa's east-facing coasts.

Key to the African elapid subfamilies

1a Tail shaped like an oar; living in the sea or stranded on the foreshore......subfamily Hydrophiinae, sea snakes (pp. 192–93)

1b Tail not shaped like an oar; not living in the sea......subfamily Elapinae, land elapids (pp. 119–92)

Subfamily Elapinae Land elapids

There are around 42 species of land elapid in Africa, in seven genera. All African species have round pupils, lack a loreal scale and lay eggs (apart from the Rinkhals, *Hemachatus haemachatus*, a curious South African elapid). Most are large snakes, reaching 1.4m or more, and are highly dangerous, the exception being the small garter snakes (*Elapsoidea*), which are not known to have caused deaths. With care, elapids are usually identifiable by eye, using a mixture of size, location, appearance and behaviour. A number of elapids (in particular the mambas and the non-spitting cobras) have a powerful neurotoxic venom and are probably the worst snakes to be bitten by, in terms of the need to take rapid action to prevent death. The spitting cobras, which cause a lot of bites, tend to have cytotoxic venoms, often causing major long-term tissue damage; they wreak a terrible long-term toll on the health of people in the savanna. However, in broad terms, the elapids are medically less significant than the vipers, as they are largely alert species; they avoid confrontation and will move away if encountered, whereas a viper will freeze and hope to remain undetected.

Key to the African land elapid genera

1a Dorsal scales strongly keeled......*Hemachatus haemachatus*, Rinkhals (pp. 190–92)

1b Dorsal scales not strongly keeled......2

2a Rostral scale huge, laterally detached and shield-like......*Aspidelaps*, shield cobras (pp. 187–90)

2b Rostral scale not enlarged and shield-like......3

3a Black; does not spread a hood; in Egypt; midbody scale rows 23...... *Walterinnesia aegyptia*, Desert Black Snake (p. 185)

3b Does not have the above combination of characters......4

4a Midbody scales in 13 rows; adults small, less than 80cm; juveniles always with broad bands......*Elapsoidea*, African garter snakes (pp. 129–42)

4b Midbody scales in 13 or more rows; large, adults usually more than 80cm; juveniles not banded or with narrow bands......5

5a Three preoculars; head coffin-shaped, long and narrow......*Dendroaspis*, mambas (pp. 120–29)

5b One to two preoculars; head not coffin-shaped, more square......6

6a Dorsal scales in 13–17 rows; glossy black or brown above, black and yellow below; hood small; eye huge......*Pseudohaje*, tree cobras (pp. 182–84)

6b Dorsal scales in 17–27 rows; usually not black and bright yellow; hood usually large (except in Burrowing Cobra, *Naja multifasciata*); eye not huge......*Naja*, true cobras (pp. 143–81)

Mambas *Dendroaspis*

Mambas occur in sub-Saharan Africa. They are all large (over 2m), agile, slender, diurnal elapid snakes with long heads. Four species are known. Three are green and live in the forests and woodlands of West, Central and East Africa. The Black Mamba (*Dendroaspis polylepis*) is grey or olive and mostly lives in the savanna. Mambas are greatly feared, but their reputation is largely unjustified, and based to a large extent on legend; they are not aggressive and avoid confrontation. However, all mambas have neurotoxic venoms, containing unusual proteins called dendrotoxins, which cause pins and needles, stimulation of the autonomic nervous system and muscle fasciculations. The venom causes a rapidly progressing, descending paralysis, which can appear soon after the bite and progress to fatal respiratory paralysis. As with the non-spitting cobras, early symptoms may include ptosis, double vision, and inability to open the mouth, clench the jaw, protrude the tongue or swallow. Thus a mamba bite is a medical emergency and needs immediate rapid transport to hospital and prompt and thorough medical treatment.

The three green tree mambas are secretive, tend to stay up in trees and cause very few bites. However, the Black Mamba is often on the ground, frequently lives in areas with extensive agriculture, is confident in defence and can strike rapidly, to a great distance. Trying to kill a Black Mamba is a very risky thing to do. In some countries, Black Mambas cause a great many bites and they may be the leading cause of snakebite death. Antivenom is available, and may be needed in large quantities. Although the benefits of first aid in snakebite are much debated, a mamba bite is one situation where pressure bandaging, immobilisation therapy using a local pressure pad and artificial respiration may be life-saving as the victim is transported to hospital.

Key to the genus *Dendroaspis*, mambas

1a Olive, grey or brown; mouth lining dark; midbody scales in 21–25 rows; widespread in savanna......*Dendroaspis polylepis*, Black Mamba (p. 126)

1a Mostly or totally green; mouth lining pink; midbody scales in 13–19 rows; usually in forest and woodland......2

2a Uniform light green; lip scales immaculate; midbody scale rows 17–19; in eastern and south-eastern Africa......*Dendroaspis angusticeps*, Eastern Green Mamba (p. 121)

2b Darker green; lip scales finely black-edged; midbody scale rows 13–17; in Central and West Africa......3

3a Midbody scale rows 15–17; occurs west from western Kenya to Ghana...... *Dendroaspis jamesoni*, Jameson's Mamba (p. 123)

3b Midbody scale rows 13; occurs west from Benin to Senegambia......*Dendroaspis viridis*, West African Green Mamba (p. 125)

Eastern Green Mamba *Dendroaspis angusticeps*

Identification: A distinctive, vivid green mamba of the low-altitude forests and woodlands of the eastern and south-eastern coastal plains. It is a large, slender tree snake with a long head, and a relatively small eye with a round pupil; the iris is yellow. The tail is long and thin, 20–23% of total length. The scales are smooth; in 17–19 (sometimes 21) rows at midbody; ventrals 201–232; subcaudals 99–126. Maximum size 2.3m (possibly more), average 1.5–2m; hatchlings 30–40cm. The colour is bright uniform green above, sometimes with a sprinkling of yellow scales, and pale green below. Juveniles less than 60cm long are bluish green.

Dendroaspis angusticeps

Habitat and Distribution: Coastal bush, thicket and forest, woodland, moist savanna, evergreen hill forest and well-vegetated plantations, mostly at low altitude (less than 500m), but up to 1,700m in parts of its range. This species is tolerant of coastal agriculture and light urbanisation in eastern and southern Africa, often occurring in cashew nut and coconut plantations, in mango trees and in bushes and garden hedges, or anywhere thick enough for the snake to rest undetected. It occurs from the Somali border and the Boni Forest south along the entire East African coastal plain to the Ruvuma River, inland to the Usambaras, along the Rufiji River system and across south-eastern Tanzania to Tunduru. There are sporadic records in Malawi. Not yet recorded from northern Mozambique, but it is probably present along the coast. It reappears in southern Malawi and central and southern Mozambique, extending west into the eastern highlands of Zimbabwe. In South Africa it occurs in the lowland forests along the Kwazulu-Natal coastline, just reaching the eastern edge of the Eastern Cape. In Kenya there are isolated populations in the Nyambene Hills, Meru National Park and around Kibwezi; in Tanzania it is known from the forests around Mount Meru

Eastern Green Mamba, *Dendroaspis angusticeps*.
Left: South Africa (Tyrone Ping). **Centre:** Tanzania (Stephen Spawls). **Right:** Tanzania (Elvira Wolfer).

Eastern Green Mamba, *Dendroaspis angusticeps.*
Left: eating a bird (Bill Roland Schroeder). **Right:** hatchling (Johan Marais).

and southern Mount Kilimanjaro, extending east to the Kitobo Forest in Kenya; it might be more widespread in the northern Tanzanian hill forests.

Natural History: A fast-moving, diurnal, secretive tree snake that climbs expertly but will descend to the ground, and is sometimes seen crossing roads. It may climb very high in trees. It sleeps at night in the branches, not seeking a hole to sleep in but coiled up in a thick patch, and will also shelter in and under makuti (coconut thatch) roofs. Individuals may emerge to sunbathe in the early morning; anecdotal evidence indicates they are most active in the morning. This is an even-natured snake, not known to spread a modified hood or threaten with an open mouth like the Black Mamba (*Dendroaspis polylepis*). It is not aggressive; if threatened it tries to escape, and rarely tries to bite. Male combat has been observed in South Africa in June. Mating has been observed in April, May and July in southern Tanzania. It lays up to 17 eggs, roughly 30 × 60mm; these take two to three months to hatch. Hatchlings have been collected between December and February in southern Tanzania, and in June and July in Watamu, Kenya. This species feeds mostly on birds and their nestlings, rodents and bats, and occasionally lizards. Tanzanian snakes took a Lilac-breasted Roller (*Coracias caudatus*) and a Southern Blue-eared Glossy Starling (*Lamprotornis elisabeth*). Ground-dwelling rodents have been found in their stomachs, indicating they will descend to feed or hunt at low levels. Captive specimens will take chameleons. A recent study around Watamu indicates that Eastern Green Mambas spend much time waiting in ambush, but they will also forage, searching in trees, investigating holes and birds' nests. This snake has been found in huge concentrations within parts of its range (especially coastal Kenya and south-east Tanzania); concentrations of two to three snakes per hectare (i.e. 200–300 per square kilometre) have been noted.

Medical Significance: A dangerous snake, with a potentially deadly neurotoxic venom, but being alert, tree-dwelling, diurnal and secretive, it rarely comes into contact with humans. However, bites on the coast of East and south-east Africa occasionally occur as a result of the snake sheltering in thatched roofs or hiding in fruit and nut trees;

occasional bites to walkers who have trodden on or near a snake are also known. Antivenom is available. The venom contains unusual neurotoxins called dendrotoxins, which cause pins and needles, tingling and numbness, muscular contractions and a progressive paralysis. Pain and swelling, sometimes extensive, are also noted, as is dizziness, nausea, difficulty in breathing and swallowing, and swollen lymph glands. Peripheral necrosis has also been recorded, but this may be connected with bites on digits and mismanagement of the wounds. Occasional cases present with no neurotoxic symptoms; severe local swelling and blistering have been described, as well as mild coagulopathy and bleeding. Bites are rare but may need rapid hospital treatment with antivenom and respiratory support, although very few deaths are recorded. Some victims have noted sudden falls in blood pressure days after the bite.

Jameson's Mamba *Dendroaspis jamesoni*

Identification: A large, slender mamba of the Central African forests, with finely black-edged head scales. The head is long and the eye small, with a round pupil. The tail is long and thin, 20–25% of total length. The scales are smooth; in 15–17 rows at midbody; ventrals 202–227; subcaudals 94–113. Maximum size 2.64m (possibly more), large specimens usually males; average 1.5–2.2m; hatchling size unknown, but probably around 30cm. The colour is dull green above, mottled with black and yellow-green, pale green below, and yellow on the neck and throat. In

Dendroaspis jamesoni

some specimens (particularly juveniles) the black and yellow dorsal scales form narrow bands. The scales on the head and body are narrowly edged with black. Specimens from Uganda and Kenya (subspecies *kaimosae*) have a black tail; those from the centre and west of the range have a yellow tail and the tail scales are black-edged, giving a netting effect; in some Central African snakes the median dorsal scales are very dark, giving the impression of a black vertebral stripe.

Habitat and Distribution: Forest, woodland, swamp forest, forest–savanna mosaic, thicket and deforested areas, from sea level to about 2,200m (up into montane woodland, although observations from the Albertine Rift suggest it doesn't like cold forests). It will persist in areas after forest has been felled, providing there are still thickets and trees to hide in, and will move across open country to reach

Jameson's Mamba, *Dendroaspis jamesoni.*
Left: Angola (Bill Branch). **Centre:** Kenya (Bio-Ken Archive). **Right:** hatchling, Uganda (Mike Perry).

Jameson's Mamba, *Dendroaspis jamesoni*.
Left: captive (Matthijs Kuijpers). **Centre:** DR Congo (Eli Greenbaum). **Right:** Gabon (Olivier Pauwels).

isolated tree clumps and thickets. It is quite often found around buildings, in and around towns within the forest, and in parks and farms. Being a secretive green tree snake, it often clings on in well-inhabited areas without being seen. It occurs westwards from the western side of the Albertine Rift across the great forests of Central Africa to Nigeria and southern Benin, south to north-west Angola. This species has also been sporadically recorded in Tanzania from Mahale Peninsula and the Minziro Forest, from the Kakamega Forest in Kenya, and in Uganda from the shore of Lake Victoria, north to Lake Kyoga, and west to Lake Edward and Budongo and Kibale Forests. There are scattered records in Rwanda and Burundi, and it is also known from the Imatong Mountains, South Sudan. West of Ghana it is replaced by the West African Green Mamba (*Dendroaspis viridis*).

Natural History: A fast-moving diurnal tree snake that climbs expertly but will descend to the ground. It is not aggressive; if threatened it tries to escape, but occasional specimens may flatten the neck to a narrow hood. It will reluctantly strike if provoked. However, some snake keepers have described this snake as very nervous, bad-tempered and willing to bite. It may shelter in hollow trees, cracks and tangles of vegetation. Male–male combat and mating were observed in Nigeria in the dry season, December to January; mating was observed in Kakamega Forest in Kenya in September. Females lay eggs; clutches of 5–15 have been recorded in Cameroon, laid in March, and 7–16 in Nigeria, laid mostly in April, May and June; females collected in June in Uganda showed no yolk deposition, but those collected in November contained developing eggs. The diet in Kenya is known to include rodents, birds and arboreal lizards. Studies in Nigeria found that the adults largely ate birds and some rodents, while juveniles also took cold-blooded prey; known prey items included toads, agamas, several bird species (including woodpeckers, cisticolas and doves), and squirrels, shrews and mice. Captive specimens readily take rodents but are nervous feeders. A mark–recapture study in Nigeria found this snake was sedentary, moving only a short distance between captures.

Medical Significance: A dangerous snake, with a potentially deadly neurotoxic venom that is also cardiotoxic. However, being alert, tree-dwelling, diurnal and secretive, this species rarely comes into contact with humans, although those who farm or work in forests are at risk, and it is sometimes active on the ground during the day. Local and extended swelling and respiratory paralysis have been recorded following a bite. A bite from a single fang of a juvenile snake in Uganda caused considerable facial pain, chills and sweating, followed by slurring of speech and inability to breathe. A man collecting firewood in Nasho, Rwanda, was bitten by a 'green snake with a black tail', probably this snake, and died in three hours. A bite victim in Nigeria with two fang marks died four hours after a bite; the symptoms included vomiting, abdominal pain, blurred vision, involuntary urination and progressive articulation problems, leading to inability to speak. Finally, breathing stopped. The correct antivenom was unavailable. Few cases are reported in the literature. Bites will need rapid hospital treatment with antivenom and respiratory support.

West African Green Mamba *Dendroaspis viridis*

Identification: A large, slender tree snake with a long head and a small eye. The pupil is round and the iris is yellow. The tail is long and thin, 20–23% of total length. The dorsal scales are smooth, and often startlingly huge; in 13 rows at midbody; ventrals 211–225 (higher counts often females); subcaudals 111–125. Maximum size 2.4m (possibly more), average 1.4–2.1m; hatchlings 35–45cm. The body is various shades of green above, or sometimes brown, yellow or blue-green. The huge, elongate dorsal scales may be black- or yellow-edged, giving

Dendroaspis viridis

the impression of oblique bars. The head scales are usually finely black-edged. The scales on the tail are usually strongly black-edged, giving a net effect. The underside is shades of green but usually pale; sometimes the ventrals are strongly black-edged, especially in juveniles.

Habitat and Distribution: Mostly in forest, thicket and woodland, but sometimes in forest–savanna mosaic, secondary forest, abandoned plantations or in riverine woodland in savanna. Usually lives at low altitude, but up to 1,400m in parts of its range. It is quite tolerant of agriculture in places, especially crops like cocoa, or even urban areas so long as parks, ornamental trees or riverine forest remain. Within forest it favours clearings. It is sporadically recorded from southern Benin, Togo and the hill country east of Lake Volta in Ghana, and found more or less continuously in the forests and fringing savanna from southern Ghana westwards to southern Senegal and the Gambia.

West African Green Mamba, *Dendroaspis viridis*.
Left: Sierra Leone (Bill Branch). **Right:** Ghana (Stephen Spawls).

West African Green Mamba, *Dendroaspis viridis*.
Left: juvenile (Paul Freed). **Right:** captive (Mark Largel).

Natural History: A fast-moving, secretive diurnal tree snake that climbs expertly but will descend to the ground. Not aggressive; if threatened it will try to escape, and it rarely tries to bite. It may climb very high in trees. It sleeps at night in the branches, not seeking a hole to sleep in but coiled up in a thick patch, or sometimes on the tops of palm trees or out on a limb. Individuals may emerge to sunbathe in the early morning. This species has a reputation as a nervous snake in captivity. It lays 6–14 eggs. It hunts in trees but also on the ground. The diet largely comprises small rodents, but it also takes birds and bats, and possibly squirrels; one captive individual swallowed another.

Medical Significance: A dangerous snake, with a potentially deadly neurotoxic venom that has a toxicity of 0.5mg/kg, and a wet venom yield averaging 1,200mg. However, being alert, tree-dwelling, diurnal and secretive, it rarely comes into contact with humans; it also avoids confrontation, and few bites are recorded. In general, mamba venoms cause a rapidly progressing, descending paralysis, which can appear soon after the bite and advance to fatal respiratory paralysis. A bite in Ivory Coast led to respiratory paralysis, which responded to antivenom. However, the zoologist George Cansdale noted only three bite cases in 14 years in Ghana; one received antivenom, but all recovered. A recent bite victim in Liberia suffered symptoms of local envenoming: considerable local pain, muscular pain in the head, neck, chest and limbs, tender skin, and he became unable to move his limbs. He received eight ampoules (80ml) of South African antivenom, which complicated the symptomology as he was asthmatic, but recovery was complete, although his heart rate was elevated for 14 days. This antivenom is not actually raised using venom from this snake. Bites from this snake are likely to need rapid hospital treatment with antivenom and respiratory support.

Black Mamba *Dendroaspis polylepis*

Identification: A long, slender, fast-moving snake, with a long, narrow 'coffin-shaped' head, a fairly pronounced brow ridge and a medium-sized eye with a round pupil. The inside of the mouth is bluish black. The body is cylindrical and the tail long and thin, 17–25% of total length. The scales are in 23–25 (rarely 21) rows at midbody; ventrals 239–281; subcaudals 105–132. Maximum size probably about 3.5m (unsubstantiated reports of larger specimens, up to 4.3m, exist); adults average 2.2–2.7m; hatchlings 45–60cm. The juveniles grow rapidly and may reach 1.5–2m in a year. The colour is olive, brownish, yellow-brown or grey, or sometimes khaki or olive-green, but almost never black as the name suggests; this may instead be based on the black mouth lining. Juveniles are greeny-grey. The scales are smooth and have a distinct purplish bloom in some adult specimens. The belly is cream, ivory or pale green. The back half of the snake is often distinctly speckled with black on the flanks, especially in animals from drier areas.

Dendroaspis polylepis

Some snakes have rows of lighter and darker scales towards the tail, giving the impression of oblique lateral bars of grey and yellow.

Habitat and Distribution: In coastal bush, woodland, moist and dry savanna and occasionally semi-desert, from sea level to about 1,600m; very rarely above this altitude. It is most common in well-wooded savanna or riverine forest, especially where there are rocky hills and large trees. It is widespread in the southern third of Africa, northwards from eastern South Africa up to Kenya (where it is absent from the central highlands and much of the dry north) and southern Somalia. There are five apparently isolated populations elsewhere: one between South Sudan and southern Ethiopia; one centred on northern Ethiopia and western Eritrea; one in eastern Ethiopia and north-west Somalia; one between north-east Cameroon and Central African Republic; and one between Senegambia and south-western Burkina Faso. It might be more widely distributed or even continuous across the West and Central African savanna, but rare. Snakes matching its description are reported from north-west Nigeria. An experienced snake collector in Burkina Faso was killed by a Black Mamba; he had never encountered one before and assumed it was harmless.

Natural History: A large, fast-moving, alert, agile diurnal snake, equally at home on the ground, climbing rocks or in trees. It often moves with the head and neck raised high. It usually moves away if approached, but if threatened it may rear up, flatten the neck into a narrow hood, hiss loudly and open the mouth to display the black interior, while shaking the head from side to side. If further antagonised, it may respond with a long upwards strike. Black Mambas shelter,

Black Mamba, *Dendroaspis polylepis*.
Left: South Africa (Johan Marais). **Right:** Kenya (Stephen Spawls).

Black Mamba, *Dendroaspis polylepsis*.
Left top: hood spreading, Kenya (Stephen Spawls). **Right top and left bottom:** Kenya (Stephen Spawls).
Right bottom: South Africa (Johan Marais).

often for several months, in holes, termitaria, rock fissures and tree cracks, Hamerkop (*Scopus umbretta*) nests, and even in tree beehives and roof spaces. In South Africa this species frequently basked between 7 a.m. and 10 a.m. and from 2 p.m. to 4 p.m. It often shares its refuges with other snakes, including cobras and pythons (*Python*); at Kimana in Kenya two Black Mambas shared their hole with two Egyptian Cobras (*Naja haje*). Black Mambas lay 6–18 eggs (although 12–18 is more usual) measuring approximately 65 × 30mm, which usually hatch after two to three months, at the beginning of the rainy season. This species takes a variety of mammals of all sizes, from mice to squirrels, bats, bushbabies, mongooses and hyraxes; it may also eat birds (especially nestlings; one ate a Peregrine Falcon, *Falco peregrinus*, chick) and snakes. One specimen was observed eating winged termites as they left the nest. Males indulge in combat, neck-wrestling with bodies intertwined; this has been misidentified as courtship. These snakes are often mobbed by birds and also by squirrels and Vervet Monkeys (*Chlorocebus pygerythrus*).

Medical Significance: This is a very dangerous snake, and probably the worst snake to be bitten by in Africa in terms of the need to take rapid action to prevent death. Although it is diurnal and alert, and tends to avoid confrontation, it is nevertheless often found around farms and homesteads in the savannas of Africa, it sometimes shelters in buildings and it is confident in defence. If it feels it is cornered, it will try to defend itself by biting and it can strike a long way. Trying to kill a Black Mamba is a very risky undertaking. The venom is a potent neurotoxin, designed to rapidly knock down relatively large mammals like hyraxes and squirrels within minutes, so it quickly affects humans. It causes nausea, vomiting, sweating, diarrhoea, thick respiratory tract secretions and fasciculations. A rapid respiratory paralysis may develop within an hour. The venom yield can be many times the lethal dose. Any Black Mamba bite is a medical emergency; rapid transport to an advanced hospital will be needed, followed by treatment with antivenom and respiratory support. A number of well-documented cases exist. Occasional 'dry' bites are known. A Black Mamba bite is one situation where immobilisation therapy using a local pressure pad and artificial respiration may be life-saving as the victim is transported to hospital.

African garter snakes *Elapsoidea*

A genus of small (most are less than 70cm in length), ground-dwelling, burrowing venomous elapids (not to be confused with the harmless American garter snakes, genus *Thamnophis*). Ten species occur in sub-Saharan Africa, but in some areas they are very hard to tell apart; it has been jokingly said that there is only a single species of *Elapsoidea* but a lot of variation. In several reviews by eminent herpetologists, museum specimens have been reassigned from one species to another; the situation in parts of the Sahel and south-central Africa is particularly unclear. When identifying a garter snake, the locality will give valuable clues; bear this in mind when using the key.

Garter snakes have small, blunt heads and short tails. Nearly all have vivid dorsal bands when juvenile, but in most species the band either fades or darkens in the centre as they grow; the banding may serve to stop large garter snakes eating juveniles. Garter snakes have front fangs, but these are short and the snakes are usually reluctant to bite. When disturbed, they often inflate and flatten the body, throwing the bands into prominence, and also elevate the body (either the head and neck, or the middle) and hiss; sometimes they flatten and roll the tail into a vertical plane. They are active on the ground, especially during the rainy season, so are liable to be trodden on. However, they are of no real medical significance (although domestic dogs have been killed by garter snakes).

A curious suite of symptoms from garter snake bites includes pain (sometimes rapid and sometimes intense), nausea, vomiting, swelling, lymphangitis, stiffness of the bitten limb, nasal congestion, blurred vision and loss of consciousness; some of these may be due to something other than the actual venom. However, no fatalities have been recorded from African garter snake bites and no generic antivenom is available; a bite should be treated symptomatically.

Key to the genus *Elapsoidea*, African garter snakes

1a Known only from Somalia; unbanded......2
1b Known elsewhere in Africa; usually banded......3

2a Ventrals 136–141......*Elapsoidea chelazziorum*, Southern Somali Garter Snake (p. 132)
2b Ventrals 165–180......*Elapsoidea broadleyi*, Broadley's Garter Snake (p. 131)

3a Six upper labials; underside uniformly dark; in western West Africa......*Elapsoidea trapei*, Trape's Garter Snake (p. 142)
3b Seven upper labials......4

4a Snout elongate; very variable; all records south of 22°S......*Elapsoidea sundevalli*, Sundevall's Garter Snake (p. 140)
4b Snout rounded; virtually all records north of 20°S (except *Elapsoidea boulengeri*, which extends to 27°S)......5

5a Juvenile head orange; only in hill forest of north-east Tanzania and the Shimba Hills in Kenya......*Elapsoidea nigra*, Usambara Garter Snake (p. 137)

5b Juvenile head not orange; not in hill forests of north-east Tanzania or Shimba Hills of Kenya......6

6a In West and Central African savannas, largely north of 5°N......7

6b Largely south of 5°N......8

7a Eight to 17 body bands; occurs from South Sudan west to northern Cameroon; has an upper labial in contact with the prefrontal......*Elapsoidea laticincta*, Central African Garter Snake (p. 134)

7b Ten to 21 body bands; occurs from Central African Republic west to Senegambia (with a population in northern Angola); no contact between prefrontal and upper labial......*Elapsoidea semiannulata* (subspecies *moebiusi*), Northern Half-banded Garter Snake (p. 138)

8a In East Africa, mostly north of 5°S; bands retained throughout life......*Elapsoidea loveridgei*, East African Garter Snake (p. 135)

8b In southern and south-eastern Africa, mostly south of 5°S; bands not retained throughout life......9

9a Dark bands and pale bands are almost equal in width......*Elapsoidea guentheri*, Günther's Garter Snake (p. 133)

9b Dark bands usually broader than pale bands......10

10a Underside white......*Elapsoidea semiannulata* (subspecies *semiannulata*), Southern Half-banded Garter Snake (p. 138)

10b Underside dark grey or brown......*Elapsoidea boulengeri*, Boulenger's Garter Snake (p. 130)

Boulenger's Garter Snake (Zambezi Garter Snake)
Elapsoidea boulengeri

Identification: A small, glossy, banded snake of the savannas of south-eastern Africa. The head is short, slightly broader than the neck, and the eyes are set well forward, with round pupils. The body is cylindrical and the tail fairly short, 6–9% of total length. The scales are smooth; in 13 rows at midbody; ventrals 140–163; subcaudals 18–27 in males, 14–22 in females. Maximum size 76.6cm, average 40–70cm; hatchlings 12–14cm. Juveniles have 8–17 white, cream or yellow bands on a black or chocolate-brown body; the head is usually white

Elapsoidea boulengeri

with a black Y-shaped central mark. As the snake grows the bands fade to grey with white edging, then to a pair of white irregular rings, and finally they fade completely, leaving a uniformly dark grey or black adult. Sometimes a few scattered white scales remain. The chin and throat are white; the rest of the belly is dark grey.

Habitat and Distribution: In moist savanna, from sea level to 1,500m, or possibly higher. It occurs from south-eastern Tanzania south through Mozambique, much of southern Zambia and Malawi, and throughout Zimbabwe; it just enters northern Botswana and the Caprivi Strip in Namibia. It also occurs along South Africa's northern border. There are isolated records from western Tanzania (from Kibondo south to the western side of Lake Rukwa), eastern DR Congo

Boulenger's Garter Snake, *Elapsoidea boulengeri*.
Left: South Africa (Bill Branch). **Right:** juvenile, Malawi (Gary Brown).

(Tanganyika Province) and on the south-east Botswana border. A report from the Omatako Canal in northern Namibia is probably misidentified.

Natural History: A burrowing snake, living in holes, it emerges on warm, wet nights to look for prey or a mate. Slow-moving and inoffensive, it doesn't usually attempt to bite if picked up, but may hiss, inflate the body and jerk convulsively. Clutches of 4–8 large eggs have been recorded, in January and February in southern Tanzania; a hatchling was collected in June. The diet comprises other snakes (including its own species), lizards, small rodents and frogs. Zimbabwean snakes had eaten a rubber frog and a rain frog, both of which have poisonous skin secretions; Tanzanian specimens have been recorded eating snout-burrowers and squeakers.

Medical Significance: Low. The venom is not life-threatening, and neurological symptoms seem to be absent from reported cases. One recorded bite caused swelling of the bitten hand, immediate pain and transient nasal congestion; another showed pain and nasal congestion but no swelling.

Broadley's Garter Snake *Elapsoidea broadleyi*

Identification: A small, glossy snake with a short head that is slightly broader than the neck, and eyes that are set well forward, with round pupils. The body is cylindrical and the tail very short, 6–7% of total length. The scales are smooth; in 13 rows at midbody; ventrals 165–180; subcaudals 22–23 in males, 18 in the only known female. Maximum size 76.6cm, average 40–70cm; hatchlings 12–14cm. This species is known solely from three preserved specimens and a sight record; in preservative it is brown above and white below, without any trace of banding; the sight record was uniform blackish brown.

Elapsoidea broadleyi

Habitat and Distribution: Recorded from Jilib and Afmadow, and a specimen was photographed near Kismayu, in the dry low-altitude savanna of southern Somalia.

Broadley's Garter Snake, *Elapsoidea broadleyi.*
Below: Somalia (Mike McLaren).

Natural History: Unknown. It has never knowingly been seen alive, but is presumably similar to other garter snakes (i.e lives on the ground or in holes, nocturnal, lays eggs, and eats reptiles and amphibians).

Medical Significance: Its tiny range (at present) and presumably typical garter snake venom indicate it is of no real medical significance, although within its range many people don't have closed shoes and are liable to tread on it at night in the rainy season.

Southern Somali Garter Snake *Elapsoidea chelazziorum*

Identification: A small, glossy snake, known only from two specimens collected in Afgooye (originally Afgoi), near Mogadishu, Somalia, in the late 1970s. The head is short, slightly broader than the neck, and the eyes are set well forward, with round pupils. The body is cylindrical and the tail very short, 7–8% of total length. The scales are smooth; in 13 rows at midbody; ventrals 136–141; subcaudals 17–21. The two known specimens were 18cm and 41.5cm in total length. The adult was dark greyish brown, with the front of the head, the upper lips and the throat

Elapsoidea chelazziorum

off-white with brownish spots; the belly was dark grey. The juvenile was black above and the head and throat were off-white above, with scattered tiny brown dots; the belly was light grey. Unusually, neither individual was banded.

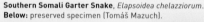

Southern Somali Garter Snake, *Elapsoidea chelazziorum.*
Below: preserved specimen (Tomáš Mazuch).

Habitat and Distribution: Taken in an area of sandy soil, in dry open woodland typical of consolidated coastal dunes, at Afgooye, in the valley of the Shebelle River, in the dry low-altitude savanna of southern Somalia.

Natural History: Unknown. It is presumably similar to other garter snakes (i.e. living on the ground or in holes, nocturnal, lays eggs, and eats reptiles and amphibians). One specimen was active at night after the first rainstorm of the season.

Medical Significance: Its tiny range (at present) and presumably typical garter snake venom indicate it is of no real medical significance, although within its range many people don't have closed shoes and are liable to tread on it at night in the rainy season.

Günther's Garter Snake *Elapsoidea guentheri*

Identification: A small, glossy snake, with equally spaced bands, of the savannas of south-central Africa. The head is short and slightly distinct from the neck. The scales are smooth; in 13 rows at midbody; ventrals 131–156; subcaudals 15–26. Maximum size about 62cm, average 30–50cm; hatchling size unknown. Juveniles are black with 14–28 broad, pale (white or grey-white) cross-bands; the bands are the same width as the spaces between. The pale bands darken in the centre as the snake grows, and usually disappear, so that adults appear grey or black. However, some adults are grey with white-edged black bands, and others retain a series of thin white bands, where the white-edged scales that marked the outer edges of the bands are all that remain of the bands.

Elapsoidea guentheri

Habitat and Distribution: Woodland, moist and dry savanna and forest–savanna mosaic, from sea level to 1,500m or slightly higher. It occurs from central DR Congo south into the miombo woodland of northern Angola, central Zambia and Zimbabwe, just getting into south-west Tanzania, and there is an isolated record from Gorongosa, Mozambique.

Natural History: Terrestrial and nocturnal, and fairly slow-moving. It hides in holes, under ground cover or in logs during the day, and is believed to spend much of its active time crawling about in holes, but will prowl on the surface

Günther's Garter Snake, *Elapsoidea guentheri*. **Left and right:** DR Congo (Colin Tilbury).

Günther's Garter Snake, *Elapsoidea guentheri*.
Below: DR Congo (Colin Tilbury).

at night, especially on damp nights after rain, and is sometimes active in the dry season. It is totally inoffensive, and many specimens will let themselves be handled freely, making no attempt to bite. However, if teased or molested it may flatten and inflate the body, showing the bands prominently, and may lift the front half of the body off the ground and jerk from side to side. It is likely to bite only if restrained. Females lay up to 10 eggs. This species eats snakes, lizards and amphibians. A small snake from DR Congo had insect larvae in its stomach, while a Zambian animal had the remains of a scorpion.

Medical Significance: A snake with a large range that lives in a well-populated area with lots of agriculture, but no bite cases are documented. No antivenom is available and it should not be necessary; treat symptomatically.

Central African Garter Snake *Elapsoidea laticincta*

Identification: A small, glossy snake with a short head, fairly large eyes and round pupils. The body is cylindrical and the tail fairly short, 6–9% of total length. The scales are smooth; in 13 rows at midbody; ventrals 139–151; subcaudals 13–25 (20 or more in males). Maximum size about 56cm, average 25–40cm; hatchling size unknown, but probably about 12cm. Young specimens have 8–19 pale brown or reddish-brown bands on a black body; the bands are about half as wide as the black spaces between. The belly is brownish yellow or dull orange. As the

Elapsoidea laticincta

snake grows, the bands darken in the centre, but the scales at the outer edges of the bands remain pale, so that adults appear to be black with a series of fine white double rings along the body; occasional specimens lose all their markings.

Habitat and Distribution: In moist and dry savanna, woodland and forest–savanna mosaic, between 300m and over 1,000m. It occurs in northern Cameroon and southern Chad, eastwards to the extreme south of Sudan, southwards through most of South Sudan to Nimule, on the border with Uganda, and in north-east DR Congo.

Natural History: Poorly known. It is terrestrial and fairly slow-moving, and nocturnal, emerging at dusk; it hides in holes or under ground cover during the day. It will prowl on the surface at night, especially on damp nights, after rain,

Central African Garter Snake, *Elapsoidea laticincta*.
Left: Central African Republic (Stephen Spawls). **Right:** preserved juvenile, South Sudan (Stephen Spawls).

and is sometimes active in the dry season. This species is totally inoffensive, and many specimens will let themselves be handled freely, making no attempt to bite. However, if teased or molested it may flatten and inflate the body, showing the bands prominently, and may lift the front half of the body off the ground and jerk from side to side. It is likely to bite only if restrained. Females lay eggs, but there is no information on size and number. The diet includes snakes, smooth-bodied lizards and frogs.

Medical Significance: A snake with a large range that lives in a well-populated area with lots of agriculture, but no bite cases are documented. No antivenom is available and it should not be necessary; treat symptomatically.

East African Garter Snake *Elapsoidea loveridgei*

Identification: A small, glossy garter snake from the highlands of eastern Africa, in which the adults retain the bands on the body. The head is short and the eyes set well forward, with round pupils. The body is cylindrical and the tail fairly short, 6–9% of total length. The scales are smooth; in 13 rows at midbody. There are three subspecies. The typical, eastern form (*Elapsoidea loveridgei loveridgei*) occurs in central Kenya (east of the Gregory Rift Valley) and northern Tanzania, and has 14–27 bands on the body, 150–164 ventrals and 19–30 subcaudals. The many-banded

Elapsoidea loveridgei

central form (*E. loveridgei multicincta*) occurs in western Kenya, parts of north-west Tanzania, most of Uganda (not the south-west), Ethiopia and South Sudan; it has 23–40 bands or pairs of transverse white lines, 151–171 ventrals and 17–28 subcaudals. Collet's East African Garter Snake (*E. l. colleti*) occurs in south-west Uganda, possibly northern Burundi (although there are only anecdotal records) and eastern DR Congo; it has 18–26 white bands, 159–172 ventrals and 17–26 subcaudals, and seems in parts of its range to be associated with the mountains of the Albertine Rift. Maximum size about 65cm, average 30–55cm; hatchling size

East African Garter Snake, *Elapsoidea loveridgei.*
Left top: Tanzania (Michele Menegon). **Right top:** Tanzania (Stephen Spawls).
Left bottom: Kenya (Konrad Mebert). **Right bottom:** Tanzania (Stephen Spawls).

East African Garter Snake,
Elapsoidea loveridgei.
Below: Tanzania (Mike McLaren).

unknown, but probably about 12–15cm. Juveniles are vividly marked with 19–36 narrow white, red or pinkish bands, or grey-brown bands with white edges, on the grey or black body; occasional individuals have yellow bands. Unlike in many other garter snakes, the bands usually persist in the adults and the band colour may vary a lot; some adults have vivid pink or red bands (particularly those from east of the Gregory Rift Valley in Kenya and northern Tanzania), others may have white-edged black bands, and in others the centre of the band darkens, leaving two fine white bands. Very rarely, specimens may become totally dark grey or black.

Habitat and Distribution: Mid-altitude woodland, moist savanna and grassland, from 300m up to 2,200m. It occurs from Kijabe and Nairobi, Kenya, south into the Crater Highlands and the vicinity of Kilimanjaro, Tanzania, and west into the Serengeti. It is widespread in high western Kenya, the northern shore of Lake Victoria in Uganda, west from Lake Victoria to the border, and south into Rwanda and northern Burundi. There are apparently isolated populations on Mount Marsabit, north and west of

Mount Kenya and the Taita Hills in Kenya, and in Sekenke and Ruaha National Park in central Tanzania. It is known from four somewhat disparate localities in southern and central Ethiopia, the Imatong Mountains in South Sudan, and a handful of localities in eastern and north-eastern DR Congo. A Somali record originally attributed to this species has been described as Broadley's Garter Snake (*Elapsoidea broadleyi*).

Natural History: Terrestrial and fairly slow-moving. It is nocturnal, emerging at dusk; individuals hide in holes or under ground cover during the day. It is believed to spend much of its active time in holes, but will prowl on the surface at night, especially on damp nights after rain, which acts as an activity trigger. A number were found dead on the road just south of Nairobi in December following a heavy, unseasonable storm the previous night. It may emerge in the day; a Nairobi specimen was crawling through the grass in the afternoon, following a rainstorm. This species is inoffensive; many specimens will not attempt to bite when handled gently, but if molested may flatten and inflate the body, showing the bands prominently, and may lift the front half of the body off the ground and jerk from side to side. If restrained it may try to bite. It lays 2–6 eggs; a Ugandan female was gravid in September; hatchlings were collected in Nairobi in April, May and July. The diet includes small snakes, lizards, frogs, reptile eggs and rodents; a beetle larva and a centipede have also been recorded. A captive specimen bitten by a large Black-necked Spitting Cobra (*Naja nigricollis*) survived despite internal bleeding.

Medical Significance: A snake with a large range that lives in a well-populated area with lots of agriculture, so liable to be trodden on. A bite in Nairobi caused local pain, which became intense when the bitten limb was moved, as well as swelling and pain in the lymph nodes; another bite was described as causing intense local tingling pain.

Usambara Garter Snake *Elapsoidea nigra*

Identification: A small, glossy snake with a short, blunt head, and eyes that are set well forward, with a yellow iris and round pupils. The body is cylindrical and the tail short, 5–7% of total length. The scales are smooth; in 13 rows at midbody; 151–169 ventrals; 13–24 paired subcaudals (males usually more than 18, females fewer than 18). Maximum size about 60cm, average 30–50cm; hatchling size unknown. Juveniles are beautiful, with an orange head and the first three or four body bands orange with black in between; the rest of the bands are grey

Elapsoidea nigra

with white edging. There is a total of 18–24 bands on the body and 2–3 on the tail; the bands are the same width as the spaces between. The pale bands darken as the snake grows, so that adults simply show a series of thin white bands, where the white-edged scales that marked their outer edges are all that remain; in some individuals the bands seem to disappear. However, if a uniformly dark adult is molested, it inflates the body, exposing the concealed white scale tips.

Usambara Garter Snake, *Elapsoidea nigra.*
Left: Tanzania (Lorenzo Vinciguerra). **Right:** Tanzania (Stephen Spawls).

Habitat and Distribution: An East African endemic. Mostly in high evergreen forest, but possibly in moist savanna as well, at 200–1,900m. It is confined to north-east Tanzania and extreme south-east Kenya; it is known from the Shimba Hills in Kenya, and Tanga, the North Pare Mountains, and the Usambara, Nguru and Uluguru Mountains in Tanzania. It is at conservation risk as it inhabits a restricted area of forest, but it might cope with deforestation.

Natural History: Poorly known. It is terrestrial, slow-moving and burrowing, living and hunting in holes, soft soil and leaf litter, and taking shelter in vegetation, rotten logs and under rocks, etc. It is sometimes active in the day following storms, but mostly hunts at night, emerging at dusk, especially on damp nights after rain; it is sometimes active in the dry season. This snake is inoffensive and gentle, but if molested it hisses, flattens the body and lashes about, sometimes raising coils off the ground. It is likely to bite only if restrained forcibly. Females lay 2–5 eggs, roughly 10 × 40mm; females with eggs in their oviducts have been captured mostly between October and December. This species eats mostly, or perhaps exclusively, caecilians (legless, worm-like amphibians).

Medical Significance: Nothing is known of its venom. Its tiny range and presumably typical garter snake venom indicate it is of no real medical significance, although within its range many people don't have closed shoes and are liable to tread on it at night in the rainy season.

Half-banded Garter Snake *Elapsoidea semiannulata*

Identification: The only garter snake in most of West Africa. A small, glossy snake, neither fat nor thin, with a short head and eyes set well forward, these with round pupils. The body is cylindrical and the tail fairly short. The scales are smooth; in 13 rows at midbody. There are two subspecies: in the Northern Half-banded Garter Snake (*Elapsoidea semiannulata moebiusi*) the ventrals number 148–167 and subcaudals 15–28; in the Southern Half-banded Garter Snake (*E. s. semiannulata*) the ventrals number 136–151 and subcaudals 13–28. Maximum size about 70cm in the northern form and 60cm in the southern form; average 30–50cm; hatchlings 14–16cm. Juveniles are vividly marked with 10–21 narrow white or yellow bands on the grey or black body. The white bands are half, or less than, the thickness of the black interspaces. When the snake reaches a length of 20cm or so the bands become grey in the centre and then darken, until all that is left are pairs of fine white lines; in large adults these lines may disappear. The underside is white, or brown in some Angolan animals.

Elapsoidea semiannulata

Habitat and Distribution: In moist and dry savanna, even semi-desert, and enters forest clearings and forest fringes in Central Africa. In south-central Africa it is mostly in savanna and woodland, but is also known from grassland. It is most common at low altitude, between sea level and 1,800m. The northern subspecies occurs across the West African savanna from Senegal and extreme southern Mauritania east to Central African Republic and South Sudan, and south into northern Republic of the Congo, DR Congo and Gabon; the southern form is in Angola, Namibia, northern Botswana and western Zambia. There are isolated records from eastern DR Congo and northern Republic of the Congo.

Natural History: Terrestrial and fairly slow-moving. This species is nocturnal, emerging at dusk. It hides in holes or under ground cover during the day, and is believed to spend much of its active time crawling about in holes, but will prowl on the surface at night, especially on damp nights after rain, and is sometimes active in the dry season. It is totally inoffensive, and many specimens will let themselves be handled freely, making no attempt to bite. However, if teased or molested it may flatten and inflate the body, showing the bands prominently, and may lift the front half of the body off the ground and jerk from side to side. It is likely to bite only if restrained. Females lay 2–8 eggs. In West Africa juveniles appear in January, indicating dry season hatching. The diet includes small snakes (it has been recorded eating burrowing asps, *Atractaspis*, in West Africa), smooth-bodied and burrowing lizards, worm lizards (amphisbaenids) and reptile eggs; it is also known to take frogs, including species with poisonous skin secretions, and termites. It probably also eats rodents.

Medical Significance: A snake with a large range that lives in a well-populated area with lots of agriculture, but no bite cases are documented. No antivenom is available and it should not be necessary; treat symptomatically.

Half-banded Garter Snake, *Elapsoidea semiannulata*. **Left:** Ivory Coast (Mark-Oliver Rödel). **Centre:** Nigeria (Gerald Dunger). **Right:** adult flattening coils and hiding head (Stephen Spawls).

Sundevall's Garter Snake (Southern African Garter Snake)
Elapsoidea sundevalli

Identification: A very variable garter snake of southern Africa, with five known subspecies. It is quite a large snake; one subspecies reaches 1m, another 1.4m. The head is short and the eyes are set well forward, with round pupils. The body is cylindrical and the scales smooth; in 13 rows at midbody; ventrals 152–181; subcaudals 13–33. The juveniles (all subspecies) are strongly marked with almost equal-sized light and dark bands, the light bands being cream, rufous or pinkish, or occasionally quite red, and the dark bands black

Elapsoidea sundevalli

or chocolate brown, often with the dark neck band extending forward to form an arrow shape on the head. Only in adults of Sundevall's Garter Snake (*Elapsoidea sundevalli sundevalli*) are the bands usually visible; the other four subspecies tend to lose all trace of banding early. The maximum size, number of bands and adult colouration for the five subspecies are as follows: Sundevall's Garter Snake (*E. s. sundevalli*), 1m, 21–38 bands, these darkening centrally, leaving only the white outer scales (occasionally they disappear, leaving uniform grey adults), and the upper lips may be yellow. Kalahari Garter Snake (*E. s. fitzsimonsi*), 76cm, 20–24 bands, belly and lowest row of scales on the flanks white, the bands usually disappearing when the snake gets to about 35cm, adults slaty-grey with a white belly. Highveld Garter Snake (*E. s. media*), 63cm, 18–26 bands, these fading in specimens over 32cm to pairs of fine white lines or disappearing, leaving uniform dark brown or grey adults with a paler

Sundavell's Garter Snake, *Elapsoidea sundevalli sundevalli*. **Left top:** adult, South Africa (Johan Marais).
Centre top: juvenile, South Africa (Johan Marais).
Kalahari Garter Snake, *Elapsoidea sundevalli fitzsimonsi*. **Right top:** Botswana (Stephen Spawls).
Highveld Garter Snake, *Elapsoidea sundevalli media*. **Left bottom:** South Africa (Bill Branch).
Long-tailed Garter Snake, *Elapsoidea sundevalli longicauda*. **Right bottom:** South Africa (Johan Marais).

De Coster's Garter Snake, *Elapsoidea sundevalli decosteri*. **Left and right:** South Africa (Tyrone Ping).

belly. Long-tailed Garter Snake (*E. s. longicauda*), 1.38m (the largest form in the genus), 19–23 bands, these disappearing when the snake reaches 33–45cm, adults black or grey, with a pale belly and upper lip scales. De Coster's Garter Snake (*E. s. decosteri*), 74cm, 22–26 bands, these darkening gradually to pairs of pale white lines and disappearing when the snake reaches around 30cm, leaving a grey snake with a white belly.

Habitat and Distribution: Endemic to southern Africa. This species has radiated to occupy a wide range of habitats, from coastal woodland and dune thicket to high-altitude grassland and the slopes of the Drakensberg Mountains, moist and dry savanna, semi-desert and rocky hill country, from sea level to 1,600m. The different subspecies may be usefully identified by their locality. Sundevall's Garter Snake occurs in southern and western Kwazulu-Natal, south-west Eswatini (Swaziland), south-east Mpumalanga and possibly the extreme Eastern Cape. The Highveld Garter Snake occurs from the west of the Northern Cape through the Free State to northern North West Province, most of Gauteng, and northern Mpumalanga and southern Limpopo. The Kalahari Garter Snake occurs in central and western Botswana and north-central Namibia. The Long-tailed Garter Snake is found in northern Limpopo, south-west Zimbabwe and southern Mozambique. De Coster's Garter Snake occurs in north-west Eswatini, extreme southern Mozambique and northern Kwazulu-Natal.

Natural History: A fairly slow-moving, nocturnal terrestrial snake. It hides in holes, under ground cover, in termitaria or in logs during the day, and is believed to spend much of its active time crawling about in holes, but will prowl on the surface at night, especially on damp nights after rain. It is sometimes active in the dry season. This species is totally inoffensive, and many specimens will let themselves be handled freely, making no attempt to bite. However, if teased or molested it may flatten and inflate the body, showing the bands prominently, and may lift the front half of the body off the ground and jerk from side to side. It is likely to bite only if restrained. Clutches of up to 10 eggs have been recorded, roughly 20 × 10mm. Its prey includes snakes, lizards and amphibians. One Long-tailed Garter Snake had eaten a golden mole; Sundevall's Garter Snake appears to eat mainly lizard eggs, but occasionally takes amphibians.

Medical Significance: Nothing known, though its venom is presumably a neurotoxin of the elapid type. A few bite cases have been recorded. One victim suffered only local swelling and slight pain that resolved in two days. Neurological symptoms were absent. Other victims have suffered nausea, vertigo, vomiting, weakness and swelling at the bite site; most symptoms resolved over four days. No antivenom is available. Treat conservatively by painkillers, elevation of the bitten limb and antibiotic cover. A bite from an adult Long-tailed Garter Snake might require medical intervention, as it is a large snake.

Trape's Garter Snake *Elapsoidea trapei*

Identification: A small, glossy snake, described in 1999 from Senegal and nearby areas in West Africa. The head is short, slightly broader than the neck, and the eyes set well forward, with round pupils. The body is cylindrical and the tail very short. The scales are smooth; in 13 rows at midbody; ventrals 155–170; subcaudals 18–27. Maximum size about 68cm, average 25–60cm; hatchling size unknown, but probably around 15cm. The colour is black, with or without white or grey-white cross-bars or bands. Juveniles are strongly banded black and white. This species

Elapsoidea trapei

differs from the northern subspecies of the Half-Banded Garter Snake (*Elapsoidea semiannulata moebiusi*) in that the ventral colouration is always grey-black, and from other garter snakes in possessing six upper labials; all the rest have seven.

Habitat and Distribution: In moist and dry savanna, at low altitude. It is known from Dakar in Senegal, eastern Senegal and extreme southern Mauritania, and is probably also in north-west Guinea.

Natural History: Unknown. It is presumably similar to other garter snakes (i.e. lives on the ground or in holes, is nocturnal and lays eggs). The diet is known to include snakes, lizards and centipedes.

Trape's Garter Snake, *Elapsoidea trapei*.
Below: Senegal (Wolfgang Wuster).

Medical Significance: Its small range (at present) and presumably typical garter snake venom indicate it is of no real medical significance, although within its range many people don't have closed shoes and are liable to tread on it at night in the rainy season.

True cobras *Naja*

Famous for their ability to spread a hood, and worshipped and feared in the ancient kingdoms of Egypt, the true cobras are a genus of over 30 species of highly venomous and medically significant snakes, some very large, found in Africa and Asia. For a long time, only four species were known from Africa: the Cape Cobra (*Naja nivea*), Black-necked Spitting Cobra (*N. nigricollis*), Egyptian Cobra (*N. haje*) and Forest Cobra (*N. melanoleuca*). Two water cobras (*Boulengerina*) and a Burrowing Cobra (*Paranaja*) were also known. However, ongoing taxonomic work, greatly aided by analysis of their DNA, has dramatically split the genus. As well as showing that *Boulengerina* and *Paranaja* belong in *Naja*, there are now seven species of spitting cobra, and the Egyptian Cobra has been split into five species (four in Africa) and the Forest Cobra into five; only the Cape Cobra has not been broken up. More splits may be on the cards.

The cobras are a major snakebite hazard in Africa. No matter where you travel in Africa (except true desert and some highland areas over 2,400m), you will find at least one species of cobra, and often more. Due to their secretive behaviour and nocturnal habits, cobras often persist in urbanised areas. They are active on the ground at night. Many are large, are willing to bite if they feel threatened, and deliver a large dose of venom. The spitting cobras have a cytotoxic venom that has hideous long-term local necrotic effects, often leading to complications that greatly reduce the quality of life, and may lead to death; see our comments in the introduction to the group ('the spitting cobra clade', pp. 167–81). They are also quick to defend themselves by spitting venom, which, if it lands in the eyes, is horribly painful, and if untreated can lead to long-term damage and blindness. The non-spitting cobras (Egyptian cobra complex, Cape Cobra and forest cobra complex) largely have a potent neurotoxic venom that can kill quickly. Necrosis is rare but there may be some swelling. Classic neurotoxic symptoms may appear as early as 30 minutes after the bite and can evolve to fatal respiratory paralysis within 2–16 hours. Symptomology often starts with ptosis, double vision, then weakness of the muscles controlled by the cranial nerves. The victim cannot open the mouth, stick out the tongue, clench the jaws, swallow, protect the airway from secretions or flex the neck, and eventually cannot breathe.

A bite from a cobra is a medical emergency and victims should be transported rapidly to hospital, particularly if the bite is from one of the Egyptian or forest cobras. As with the Black Mamba (*Dendroaspis polylepis*), a bite from a forest or Egyptian cobra is one situation where immobilisation therapy using a local pressure pad (see Figures 48–50, p. 299) and artificial respiration may be life-saving as the victim is transported to hospital.

Identification of cobras to species level is often possible; look at our pictures. The locality gives clues (there are rarely more than two or three species in an area), and if the snake is seen clearly, the size, colour and the patterns/bands on the neck will help. Take a photograph if it can be done safely. We have provided a technical key, although this can really only be used with a dead snake.

Key to the genus *Naja*, true cobras

Note: although we are fairly confident with this key, some black specimens of *Naja melanoleuca* and *Naja subfulva* from Central Africa (see couplet 13) are virtually impossible to distinguish, even in the laboratory.

1a Fewer than 40 subcaudal scales; does not spread a hood......*Naja multifasciata*, Burrowing Cobra (p. 155)

1b More than 40 subcaudals; spreads a hood when threatened......2

2a Dorsal scales from spine down to ventral scales in an oblique but straight line; always associated with watercourses in Central Africa......3

2b Dorsal scales from spine down to ventral scales change direction, curving or kinking on the lower flanks; not necessarily associated with watercourses in Central Africa......4

3a Midbody scale rows 21–25; where present, bands black on brown......*Naja annulata*, Banded Water Cobra (p. 156)

3b Midbody scale rows 17; bands yellow on dark brown or black......*Naja christyi*, Congo Water Cobra (p. 158)

4a Upper labials separated from the eye by subocular scales......5

4b At least one upper labial in contact with the eye......8

5a Dorsal scales usually in 25–27 rows on the neck and 21–23 at midbody; found only in West African savannas between north-west Nigeria and Senegal......*Naja senegalensis*, Senegal Cobra (p. 148)

5b Dorsal scales usually in 23 or fewer rows on the neck and 21 or fewer at midbody; largely not in West African savannas between north-west Nigeria and Senegal......6

6a Dorsal scales usually in 21–23 scale rows on the neck and 21 (sometimes 19) at mid-body; always north of 10°S......*Naja haje*, Egyptian Cobra (p. 146)

6b Dorsal scales usually in 14–21 scale rows on the neck and usually 15–19 at midbody; always south of 10°S......7

7a Dorsal scale rows usually 19 at midbody; on the eastern side of southern Africa......*Naja annulifera*, Snouted Cobra (p. 150)

7b Dorsal scale rows usually 17 at midbody; on the central and western side of southern Africa......*Naja anchietae*, Anchieta's Cobra (p. 151)

8a Sixth upper labial largest and in contact with the postoculars; one anterior temporal; do not spit venom......9

8b Sixth upper labial not the largest and not in contact with the postoculars; two or three anterior temporals; all capable of spitting venom......14

9a Rostral scale about as broad as deep; dorsal colour usually shades of yellow or brown; upper labials never black-edged; always south of 18°S and west of 30°E, in southern Africa......*Naja nivea*, Cape Cobra (p. 153)

9b Rostral scale much broader than deep; dorsal colour usually olive, brown, black or banded yellow and black; upper labials nearly always black-edged; always north of 18°S except in South Africa, where always east of 30°E......10

10a Confined to São Tomé Island; ventral scales mostly black......*Naja peroescobari*, São Tomé Forest Cobra (p. 166)

10b On mainland Africa or adjacent islands; ventral scales not always mostly black......11

11a Always strongly banded in black and yellow bands of almost equal width; in West African savannas from south-west Chad west to Senegal......*Naja savannula*, Banded Forest Cobra (p. 164)

11b Not banded black and yellow, or with fairly narrow yellow bands; mostly in forest and woodland outside of the West African savannas......12

12a Dorsal colour black; in forest west of Benin; midbody scale rows usually 17, sometimes 19......*Naja guineensis*, West African Forest Cobra (p. 165)

12b Dorsal colour brown or black; in forest, woodland and moist savanna east of central Ghana; midbody scale rows usually 19, range 17–21......13

13a Dorsal colour always black; ventral banding usually extends over anterior third of the belly; posterior venter black; largely confined to Central African forest areas......*Naja melanoleuca*, Central African Forest Cobra (p. 162)

13b Dorsal colour brown or black above; ventral banding usually extends over 15–20% of belly, sometimes indistinct; remainder of venter sometimes light; widespread in forest, woodland, riverine forest and moist savanna in central, eastern and southern Africa......*Naja subfulva*, Eastern Forest Cobra (p. 160)

14a Adults usually less than 1.5m; dorsal surface of entire body usually brown, maroon, rufous, red, orange or yellow; unmarked dorsally except 1–3 dark rings on the neck (often faded in adults); always north of 5°S......15

14b Adults not necessarily less than 1.5m; dorsal surface usually black, brown, olive or grey; throat bars do not form rings; not always north of 5°S......17

15a Confined to West African savannas and semi-desert west of Cameroon; ventrals 160–195; no black 'teardrop' marking under eye......*Naja katiensis*, West African Brown Spitting Cobra (p. 177)

15b All specimens (apart from those from Aïr Massif, Niger) always east of Cameroon and the eastern Chad–Niger border; ventrals 192–228; usually has a black 'teardrop' marking under eye......16

16a Dorsal colour brown, maroon, grey or olive, with 2–3 bands on the neck; mostly in north-east Africa north of 10°N......*Naja nubiae*, Nubian Spitting Cobra (p. 180)

16b Dorsal colour usually orange, pink or red, sometimes pinky-grey, red-brown, yellow or steel grey, with 1–2 bands on the neck; in eastern north-east Africa, south of 10°N......*Naja pallida*, Red Spitting Cobra (p. 178)

17a Never black; underside yellow with a single broad, deep brown throat bar; never pink on the neck; reaches 2.7m in length......*Naja ashei*, Ashe's Spitting Cobra (p. 170)

17b Dorsal colour brown, grey or black; never has a single brown, deep throat bar; usually less than 2m in length......18

18a Midbody scale rows 23–25 (rarely 21 or 27); dorsal colour grey, olive or brown; dorsal scales usually dark-edged; black throat bars almost never more than seven scales deep......*Naja mossambica*, Mozambique Spitting Cobra (p. 175)

18b Midbody scale rows 17–21 (rarely 23); dorsal colour grey, black, olive-brown or finely banded; dorsal scales not dark-edged; a black throat bar at least 12 scales deep almost always present......19

19a Usually numerous fine, zebra-like cross-bands on body and tail......*Naja nigricincta nigricincta*, Zebra Cobra (p. 172)

19b Dorsum uniform, not finely banded......20

20a Ventrals 175–219......*Naja nigricollis*, Black-necked Spitting Cobra (p. 168)

20b Ventrals 221–228......*Naja nigricincta woodi*, South-western Black Spitting Cobra (p. 174)

The Egyptian cobra (*Naja haje*) complex

The Egyptian Cobra was long regarded as a single species, but taxonomic work has revealed that there are five species in the group, with four in Africa: the 'typical' Egyptian Cobra (*Naja haje*), the Snouted Cobra (*N. annulifera*), Anchieta's Cobra (*N. anchietae*) and, most recently, the Senegal Cobra (*N. senegalensis*). Extralimitally, an Arabian form is known, the Arabian Cobra (*N. arabica*). Whether the two southern African forms are truly two separate species or merely a cline is uncertain; intermediates exist and the exact identity of some Zambian specimens is debated.

Egyptian Cobra *Naja haje*

Identification: A large, thick-bodied cobra with a broad head, distinct from the neck, a fairly large eye and a round pupil. Members of the Egyptian cobra complex have a subocular scale row, separating the eye from the upper labials. The body is cylindrical and the tail fairly long, 15–18% of total length. The scales are smooth; in 21 (very occasionally 19) rows at midbody; ventrals 191–222; subcaudals paired, 53–68. Maximum size 2.5m (possibly larger), average 1.3–1.8m; hatchlings 25–30cm. The colour is astonishingly variable. Most often it is some

Naja haje

Egyptian Cobra, *Naja haje*.
Left top: Kenya (Stephen Spawls). **Right top:** juvenile, Morocco (Konrad Mebert).
Left bottom: Uganda (Stephen Spawls). **Right bottom:** Morocco (Konrad Mebert).

Egyptian Cobra, *Naja haje.*
Left top: Nigeria (Gerald Dunger). **Right top:** West Ethiopia (Stephen Spawls).
Left bottom: Tanzania (Stephen Spawls). **Right bottom:** Eritrea (Tony Kamphorst).

shade of brown or tan, but may be black (especially in Morocco and parts of West Africa), rufous (upper Nile basin), red-brown, yellow or grey. It is usually light brown, yellow or cream underneath, often with a broad grey or brown throat bar, 8–20 scales deep, which is visible when the hood is spread. There are often irregular brown or yellow blotches and speckles on the neck, back and underside. Some brown specimens are patterned with yellow scales on the back, and the light ventral colour may extend onto the flanks in irregular blotches; some East African and Ethiopian snakes have light bands on the posterior body. Juveniles may be yellow, grey, orange or reddish, with a distinctive dark ring on the neck or a black head and neck, sometimes with a cream or yellow collar. Some juveniles have fine dark back bands; others have yellow bands, with a fine dark ring in the centre of the band.

Habitat and Distribution: Mostly in semi-desert, moist and dry savanna, woodland (never forest), grassland and agricultural lands. Around the Sahara it is not in true desert but instead in vegetated wadis, agriculture, etc. In some areas (e.g. the Ethiopian Rift Valley) it is common in degraded badlands. It is found from sea level to around 1,800m. This species occurs from Western Sahara eastwards across North Africa (although it is absent from the north Moroccan and Algerian coasts, and records are lacking for parts of Algeria) to Egypt, and southwards down the Nile. It then occurs westwards across the Sahel and savanna to at least Mali (records from further west are probably the Senegal Cobra, *Naja senegalensis*). Its range extends south into northern Uganda, south-central Kenya and northern Tanzania, and also into Eritrea, northern Somalia and the central Ethiopian Rift Valley. It is often uncommon, or curiously absent, from pockets in its range. It may have been displaced from some areas by more successful snakes, or it might be present and simply overlooked; in other restricted areas it can be surprisingly common.

Natural History: Mostly terrestrial. Clumsy but quick-moving, it is active by day and night; it often basks in the day, but will hunt at any time. It often lives in termite holes, rock fissures, hollow trees or holes; in Kenya it is known to share its refuge with Black Mambas (*Dendroaspis polylepis*). It sometimes climbs trees. This snake is quick to rear up (usually only to 30cm or so, but sometimes it rears quite high) and spread a broad hood when threatened; if further provoked it will hiss loudly, rush and strike at the aggressor. It sometimes shams (feigns) death. Females lay up to 23 eggs, averaging 50–60 × 30–35mm; a Kenyan snake laid 23 eggs in early March, and these hatched at the end of May. This species eats a wide range of prey; it is fond of toads but will take eggs (a notorious chicken-run raider), small mammals and other snakes.

Medical Significance: This is a highly dangerous snake. It is large, and thus delivers a large dose of potent neurotoxic venom if it bites; it will rear up if molested and may cause a bite high up on the body. It is often found in agricultural areas; it raids domestic fowl runs, and may come around habitation looking for water or food. If it feels threatened it will try to bite from a short rush. Bites are reported from the Nile Valley and parts of the West African Sahel and savanna, although there are relatively few documented cases. In West Africa, this is a species favoured by snake charmers, as these snakes are alert and willing to remain reared up for a while; consequently the charmers (and members of their audiences) are often bitten. Recent work indicates that East African examples of this snake have an extremely toxic venom. Bites may cause some local swelling, but necrosis does not develop. Classic neurotoxic symptoms appear as early as 30 minutes after the bite and can evolve to the point of fatal respiratory paralysis within 2–16 hours.The neurotoxic venom causes progressive descending paralysis, starting with ptosis and paralysis of the neck and head muscles, and proceeding to paralysis of the lungs. A bite is a medical emergency and will probably need treatment with a respirator and antivenom in a well-equipped hospital. A bite from this snake is one situation where pressure bandaging, immobilisation therapy and artificial respiration may be of benefit as the victim is transported to hospital.

Taxonomic Notes: Some authorities regarded the Moroccan and Western Saharan populations of this snake as a separate subspecies, or even species, *Naja (haje) legionis*, but molecular analysis indicates that they cluster within the typical species, *Naja haje*.

Senegal Cobra *Naja senegalensis*

Identification: A recently described large, thick-bodied brown cobra, in the savanna of West Africa. The head is broad and distinct from the neck, the eye is fairly large with a round pupil, and a subocular scale separates the eye from the upper labials. The body is cylindrical and the tail fairly long, 15–19% of total length (usually longer in males). The scales are smooth; in 21 (very occasionally 23) rows at midbody and 25–27 rows on the neck (and thus it is distinguished from other members of the Egyptian cobra complex, which nearly always have 23 or

Naja senegalensis

Senegal Cobra, *Naja senegalensis.*
Left top: Senegal (Jean-Francois Trape). **Right top:** Senegal (Laurent Chirio).
Left bottom: juvenile (Jean-Francois Trape). **Right bottom:** underside (Jean-Francois Trape).

fewer); ventrals 205–225; subcaudals 56–66. Maximum size 2.45m (possibly larger), average 1.3–1.8m; hatchlings probably 20–30cm. The colour is uniformly dark brown above, sometimes with light (yellow or reddish) speckling; the underside is yellow-brown, with darker speckling (especially towards the middle of the ventral scales), but the throat and neck are usually uniformly dark. The sides of the head are usually uniform brown, and this is a good field character. Juveniles are more grey to greyish brown, sometimes with faint fine cross-bars, and with a black head and neck; there is usually a distinctive pale mark on the nape and this occasionally persists in adults. The juvenile underside is unmarked (except the black head and neck barring).

Habitat and Distribution: In the Guinea and Sudan savanna of West Africa, between sea level and 600m. Local information indicates that in some areas it is often found near water sources; one was caught in a fisherman's net. It occurs from south-western Niger, eastern Nigeria and northern Benin westwards to Senegal.

Natural History: Presumably similar to the Egyptian Cobra (*Naja haje*); that is, terrestrial, quick-moving, active by day and night, and shelters in holes, rock fissures, hollow trees, etc. It presumably lays eggs. Known prey items include the Desert Toad (*Sclerophrys xeros*) and the Western Beaked Snake (*Rhamphiophis oxyrhynchus*); it probably takes a wide range of small vertebrates.

Medical Significance: Presumably similar to the Egyptian Cobra, so likely to be a major snakebite hazard within its range, being large and liable to deliver a dose of potent neurotoxic venom if molested. However, few bites are reported; a recent one from Nigeria showed a typical suite of neurotoxic symptoms. The neurotoxic venom probably causes progressive descending paralysis, starting with ptosis and paralysis of the neck and head muscles, and proceeding to paralysis of the lungs. A bite is a medical emergency and will probably need treatment with a respirator and antivenom in a well-equipped hospital. A bite from this snake is one situation where pressure pad immobilisation and artificial respiration may be life-saving as the victim is transported to hospital.

Snouted Cobra *Naja annulifera*

Identification: A large brown, grey or banded cobra of the savannas of south-east Africa. The head is broad and distinct from the neck, the eye is fairly large with a round pupil, and a subocular scale separates the eye from the upper labials. The body is cylindrical and the tail fairly long, 14–18% of total length (usually longer in males). The scales are smooth; in 19 (very occasionally 21) rows at midbody; ventrals 175–208; subcaudals 50–67. Maximum size about 2.5m (possibly larger; anecdotal reports of 3m specimens exist), average 1.3–1.8m; hatchlings

Naja annulifera

probably 22–34cm. The colour is usually shades of brown, rufous-brown, grey, tan or sometimes black. A banded morph, with brown and yellow bands, occurs throughout its range; the juveniles are unbanded but the adults develop bands at a length of 60cm or so. The underside is dull yellow, mottled or barred with dirty brown to a greater or larger extent.

Habitat and Distribution: A snake of dry and moist savanna and woodland ('bushveld'), between sea level and 1,600m. It occurs from central Zambia south through most of Zimbabwe, southern Malawi and the southern half of Mozambique to eastern Botswana and northern South Africa, as far south as parts of coastal Kwazulu-Natal. There are isolated northern records from Mpika in Zambia, Makuzi Lodge and Likoma Island in Malawi, and Livingstone in southern Zambia.

Natural History: A large terrestrial snake (although it will climb rocks), active by day and night – often by day in wild regions and by night in inhabited areas. It tends to have a permanent home in termite mounds, holes or rock fissures, where

Snouted Cobra, *Naja annulifera*. **Left and right:** Eswatini (Swaziland) (Richard Boycott).

Snouted Cobra, *Naja annulifera*. **Left:** South Africa (Luke Verburgt). **Right:** Botswana (Stephen Spawls).

it may live for years if undisturbed. In the cooler months it often basks near its home, especially in the morning. It is relatively quick-moving and, although not aggressive, if confronted and unable to escape it will rear up fairly high and spread a broad hood; if further antagonised it may hiss loudly and rush forward to bite. Occasional individuals may sham (feign) death. This snake lays 8–33 eggs (roughly 30 × 50mm), in early summer in southern Africa. The diet is mostly small vertebrates, such as toads, birds, rodents and other snakes, including Puff Adders (*Bitis arietans*). It often raids chicken runs and can be a nuisance to poultry farmers.

Medical Significance: A dangerous snake; widespread in agricultural country, active by night and day and on the ground, will bite if inadvertently cornered, and rears up, so may deliver a bite high on the body. The neurotoxic venom causes progressive descending paralysis, starting with ptosis and paralysis of the neck and head muscles, and proceeding to paralysis of the respiratory muscles. A bite is a medical emergency and will probably need treatment with a respirator and antivenom in a well-equipped hospital. A bite from this snake is one situation where pressure pad immobilisation and artificial respiration may be life-saving as the victim is transported to hospital.

Taxonomic Notes: A member of the Egyptian cobra (*Naja haje*) complex, originally seen as a subspecies of that snake, but elevated to full species level by Don Broadley in 1995.

Anchieta's Cobra *Naja anchietae*

Identification: A large brown or banded cobra of the savannas of south-west Africa. The head is broad and distinct from the neck, the eye is fairly large with a round pupil, and a subocular scale separates the eye from the upper labials. The snout is quite noticeably pointed. The body is cylindrical and the tail fairly long, 15–19% of total length (usually longer in males). The scales are smooth; in 17 (very occasionally 15 or 19) rows at midbody; ventrals 171–200; subcaudals 49–66. Maximum size about 2.3m (possibly larger), average 1.3–1.8m; hatchlings probably 22–34cm. Both uniform and banded morphs occur. The uniform morph is usually grey or brown, although some individuals are purple-brown or almost black; juveniles are usually yellow or yellow-brown. The underside is a lighter cream, light grey, brown or tan, often with darker mottling that increases towards the tail. There is a deep, dark throat bar, which is conspicuous and forms a ring in juveniles, but fades in adults. The banded morph is yellow and brown; the yellow bands are usually narrower than the brown bands (but not always).

Naja anchietae

Habitat and Distribution: Semi-desert, dry and moist savanna and woodland, from sea level to about 1,600m, or possibly higher. It occurs from the northern Kalahari in Botswana and central Namibia, north into most of Angola and south-west Zambia. There are isolated records from the Lake Bangweulu area in Zambia and southern Haut-Katanga in DR Congo.

Natural History: Similar to the previous species. A large terrestrial snake (although it will climb rocks), active by day and night – often by day in wild regions and by night in inhabited areas. Juveniles seem to be more diurnal than adults. It is crepuscular at some times of the year. It tends to have a permanent home in termite mounds, holes or rock fissures, where it may live for years if undisturbed. In the cooler months it often basks near its home, especially in the morning. It tends to be more common in well-wooded areas and near water sources, but will venture into arid country. It is relatively quick-moving and, although not aggressive, if confronted and unable to escape it will rear up fairly high and spread a broad hood; if further antagonised it may hiss loudly and rush forward to bite. Occasional individuals may sham (feign) death. This species is reported as laying up to 60 eggs (roughly 30 × 50mm), in early summer in southern Africa. The diet is mostly small vertebrates, including toads, birds, rodents and other snakes. It often raids chicken runs and can be a nuisance to poultry farmers.

Medical Significance: A dangerous snake; active by night and day and on the ground, will bite if inadvertently cornered, and rears up, so may deliver a bite

Anchieta's Cobra, *Naja anchietae*.
Left: Namibia (Francois Theart). **Centre:** Namibia (Tyrone Ping). **Right:** juvenile (Matthijs Kuijpers).

high on the body. The neurotoxic venom causes progressive descending paralysis, starting with ptosis and paralysis of the neck and head muscles, and proceeding to paralysis of the lungs. A bite is a medical emergency and will probably need treatment with a respirator and antivenom in a well-equipped hospital. A bite from this snake is one situation where pressure bandaging, immobilisation therapy and artificial respiration may be of benefit as the victim is transported to hospital.

Taxonomic Notes: A member of the Egyptian cobra (*Naja haje*) complex, originally seen as a subspecies of that snake, then as a subspecies of the Snouted Cobra (*Naja annulifera*); it was elevated to full species level in 2004.

Cape Cobra *Naja nivea*

Identification: An attractive, alert cobra, often yellow or orange, endemic to the dry country of south-western Africa. It has a broad head and medium-sized eye with a round pupil. The body is cylindrical and the tail is 14–17% of total length. The scales are smooth and glossy; in 19–21 rows at midbody; ventrals 193–227; subcaudals 50–68. The cloacal scale is undivided. Maximum size recorded 2.3m (an unusually large Western Cape specimen), average 1.2–1.5m; hatchlings 25–40cm. Several distinct colour forms exist: one is bright yellow above, sometimes orange or reddish; another is warm brown, often speckled with yellow; a third is pale brown or yellow-brown, heavily speckled both above and below with black; and a fourth is uniform shiny black. Other colour forms have also been described. Juveniles are yellow or reddish yellow, and usually have a broad black or brown neck band that fades with age.

Naja nivea

Habitat and Distribution: Predominantly a snake of dry country, and the only cobra in much of South Africa. It occupies a variety of habitats, from sea level to highlands at 2,500m; it occurs in the fynbos and savanna (bushveld) of the Cape provinces, the high grasslands of the Free State, the semi-desert of the Kalahari in Botswana and the Karoo in the Cape, and the dry rocky hills and desert of Namibia. It is often found along dry riverbeds (dongas). Endemic to southern Africa, it occurs in extreme south-western Lesotho, throughout most of the western side of South Africa, north into the Kalahari in south-west and south-central Botswana, and north-west into north-western Namibia.

Natural History: Terrestrial, but will climb trees and bushes; it is known to raid weaver bird nests in the Kalahari. It is almost entirely diurnal, emerging to bask for a while before foraging; it is occasionally crepuscular. When not active, it hides in holes, or sometimes under ground cover (brush, wood and rock piles, etc.), often remaining in the same retreat for some time. Sometimes more than one individual occupies the same refuge. It is very quick-moving and alert; it will spread a hood and look threatening at the least provocation, and if approached closely will strike without hesitation. It is also stated to be more aggressive and

Cape Cobra, *Naja nivea*. **Left top and centre top:** South Africa (Johan Marais). **Right top:** South Africa (Tyrone Ping). **Left bottom:** South Africa (Johan Marais). **Right bottom:** South Africa (Wolfgang Wuster).

irritable in the mating season, during September and October. Females lay 8–20 eggs, roughly 60 × 30mm, in December–January, in a hole or termite hill or other warm, wet place. This species eats a wide variety of prey, including rodents, birds, lizards, amphibians and snakes (including young of its own kind); it will feed on dead snakes and its habit of climbing trees to get at the eggs and young of the Sociable Weaver (*Philetairus socius*) is well documented. It is often hounded and sometimes killed in defence by Meerkats (*Suricata suricatta*) and Yellow Mongooses (*Cynictis penicillata*) working collectively; these small mammals have a high resistance to its venom. Cape Cobras are widely but erroneously believed to drink milk from cows.

Medical Significance: Often relatively common in parts of its range and active by day in farming areas, and so a major snakebite hazard in southern Africa, both to humans and stock. In the Cape provinces this snake accounts for most human snakebite fatalities, due to the rapid onset of paralysis. It has a very potent neurotoxic venom, causing progressive descending paralysis, starting with ptosis and paralysis of the neck and head muscles, and proceeding to paralysis of the lungs. A bite is a medical emergency and will probably need treatment with a respirator and antivenom in a well-equipped hospital. A bite from this snake is one situation where pressure pad immobilisation and artificial respiration may be life-saving as the victim is transported to hospital. Antivenom is available.

The forest, water and burrowing cobra clade

This clade, elucidated by molecular analysis, now contains the Burrowing Cobra (*Naja multifasciata*, originally *Paranaja*), two water cobras (*N. annulata* and *N. christyi*, originally *Boulengerina*) and the true forest cobras, originally *N. melanoleuca* but now split into five species. The Burrowing Cobra and the water cobras are not very significant medically, as they rarely come into contact with humans. However, the forest cobras are another matter; see our notes on the forest cobra (*N. melanoleuca*) complex (pp. 160–67).

Burrowing Cobra *Naja multifasciata*

Identification: A small, stout cobra with high-contrast bars on the lips, from the forests of Central Africa. It doesn't spread a hood. It has a short head and a fairly large eye with a round pupil. The sixth upper lip scale (upper labial) situated below and behind the eye is huge; the rostral is slightly prominent and it is speculated that it is used for digging. The body is cylindrical, and the tail is quite short and ends in a blunt spike. The scales are smooth and glossy; in 15 (rarely 17) rows at midbody; ventrals 150–175; subcaudals 30–39, paired. Maximum size probably

Naja multifasciata

about 80cm, usually 50–70cm; hatchling size unknown. Juveniles are distinctly marked, with the head black above and the snout cream and black. The upper lip scales are yellow or cream, heavily edged with black; the chin, throat and belly are cream; and the scales on the back are yellow or cream on the front half and black on the back half, so the snake appears finely barred black and yellow. Adults usually darken with age, those of 60cm or more becoming uniform dull brown or blackish, with the chin and throat yellow or yellow-brown, brownish below, and the upper lips black with dull brown blotches.

Habitat and Distribution: A snake of the great Central African forest, although work in Cameroon suggests it doesn't occur in dense rainforest, but is found

Burrowing Cobra, *Naja multifasciata.* **Left and right:** captive (Wolfgang Wuster).

in clearings, swamps and forest–savanna mosaic. It is found between sea level and 1,500m. It occurs from southern Cameroon, Equatorial Guinea and Gabon eastwards across the forest to Bunyakiri, in South Kivu, DR Congo, not far west of the Rwandan border, and it might just extend into northern Angola (not recorded from Cabinda). It is suggested that it fills an ecological niche in the forest that is occupied in the savanna by the garter snakes (*Elapsoidea*) – that of a small, inoffensive, terrestrial elapid, albeit a rodent-eating one.

Natural History: Little known. It is almost certainly terrestrial, and, despite the common name, burrowing behaviour is not yet reported. It moves fairly quickly and is probably diurnal, like most forest elapids. It presumably hides in holes, under ground cover or among leaf litter when inactive. A captive DR Congo specimen from South Kivu constantly tried to hide under debris in its cage, and pushed its snout into the earth. It was not aggressive and tried to bite only if restrained. A captive specimen maintained for a number of years in the UK was always calm and never tried to bite. When threatened, the DR Congo snake flattened itself to almost double its width, and raised its head in a cobra-like fashion, but without spreading a hood. It would take no prey other than small rodents; likewise, the UK captive ate only dead mice. This species might take amphibians. It presumably lays eggs, although no details are known.

Medical Significance: Unlikely to pose a threat to humans, being a small, secretive and apparently very rare species, although those who work in the forest are at risk. The venom is produced in very small amounts. However, a 12g mouse died one and a half minutes after being bitten on the flank by one of these snakes, and analysis confirms that the venom contains neurotoxins and cardiotoxins. No antivenom is available; bites should be treated symptomatically.

Banded Water Cobra *Naja annulata*

Identification: A large water-dwelling cobra of Central Africa, with several distinct colour forms. It is a large, heavy-bodied snake (weighing up to 4kg), with a broad, flat head and a medium-sized dark eye with a curious small round pupil. The body is cylindrical and the tail is long, 18–23% of total length. The scales are smooth and glossy; in 21–25 rows at midbody; ventrals 192–226; cloacal scale entire; paired subcaudals 67–78. Maximum size about 2.5m (possibly larger), average 1.4–2.2m. The size of wild hatchlings is unknown, but juveniles from Lake Tanganyika measured 43–48cm.

Naja annulata

The colour is quite variable. The animals associated with Lake Tanganyika (Storm's Water Cobra, *Naja annulata stormsi*) are shades of brown, with two or more black rings and a black tail; the number of rings and black dorsal blotches increases northwards, but totally banded specimens have also been found in the lake. Those from eastern DR Congo are banded black and warm brown or orange-brown, and are bright orange below. Further west, the black bands may have a yellow centre, and in occasional specimens the light cross-bars coalesce to form huge, dark-edged blotches down the centre of the back. Another colour form, which seems to be

Banded Water Cobra, *Naja annulata*. **Left:** subspecies *stormsi*, Tanzania (Michele Menegon). **Centre:** subspecies *stormsi*, Tanzania (Wolfgang Wuster). **Right:** Angola (Bill Branch).

smaller (usually less than 1.4m), occurs on the lower Congo River; some adults are speckled black and white while others are black or dark grey to mottled grey-brown above and ivory below, with juveniles being heavily speckled black and white; this might prove to be another species, although at present both the scalation and the DNA places it within *Naja annulata*.

Habitat and Distribution: Associated with lakes and rivers, in forest and well-wooded savanna, between sea level and 1,200m. It is usually found where there is enough waterside cover to conceal a large snake, but will venture out onto open beaches and sand bars. It occurs from Lake Tanganyika westwards across Central Africa to Gabon and the mouth of the Congo River, south to northern Angola, and north to south-western Central African Republic and southern Cameroon.

Natural History: An aquatic snake that spends most of its time in the water, hunting fish by day and by night. It may bask in the morning and later

Banded Water Cobra, *Naja annulata*.
Left top: DR Congo (Václav Gvoždik). **Right top:** dark morph, DR Congo (Eli Greenbaum).
Left bottom: dark morph juvenile (Václav Gvoždik). **Right bottom:** unusually marked (Václav Gvoždik).

afternoon. It moves somewhat ponderously on land, but is a superb, fast swimmer, spending much time underwater; it is recorded as staying down for more than 10 minutes and diving to depths of 25m. Groups of these cobras can be seen in Lake Tanganyika at a distance with their nostrils just out of the water, seemingly hanging suspended vertically. Hunting snakes also investigate recesses like rock cracks, mollusc shells and underwater holes, looking for concealed fish. They rapidly swallow small fish caught this way, without waiting for the venom to take effect. DR Congo animals are often caught in fish traps in small rivers coming downstream, leading fishermen to suggest the species may follow fish movements from large to small rivers at night, and return at dawn. When not active, it tends to hide in shoreline rock formations, in holes in banks or overhanging root clusters, or in holes of waterside trees; in such refuges it may live for long periods if not disturbed. It often makes use of man-made cover such as jetties, stone bridges and pontoons, and will also use buildings and fish-drying huts. Juveniles have been found under boats, waterside debris, logs, etc. When approached in water, this snake simply swims away, and on land will attempt to escape into water, but if cornered it will rear, spread a hood and demonstrate with an open mouth. It lays eggs; clutches of 22–24 are laid in August and September in Lake Tanganyika. A captive of the speckled form laid four elongate eggs that hatched in 70 days; the hatchlings were around 30cm. A 48cm juvenile was captured in July. Only fish have been recorded in the diet of wild snakes, but captives took amphibians and rodents.

Medical Significance: A large snake, with a potent neurotoxic venom (the lethal dose is very low compared to that of other elapids), so theoretically a risk to inhabitants of Central Africa, especially fishermen. However, there seem to be no documented bites, possibly because, being both nervous and largely tied to water where it hunts fish, this snake rarely comes into confrontational contact with humans. No specific antivenom is available; however, South African polyvalent antivenom is shown to be effective in neutralising the venom.

Congo Water Cobra *Naja christyi*

Identification: A fairly large, semi-aquatic cobra of the lower Congo River. The head is short and subtriangular, and the eye small and set forward, with a small round pupil. The body is cylindrical and the tail is fairly long, 18–20% of total length. The scales are smooth and glossy; in 17 rows at midbody; ventrals 206–221; cloacal scale entire; paired subcaudals 67–73. Maximum size about 1.4m, average probably 80–120cm; hatchlings around 30cm. The colour is a rich brownish maroon above; some specimens have low-contrast

Naja christyi

blotches or bars, and some are very dark. The tail is black. On the neck are two to six yellow or cream rings, clearly visible on the underside of the hood. The lips and other head scales may have lightish blotches, and the upper labials are finely black-edged. The underside is brown, and the ventrals are often lightly edged yellow.

Congo Water Cobra, *Naja christyi*. **Left and right:** DR Congo (Václav Gvoždik).

Habitat and Distribution: Known only from the lower Congo River, on both banks, from Kinshasa–Brazzaville down to the sea. Informal records indicate it extends upriver from Kinshasa, and is probably also in the tributaries of the Congo River in the region, but this is poorly documented. It might occur in Cabinda and the Zaire Province of north-west Angola.

Natural History: Behaviour little known. This snake is semi-aquatic, spending a lot of time in water, and moves quite quickly. It might be nocturnal or diurnal. It presumably hides in riverside holes, root clusters and holes of waterside trees, etc.; juveniles have been taken under rocks on the floodplains of the lower Congo River, and in and around small ponds with rocky banks. It swims well, and if encountered in water swims away; if cornered on land it will spread a narrow hood. It lays eggs and probably eats amphibians and/or fish. Little else is known. This species was originally placed in the genus *Limnonaja*, then *Boulengerina*.

Medical Significance: A fairly large snake, with a potent neurotoxic venom (the lethal dose is low compared to that of other elapids), so theoretically a risk to inhabitants of the lower Congo River, especially fishermen. However, there seem to be few documented bites, possibly because, being both nervous and largely tied to water where it hunts fish, this snake rarely comes into confrontational contact with humans. No specific antivenom is available; however, South African polyvalent antivenom is shown to be effective in neutralising the venom. One victim experienced only a headache following a bite.

The forest cobra (*Naja melanoleuca*) complex

Regarded for a long time as a single well-defined species, but quite variable in colour, this large snake has now been split, mainly on the basis of DNA, into five species. This split largely makes zoogeographic sense, and indicates separate lineages. The São Tomé Forest Cobra (*Naja peroescobari*) is confined to that island, while the West African Forest Cobra (*N. guineensis*) occurs in the forests of West Africa, west of the Dahomey Gap. The Banded Forest Cobra (*N. savannula*) occurs in the savannas of West Africa. However, there are problems in identifying (at least visually) some specimens of the Eastern Forest Cobra (*N. subfulva*) and the Central African Forest Cobra (*N. melanoleuca*). The ranges of these two snakes overlap widely in the Albertine Rift, parts of Central Africa, and in and around Cameroon, DR Congo and the Republic of the Congo. Even the authors of the paper that split the group struggled to allocate some specimens from these areas and state that these can be differentiated only using DNA; no key is provided in their paper.

These are medically important snakes, although bites seem to be rare. They have a huge range (most of forested and wooded tropical Africa, extending into savanna in places). They are secretive, and often occur in urban and agricultural areas; they may be active by day and by night on the ground. They are quick to rear up if molested, if they perceive themselves to be in danger they will bite, and their height means a bite may be delivered high on the body. They have a potent neurotoxic venom, causing a rapid descending paralysis and respiratory depression, leading to death from suffocation within 2–16 hours of the bite. Initial symptoms include ptosis, double vision, and inability to open the mouth, clench the jaw, protrude the tongue or swallow. These cobras are large and a bite from one is a medical emergency, needing immediate hospitalisation. However, antivenom is available, and surprisingly few bites are documented; these alert species are quick to avoid confrontation. The symptomology of forest cobra bites seems to be similar for all five species, indicating that whatever else may have changed, the venom remains much the same.

Eastern Forest Cobra (Brown Forest Cobra) *Naja subfulva*

Identification: A large black or brown cobra of the African woodland and forest. A fairly thick-bodied snake, it has a large head and a large, dark eye with a round pupil. The body is cylindrical and the tail fairly long and thin, 15–20% of total length. The scales are smooth and glossy; in 19 (rarely 17 or 21) rows at midbody; ventrals 196–226; subcaudals 55–71. Maximum size about 2.7m, average 1.4–2.2m; hatchlings 25–40cm. Several colour forms exist. Those from the area around Lake Victoria and parts of the Congo basin are glossy black; the juveniles

Naja subfulva

are usually white-spotted and this may persist in some adults. The chin, throat and anterior part of the belly are white, cream or pale yellow, with broad black cross-bars and blotches, although in some specimens most of the underside is

Eastern Forest Cobra, *Naja subfulva*.
Left: Tanzania (Stephen Spawls). **Centre:** South Africa (Tyrone Ping). **Right top:** South Africa (Johan Marais).
Right bottom: Kenya (Stephen Spawls).

black. The sides of the head are strikingly marked with black and white, giving
the impression of vertical black and white bars on the lips. Those from south-
east Africa, the eastern and south-eastern coast and the savannas around the
Central African forest are usually brownish or blackish brown above, sometimes
spotted darker brown, with a darker or black tail. The underside is paler, with
a yellow or cream belly, heavily speckled with brown or black; specimens from
the southern part of the range have black tails. Mount Kenya snakes are yellow-
brown anteriorly, becoming dark or black towards the tail, and the head lacks the
black lip edging. Ventral banding is usually restricted to the anterior fifth of the
underside; this may be indistinct or faded.

Habitat and Distribution: Forest, woodland, coastal thicket, moist savanna and
grassland, from sea level to over 2,500m. It occurs northwards from north-
eastern South Africa (from Durban in Kwazulu-Natal and Kruger National Park),
through parts of Mozambique, most of Malawi, northern and eastern Zambia and
southern DR Congo, west to southern Gabon. It is also found in coastal Tanzania
and Kenya, western Kenya, most of Uganda, Rwanda, Burundi and the watershed
west of the Albertine Rift Valley in DR Congo, then across the north of the forest
to south-western Chad and Cameroon; there are some old records from the Niger
Delta. Records are lacking from north-east DR Congo. Isolated populations are
known from the Udzungwa Mountains and the Crater Highlands in Tanzania, the
forests of south-eastern Mount Kenya, the Nyambene Hills and parts of Laikipia
in Kenya, the forests of south-west Ethiopia, and the Imatong Mountains and
Torit in South Sudan.

Eastern Forest Cobra, *Naja subfulva*.
Left: juvenile, Kenya (Stephen Spawls). **Right:** Kenya (Stephen Spawls).

Natural History: Terrestrial, but a fast, graceful climber, known to ascend trees to a height of 10m or more. It is quick-moving and alert. It swims well and readily takes to the water; in some areas it eats mostly fish and could be regarded as semi-aquatic. It is active both by day and by night, mostly by day in uninhabited areas and by night in urban areas. When not active, it takes cover in holes, brush piles, hollow logs, among root clusters or in rock crevices, or in termite hills at forest fringes or in clearings; in some areas it is fond of hiding along riverbanks, in overhanging root systems or bird holes; in urban areas it will hide in junk piles or unused buildings. If molested it rears up to a considerable height and spreads a long, narrow hood; if further stressed and unable to retreat, it may rush forward and bite. Male–male combat has been observed. Females lay 15–26 eggs, roughly 30 × 60mm; incubation times of around two months have been recorded. This snake eats a wide variety of prey; it is fond of mammals (squirrels, mice, shrews and elephant shrews) and amphibians, and will also take fish (including mudskippers), other snakes, monitor lizards and other lizards, and birds and their eggs. Captive specimens can seem almost cunning in their timing of escape bids, and have been reported as trying to attack rather than escape. When restrained, it will jab the sharp tail tip into the handler. A zoo specimen of the forest cobra (*Naja melanoleuca*) complex lived 28 years in captivity.

Medical Significance: This is a dangerous snake; it is wide-ranging, secretive, large and active on the ground, often in rural areas. It will bite if cornered. It has a neurotoxic venom that causes a progressive paralysis, which can be rapid; fatalities are recorded. Antivenom is available. A bite must be rapidly hospitalised and treated with antivenom and respiratory support in hospital. As with mambas (*Dendroaspis*), a bite from this snake is one situation where pressure bandaging, immobilisation therapy and artificial respiration may be of benefit while the victim is being transported to hospital.

Central African Forest Cobra *Naja melanoleuca*

Identification: A large black cobra of the Central African forest. It has a large head and a large, dark eye with a round pupil. The body is cylindrical and the tail fairly long and thin, 15–20% of total length. The scales are smooth and glossy; in 17–21 rows at midbody (usually 19); ventrals 206–232; subcaudals 57–74. Maximum size about 2.7m, average 1.4–2.2m; hatchlings 25–40cm. The colour is usually black above, sometimes with white speckling, especially in juveniles. The underside of the throat and neck is cream or yellow, with black cross-bands extending over the first third of the underside; after about 100 ventrals the venter is usually uniform black. The cream cross-bars may extend to the back of the neck (thus forming rings), and sometimes there is an ocellate mark on the neck.

Naja melanoleuca

Habitat and Distribution: Forest and woodland, coastal thicket, moist savanna and grassland, from sea level to over 2,000m. It occurs northwards from north-west Angola through DR Congo and Gabon, north and west to southern Central African Republic, southern Cameroon, Nigeria and extreme southern Togo, and east to western Rwanda, Burundi and south-west Uganda. It might occur in Ghana; west of there it is replaced by the West African Forest Cobra (*Naja guineensis*).

Natural History: As for the Eastern Forest Cobra (*Naja subfulva*); that is, terrestrial but also climbs, fast-moving, swims well, active by day and night, lays 12–23 eggs,

Central African Forest Cobra, *Naja melanoleuca.*
Left top: DR Congo (Jean-Francois Trape). **Right top:** Cameroon (Jean-Francoise Trape).
Left bottom: Uganda (Andre Zoersel). **Right bottom:** Rwanda (Harald Hinkel).

and eats a wide range of vertebrate prey, including mammals, snakes, amphibians and fish. Cameroon specimens have been described as shamming (feigning) death, particularly if trapped.

Medical Significance: As for the previous species; a dangerous snake with a rapidly acting neurotoxic venom.

Banded Forest Cobra *Naja savannula*

Identification: A large, banded black and yellow cobra of the West African savanna. It has a large head and a large, dark eye with a round pupil. The body is cylindrical and the tail fairly long and thin, 15–20% of total length. The scales are smooth and glossy; in 19 rows at midbody; ventrals 211–226; subcaudals 67–73. Maximum size about 2.3m (possibly larger), average 1.3–2m; hatchling size unknown, but probably around 30cm. The colour is distinctive; the front half of the body is banded black and yellow. The yellow bands (rarely more than five) are usually

Naja savannula

slightly narrower than the black bands, sometimes with a fine black band in the centre of the yellow; sometimes the yellow bands are black-speckled. The ventral surface is vivid yellow anteriorly, with black cross-bars on the neck, gradually darkening towards the tail.

Habitat and Distribution: In the West African Guinea and Sudan savanna, from sea level to 1,250m, but most common below 600m. It is usually in the vicinity

Banded Forest Cobra, *Naja savannula*.
Left top: Nigeria (Gerald Dunger). **Centre top:** underside, Ghana (Barry Hughes). **Left bottom:** Guinea (Jean-Francois Trape).
Centre bottom: Ghana (Stephen Spawls). **Right:** Ivory Coast (Mark-Oliver Rödel).

of water sources. It occurs from extreme south-west Chad through northern Cameroon and Nigeria to Senegal and the Gambia, mostly between latitudes of 10 and 14°N, except in eastern Ghana, Togo and Benin, where its range approaches the coast in the Dahomey Gap.

Natural History: As for the Eastern Forest Cobra (*Naja subfulva*); that is, terrestrial but also climbs, fast-moving, swims well, active by day and night, lays eggs, and eats a wide range of vertebrate prey. A 2m specimen from Wa, north-west Ghana, had eaten an African Giant Shrew (*Crocidura olivieri*), a noxious-smelling mammal.

Medical Significance: Few bite cases are recorded, but, as for the previous species, this is a dangerous snake with a rapidly acting neurotoxic venom.

West African Forest Cobra (Black Forest Cobra)
Naja guineensis

Identification: A large black cobra of the West African forest. It has a large head and a large, dark eye with a round pupil. The body is cylindrical and the tail fairly long and thin, 15–20% of total length. The scales are smooth and glossy; in 17–19 rows at midbody; ventrals 203–221; subcaudals 60–70. A specimen of 2.6m from Tafo, Ghana, appears to be the largest measured; average size 1.4–2.2m; hatchlings 25–40cm. The colour is usually black or brownish black above, sometimes with cross-bars (or pairs of cross-bars) of speckled white scales;

Naja guineensis

juveniles may be entirely speckled. The underside is largely black with white or cream cross-bars on the throat, which may become suffused with black in large adults. Sometimes there is an ocellate mark on the neck. The lips are cream, with black edging to the scales; in large adults the lips may become entirely black. Melanistic (all-black) specimens are known.

Habitat and Distribution: Forest and woodland, coastal thicket, occasionally moist savanna and grassland, from sea level to over 2,000m. It occurs from western

West African Forest Cobra, *Naja guineensis*. **Left and right:** Guinea (Laurent Chirio).

West African Forest Cobra, *Naja guineensis*.
Left: Togo (Jean-Francois Trape). **Right:** Guinea (Johan Marais).

Togo westwards through Ghana to Liberia and Guinea, south of 10°N, except a single record (Contuboel) from Guinea Bissau.

Natural History: As for other forest cobras; that is, terrestrial but also climbs, fast-moving, swims well, active by day and night, lays eggs, and eats a wide range of vertebrate prey.

Medical Significance: As for the previous species; a dangerous snake with a rapidly acting neurotoxic venom. A child in Ghana died 20 minutes after a bite from one of these snakes.

São Tomé Forest Cobra *Naja peroescobari*

Identification: A large, black cobra from São Tomé Island. It has a large head and a large, dark eye with a round pupil. The body is cylindrical and the tail fairly long and thin, 15–20% of total length. The scales are smooth and glossy; in 19–21 rows at midbody; ventrals 204–215; subcaudals 52–70. Maximum size 2.6m, average probably 1.4–2.2m; hatchling size unknown, but probably 25–40cm. The colour is uniform glossy black above. The chin, throat and neck are white or cream, with lateral black blotching and irregular cross-bars, but any

Naja peroescobari

white on the neck is usually (but not always) confined to the first 22 ventral scales. The lips are cream, with black edging to the scales.

Habitat and Distribution: A species known only from the volcanic São Tomé Island, in the Gulf of Guinea west of Gabon. It occurs in forested areas between sea level and 1,154m, and is apparently absent from the drier north-east side of the island.

Natural History: As for other forest cobras; that is, terrestrial but also climbs, fast-moving, swims well, active by day and night and lays eggs. It is most often seen basking on roads during the daytime, and is said to prefer shaded, moist habitats. Known prey animals include a weasel and a Black Rat (*Rattus rattus*), but it probably takes a range of vertebrates, like other members of the group.

Medical Significance: No bite cases seem to have been formally documented, but it probably has, like other members of the complex, a potent neurotoxic venom capable of causing paralysis. The inhabitants of the island state that it is highly venomous and capable of causing death in hours; a local saying about this snake translates as 'man bitten, man lost', and a local tradition is that to survive a bite, the victim must immediately cut the limb off. Palm-wine tappers are reported to jump from the tree if they encounter this snake.

São Tomé Forest Cobra, *Naja peroescobari*. **Below:** São Tomé (Barry Hughes).

The spitting cobra clade

A group of elapid snakes able to squirt venom from their fangs, to a distance of 2m or 3m; if the venom enters the eyes it is intensely painful. The snakes also bite. Unusually, the use of venom for spitting is the only situation where snake venoms have a defensive purpose, deterring a potential predator. Spitting probably evolved for defence against predatory birds such as ground hornbills and Secretary Birds (*Sagittarius serpentarius*), since spitting cobras appear in the fossil record at least 15 million years ago, before humans.

For a long time, all African spitting cobras were believed to be a single, fairly variable species, and the various new taxa that were found were regarded as subspecies or varieties from 1843 until 1968, when Don Broadley elevated the Mozambique Spitting Cobra (*Naja mossambica*) to full species status. More splits followed; at least seven species are now recognised. Spitting cobras show wide variation (within the constraints of being true cobras); some are huge, 2.5m or more, and others a maximum of 1m or so. However, they form a monophyletic clade; they have a single common ancestor that lived about 15 million years ago. We have provided separate entries for the two subspecies of the South-western Black Spitting Cobra (*N. nigricincta*) as they look very different, although intergrades exist.

These are medically important snakes. Their venom rarely shows neurotoxic effects, unlike that of the forest and Egyptian cobras. It is potently cytotoxic. There is usually immediate pain after a bite, often followed by vomiting within six hours, and extensive local swelling. Over the next few days a horrible local shallow necrosis develops, involving the skin and connective tissue, often with skip lesions. This then leads to long-lasting disfigurement and complications that may cause death. In parts of West Africa, these species are a major cause of snakebite and suffering. A large proportion of bites occur inside houses while the victims are asleep. Recent evidence has uncovered a spate of bites in southern Africa, by the Mozambique Spitting Cobra, where the snake may have deliberately bitten the victim (most snakebites are in defence), possibly under the impression that they were prey. If threatened, a spitting cobra's first line of defence is to spit venom, and if this gets in the victim's eyes it can cause irreversible damage; there is intense local pain, blepharospasm, swelling and watering. In theory, if the venom is rapidly washed out with quantities of water and the eyes medically monitored, no lasting harm should result, although some corneal erosion may occur. But in reality the harsh conditions of everyday life in remote Africa often mean that victims cannot get water easily, and a frightful slew of inappropriate local remedies can sometimes leave them with permanent damage to their sight.

Black-necked Spitting Cobra *Naja nigricollis*

Identification: A large spitting cobra, widespread in the savannas of tropical Africa. It has a broad head, a cylindrical body and smooth scales; the tail is 15–20% of total length. Midbody scale count is 17–23; ventrals 175–212; subcaudals 50–70. Maximum size seems to vary; over most of Africa 1.8m is about as large as it gets, averaging 1.2–1.5m; hatchlings 33–36cm. However, savanna snakes from immediately west of the Gregory Rift Valley in Kenya and Tanzania, and in north-east Tanzania, seem to get much larger, reaching 2.3m or even more. The colour

Naja nigricollis

is variable. In most of Africa adults are black (sometimes grey or sometimes dark brown), and juveniles are grey. The upper neck is black above, and the throat is barred black and either red, orange or pink (or occasionally yellow, pale grey or white). In highland Kenya, adults may be copper-coloured, pinky-grey, rufous or rufous-brown, but most have red and black throat bars. A large all-black (above and below) colour form occurs west of the central Gregory Rift Valley in Kenya and in north-west Tanzania. Some individuals of this form have white blotching on the underside, and some juveniles have red throats.

Black-necked Spitting Cobra, *Naja nigricollis*.
Left: Angola (Warren Klein). **Right:** juvenile, Benin (Jordan Benjamin).

Black-necked Spitting Cobra, *Naja nigricollis*.
Left top: Kenya (Anthony Childs). **Centre top:** juvenile, Kenya (Anthony Childs). **Right top:** DR Congo (Johan Marais).
Centre bottom: underside, Nigeria (Gerald Dunger). **Right bottom:** Tanzania (Caspian Johnson).

Habitat and Distribution: Moist and dry savanna and woodland, but rarely semi-desert or forest, although it will colonise forest clearings and fringes and follow rivers into dry country; it is often in suburbia if there are suitable refuges and prey. This species is found from sea level to above 1,800m. It occurs east from southern Mauritania across the savannas of West and Central Africa to western Ethiopia, then south down the eastern side of Africa, south-westwards to southern Zambia, Angola and extreme northern Botswana; south of there it is replaced by the Zebra Cobra (*Naja nigricincta nigricincta*). It is known on the lower Congo River, and there is an isolated population on Jebel Marra, western Sudan.

Natural History: Terrestrial, but it climbs readily and will even attempt to escape up trees or up into rocks. It is quick-moving and alert. Large adults are mostly active at night, but will also hunt or bask during the day, especially in uninhabited areas. Juveniles are often active during the day. When not active, this snake takes cover in termitaria, holes, hollow trees, old logs or brush piles, etc. When threatened, it rears and spreads a hood and will spit if further molested, but may also spit without actually spreading a hood. Large adults can spit 3m or more. They tend to stand their ground when threatened, rather than rush forward as the forest and Egyptian cobras may do. Male–male combat has been observed, with the snakes intertwined and neck-wrestling. Females lay 5–20 eggs, roughly 25 × 40mm, often in termitaria; the larger the female, the more eggs she lays. Hatchlings have been collected in Nairobi in April and May, and in May and June in West Africa. This species eats a wide variety of food, including amphibians, various mammals, lizards, snakes and even fish; it will raid chicken runs and will take eggs and chicks. It is quick to exploit suitable habitats, aided by its secretive nocturnal habits. In Nigeria it has been observed moving into what were previously forested areas, and in urbanised areas the disappearance of vipers has allowed this snake to become common. In Kenya it is known to exploit urban areas; in the satellite suburbs around Athi River it has displaced the Egyptian Cobra (*Naja haje*), while on the southern side of Nairobi, in areas where both this snake and the Puff Adder (*Bitis arietans*) were common, the Puff Adders are long gone but the spitting cobras persist.

Medical Significance: One of Africa's most dangerous species, on account of its huge range, relative abundance and potent cytotoxic venom. It is active on the ground at night and tolerates suburbia, where it can remain undetected due to its habit of living in holes and emerging only after dark; it is often also in agricultural land. Many bites occur inside houses. There is immediate pain followed by vomiting within six hours, extensive local swelling, local blistering in 60% of cases and local tissue necrosis in 70% of envenomed cases. Necrosis usually involves only the skin and subcutaneous connective tissues. In Nigeria 20% of bite victims had permanent disability. Bite victims should be transported rapidly to hospital. Rapid death is very rare in bites from this snake, but this has the unfortunate result that victims are not taken to hospital as the venom seems to be local in its effects. As time goes by, frightful medical complications arise from the necrotic effects of the venom and these can become life-threatening; they include spontaneous haemorrhage. Antivenom is available, but treatment for any bite must also address the local tissue damage. Venom in the eye can be treated by careful flushing with large volumes of water; there is no need to inject antivenom. However, the more rapidly the eye is flushed out, the better; see our section on 'Spitting snakes: first aid for venom in the eyes' (p. 300). Anyone who has been spat at should visit a hospital or eye clinic for an assessment of the damage to the eye.

Ashe's Spitting Cobra (Large Brown Spitting Cobra)
Naja ashei

Identification: A recently described, huge brown spitting cobra of the coast and dry country of eastern Africa. The eye has a round pupil and orange-brown iris. The body is cylindrical and the tail fairly long, 15–18% of total length. The scales are smooth; in 20–23 rows at midbody; ventrals 192–207; subcaudals 55–65. Maximum size about 2.7m (maybe larger), average 1.2–1.8m; hatchlings 35–38cm. The dorsal colour is various shades of brown or olive, or sometimes grey; some specimens can be quite yellow-brown, others uniform, and

Naja ashei

others have a sprinkling of yellow scales, which may form narrow yellow dorsal bars. The underside is dull light yellow, each ventral scale edged with light brown; there is usually a deep, dark brown bar on the underside of the neck, but this may be fragmented into several bars with yellow scales between. Inland juveniles are light grey-brown; the throat bar is dark brown or black and the head deep brown. Juveniles from Kenya's north coast have a shiny black head and neck, which fades in their second year. This species is named after James Ashe, a charismatic Kenyan herpetologist.

Ashe's Sptting Cobra, *Naja ashei*.
Left top: Kenya (Stephen Spawls). **Right top:** juvenile, Kenya (Bio-Ken Archive).
Left bottom: Kenya (Stephen Spawls). **Centre bottom:** Kenya (Anthony Childs). **Right bottom:** Ethiopia (Tomáš Mazuch).

Habitat and Distribution: Dry savanna, semi-desert and coastal thicket, from sea level to about 1,300m. It occurs from southern Somalia into Kenya, south down the coast to the Tanzanian border, and west across large areas of eastern and northern Kenya (it probably occurs throughout most of dry country Kenya, although records are sporadic at present), just reaching Uganda at Amudat in the east. In Tanzania it is known only from around Ndukusi, a village near Lolkisale Mountain south-east of Arusha, so is possibly in Tarangire National Park. It should also be on the extreme north Tanzanian coast as it is recorded from Lunga-Lunga. It has also been recorded from extreme south-east South Sudan (Torit area). In Ethiopia it is known from the lower Omo River area, north to Arba Minch; there is a cluster of records from in and around Awash National Park, and also from Gode in the Ogaden.

Natural History: Terrestrial, but it may climb into bushes and low trees; a specimen on the road in Meru National Park climbed an acacia when pursued. This species is quick-moving and alert. Adults in undisturbed habitat seem to be both nocturnal and diurnal; one was found mid-morning pursuing frogs on the edge of a dam near Mount Kasigau, and when approached also climbed an acacia tree; several have been photographed active by day in Laikipia. Juveniles in the Shimba Hills and at Lunga-Lunga were active by day. When inactive, this snake shelters in holes (particularly in termite hills), old logs or under ground cover. When disturbed it rears up, spreads a broad hood and may spit twin jets of venom; it can spit up to 3m. Females lay eggs. This species eats a range of prey, taking mammals, birds, amphibians and reptiles; known prey items include a domestic cat, chickens and whole nests of chicken eggs, Red Spitting Cobras (*Naja pallida*), White-throated Savanna Monitors (*Varanus albigularis*) and Puff Adders (*Bitis arietans*). A Watamu specimen ate a Puff Adder so large it could subsequently crawl only in a straight line. They will also take carrion, including chicken heads and roadkilled snakes; one ate a dead egg-eating snake (*Dasypeltis*) that was maggot-ridden and decomposing.

Medical Significance: A dangerous species, widespread in parts of eastern Africa, and in some areas (particularly Kenya's north coast) relatively abundant. It has a potent cytotoxic venom produced in huge quantities (6ml were milked from one snake), which contains proteins not described from other African spitting cobras. This snake is active on the ground at night and tolerates suburbia, where it can remain undetected due to its habit of living in holes and emerging only after dark; it is often also in agricultural land. Bite victims should be transported rapidly to hospital. Antivenom is available, but treatment for any bite must also address the local tissue damage. Note that this snake spits readily; the venom is produced in huge quantities and is damaging to the eyes. Venom in the eye can be treated by gentle flushing with large volumes of water; there is no need to inject antivenom. However, the more rapidly the eye is flushed out, the better; see the relevant section on 'Spitting snakes: first aid for venom in the eyes' (p. 300). Anyone who has been spat at should visit a hospital or eye clinic for an assessment of the damage to the eye.

Zebra Cobra (Zebra Snake/Western Barred Spitting Cobra)
Naja nigricincta nigricincta

Identification: A relatively small, attractively banded cobra of dry Angola and Namibia. It has a fairly large head and a large, dark eye with a round pupil. The body is cylindrical and the tail fairly long and thin, 17–19% of total length. The scales are smooth and glossy; in 21 (sometimes 23) rows at midbody; ventrals 192–226; subcaudals 57–73. Maximum size about 1.9m, but this is unusually large; most adults are 0.8–1.1m; hatchlings 28–35cm. Adults are strikingly banded black and white (hence the common Namibian name 'Zebra Snake'), although

Naja nigricincta nigricincta

the ground colour may be light brown or pinkish. The neck is usually black. Angolan animals are often rufous-brown with faded cross-bars. In intergrades from central and southern Namibia the clarity of the bands fades.

Habitat and Distribution: Inhabits dry savanna, semi-desert and desert, often in rocky areas, between sea level and 1,800m, or occasionally higher (for example, in the Brandberg). It occurs from south-west Angola south through north and central Namibia (largely to the west, not quite reaching Botswana); south of

Zebra Coba, *Naja nigricincta nigricincta*.
Left top: Namibia (Bill Branch). **Right top:** Namibia (Wolfgang Wuster).
Left bottom: Angola (Bill Branch). **Right bottom:** intergrade (Francois Theart).

22°S it starts to intergrade with the South-western Black Spitting Cobra (*Naja nigricincta woodi*).

Natural History: Terrestrial, but it climbs well and often attempts to escape up trees. It is active by day and night, adults usually by night and juveniles particularly by day. Individuals shelter in holes, under ground cover or in rock fissures, tree cracks and termite mounds. When threatened it will try to escape, but if cornered it will rear up and spread a broad hood, and will spit venom to 2m or more. It may also bite. Between 10 and 24 eggs are laid. This snake feeds on small vertebrates, including mammals, frogs, lizards and other snakes, including its own kind. It is often relatively common; around Windhoek it was the second most common species relocated.

Medical Significance: A dangerous species in the northern half of Namibia; it is relatively abundant and has a potent cytotoxic venom that causes major damage to superficial tissues. This snake is active on the ground at night and tolerates suburbia, where it can remain undetected due to its habit of living in holes and emerging only after dark; it is often also in agricultural land. Bite victims should be transported rapidly to hospital. Antivenom is available, but has a limited effect; it does not neutralise some systemic symptoms and treatment for any bite must also address the local tissue damage, which does not seem to be affected by antivenom. Note that this snake spits readily; the venom is damaging to the eyes. It can be treated by gentle flushing of the eye with large volumes of water; there is no need to inject antivenom. However, the more rapidly the eye is flushed out, the better; see our section on 'Spitting snakes: first aid for venom in the eyes' (p. 300). Anyone who has been spat at should visit a hospital or eye clinic for an assessment of the damage to the eye.

South-western Black Spitting Cobra *Naja nigricincta woodi*

Identification: A black spitting cobra of dry southern Namibia and western South Africa. It has a fairly large head and a large, dark eye with a round pupil. The body is cylindrical and the tail fairly long and thin, 15–20% of total length. The scales are smooth and glossy; in 21 rows at midbody; ventrals 221–228; subcaudals 65–74. Maximum size about 1.8m, average 0.9–1.4m; hatchlings 35–38cm. Adults are shiny black above and below, and juveniles purply-grey or grey with a black throat, darkening when around 70–80cm in length. North into Namibia, the bands appear.

Naja nigricincta woodi

Habitat and Distribution: Lives in rocky desert, hills, rocky outcrops and semi-desert, extending south into the shrublands and heathlands (fynbos) of Northern Cape and Western Cape, between sea level and 1,000m. It occurs west from Kimberley in South Africa, and north from Citrusdal up into Namibia, where it meets and intergrades with the Zebra Cobra (*Naja nigricincta nigricincta*) between 22°S and 25°S.

Natural History: Terrestrial, but it climbs well and often attempts to escape up trees. It is active by day and night, adults usually by night and juveniles particularly by day. Individuals shelter in holes, under ground cover or in rock fissures, tree cracks and termite mounds; they are often seen in dry riverbeds and crossing roads. When threatened, it will try strenuously to escape; it will spread a hood and spit if totally cornered, but does so reluctantly. It can also bite. Between 10 and 20 eggs are laid. This snake feeds on small vertebrates, including mammals, frogs, lizards and other snakes.

Medical Significance: A dangerous species in the southern half of Namibia and parts of the Northern Cape and Western Cape; it has a potent cytotoxic venom causing major damage to superficial tissues. This snake is active on the ground at night. However, its rarity and remote habitat mean it doesn't often come into

South-western Black Spitting Cobra, *Naja nigricincta woodi*.
Left: South Africa (Johan Marais). **Right:** juvenile, South Africa (Richard Boycott).

contact with humans. Bite victims should be transported rapidly to hospital. Antivenom is available, but treatment for any bite must also address the local tissue damage. Venom in the eyes should be treated by gently flushing the eye with large volumes of water; there is no need to inject antivenom. However, the more rapidly the eye is flushed out, the better; see our section on 'Spitting snakes: first aid for venom in the eyes' (p. 300). Anyone who has been spat at should visit a hospital or eye clinic for an assessment of the damage to the eye.

Mozambique Spitting Cobra *Naja mossambica*

Identification: A medium-sized brown cobra of the savannas of south-eastern Africa. The head is blunt and the eye medium-sized, with a round pupil. The body is cylindrical and the tail quite long, 15–20% of total length. The scales are smooth; in 23–25 rows at midbody; ventrals 177–205; subcaudals 52–69. Maximum size usually about 1.5m (an unusual recent specimen was 1.8m), average 0.8–1.3m; hatchlings 23–25cm. The back colour is usually some shade of brown, or occasionally pinkish; juveniles may appear olive-green, while large adults may be

Naja mossambica

very dark brown or grey. The underside is pale brown, orange, pinkish or grey; on the neck, throat and anterior third of the belly there is a mixture of black bars, half-bars, blotches and spots. Some specimens have only a few small markings; others have the throat heavily mottled with black. In some individuals the ventral scales are very dark. The skin between the scales is blackish and visible, giving a 'net-like' appearance in some specimens, and the scales on the side of the head (especially the lips) are usually black-edged.

Habitat and Distribution: Occupies a wide range of habitats, including coastal forest, woodland, thicket, moist and dry savanna and semi-desert, between sea level and about 1,800m. The northernmost records are of isolated populations on Pemba and Zanzibar Islands. From there, it occurs south and west from Morogoro across

Mozambique Spitting Cobra, *Naja mossambica*.
Left: Mozambique (Mark-Oliver Rödel). **Right:** Botswana (Stephen Spawls).

Mozambique Spitting Cobra, *Naja mossambica*.
Left: Malawi (Gary Brown). **Right:** Northern Mozambique (Luke Verburgt).

south-eastern Tanzania, southern Malawi and the southern quarter of Zambia, through virtually all of Mozambique (records are sporadic in the south) and Zimbabwe, to south-east Angola, most of northern and eastern Botswana, and northern South Africa, in Kwazulu-Natal.

Natural History: Mostly terrestrial, but able to climb well; adults readily ascend low trees and even sleep in them. Adults are mostly active at night, but sometimes by day; juveniles are often active during the day, presumably to avoid competition with adults and/or to avoid being eaten by them. When not active, individuals shelter in termite hills, holes, rock fissures or under ground cover such as logs and brush piles. This species is very quick and alert; if molested it may rear up (sometimes quite high), spread a hood and spit readily (it can also spit without spreading a hood) to a distance of 2m or more. It sometimes pretends to be dead. Females lay 10–22 eggs, roughly 35 × 20mm, in December/January (summer) in southern Africa. The diet is quite varied; it is fond of amphibians but also takes lizards, rodents, other snakes (including its own kind) and even insects. It is often common; in a large snake collection from south-east Botswana it was the most abundant species.

Medical Significance: This snake represents a major snakebite hazard in south-eastern Africa; in some areas it causes the majority of serious snakebites. It is common, adults are active on the ground at night, it tolerates suburbia due to its secretive habits, and it is often present in and around villages, smallholdings and even within large towns. It will spit in defence if it feels cornered, and is liable to be trodden on. In some curiously disturbing recent cases, particularly in South Africa, the snake entered a building and apparently bit a sleeping victim unprovoked, often multiple times. Bites were experienced almost anywhere on the body, including the face and the torso. This is unusual, as most snakebites result from a snake perceiving itself to be threatened and biting in defence. Possibly a sleeping human gives off some sort of signal (warmth or smell) that this snake associates with prey, although it might be that the snake was inadvertently restrained by the sleeper. The venom is predominantly and potently cytotoxic, causing agonising pain, swelling (which may involved the entire limb and spread to the trunk), discolouration, blistering and serious local long-term tissue damage.

Occasionally, effects such as drowsiness occur. Deaths are rare, but treatment for the necrosis is often prolonged, complicated and very expensive. Antivenom is available. A bite is a medical emergency; it must be treated in a well-equipped hospital, and vigorous action to reduce the amount of tissue damage, probably involving early antivenom administration, must be taken. Venom in the eye is also a frequent result of an encounter with this snake, causing agonising pain, and, if untreated, corneal lesions and complications. The affected eye must be gently irrigated with large quantities of water. See our section on 'Spitting snakes: first aid for venom in the eyes' (p. 300). Anyone who has been spat at should visit a hospital or eye clinic for an assessment of the damage to the eye.

West African Brown Spitting Cobra *Naja katiensis*

Identification: A small red-brown cobra of the West African savanna, between 10°N and 15°N. It is moderately thick, and has a small head and a large eye with a round brown pupil. The body is cylindrical and the tail fairly long. The scales are smooth; in 21–27 rows (usually 25) at midbody; ventrals 160–195; cloacal scale entire; subcaudals 42–65. Maximum size just over 1m (possibly more; there is an unsubstantiated report of a 1.4m specimen), average 50–80cm; hatchling size unknown, but probably 14–16cm. Usually reddish brown, warm brown or

Naja katiensis

maroon, orange-brown on the flanks, and light orange-brown underneath. There is a broad dark band on the underside of the neck; this may form a ring, which may fade on large specimens.

Habitat and Distribution: Sudan savanna and semi-desert (Sahel), penetrating into the northern Guinea savanna, between sea level and about 900m. For example, in Nigeria it occurs at Zonkwa but not at Jos. It is not found in desert or dense woodland. It occurs from northern Cameroon westwards through northern Nigeria across to Senegal. Oddly, it is as yet unrecorded for Niger, but it probably occurs there as it is known immediately west in Burkina Faso.

Natural History: Mostly terrestrial, but it will climb into low bushes. This snake is fast-moving and alert. It is active both by day and night, juveniles often moving

West African Brown Spitting Cobra, *Naja katiensis*.
Left: Nigeria (Gerald Dunger). **Centre:** Ghana (Stephen Spawls). **Right:** Nigeria (Gerald Dunger).

around in the day. It shelters under logs, rocks and similar ground cover, or in holes. When disturbed it will usually try to make off, but if cornered it rears up and spreads a small, narrow hood; if further molested it will spit twin jets of venom at its adversary, to a distance of nearly 2m. It lays eggs, but no clutch details have been published. It eats mostly amphibians, but will take lizards, other snakes, birds and their eggs and rodents; insect remains were found in a Senegal snake, but these may have originated from an amphibian that the snake ate.

Medical Significance: Often common in parts of its range, active on the ground at night and liable to be trodden on, so probably a significant hazard, although there seem to be very few documented bites from this snake. An average wet venom yield of 100mg has been reported, and other cobras in this clade have predominantly cytotoxic venoms, causing swelling and local necrosis. The venom is probably produced in large quantities. As with other spitting cobras, a bite should be treated in a well-equipped hospital. Antivenom is available. Venom in the eye is also a frequent result of an encounter with this snake, causing agonising pain, and, if untreated, corneal lesions and complications. The affected eye must be gently irrigated with large quantities of water. Anyone who has been spat at should visit a hospital or eye clinic for an assessment of the damage to the eye.

Red Spitting Cobra *Naja pallida*

Identification: A relatively small, usually reddish cobra of the dry country of the Horn of Africa. It has a smallish head and a large eye with a round pupil. The body is cylindrical and the tail fairly long, 15–19% of total length. The scales are smooth; in 21–27 rows at midbody; ventrals 192–228; subcaudals 56–81. Maximum size about 1.5m (possibly larger), most adults 0.7–1.2m; hatchlings about 16–20cm. The colour is variable; specimens from northern Tanzania and eastern and southern Kenya (especially those from red soil areas) are orange or red, with

Naja pallida

a broad black throat band several (sometimes two or three) scales deep, and a black 'suture' on the scales below the eye, giving the impression of a teardrop. They are reddish below, sometimes with a white chin and throat. Specimens from other areas may be pale red, pinkish, pinky-grey, red-brown, yellow or steel grey. Most, however, have the dark throat band. In large adults, this band may fade or disappear. Red specimens become dull red-brown with increasing size.

Habitat and Distribution: Dry savanna and semi-desert, from sea level to about 1,200m. This species occurs from Djibouti and northern Somalia south throughout Somalia and the Ogaden (and back up the Ethiopian Rift Valley to Awash National Park). It is found in extreme southern Ethiopia and most of northern and eastern Kenya, although it does not reach the coast. From Tsavo West National Park in Kenya it extends south into Mkomazi, in north-east Tanzania, and from Amboseli it occurs west in Kenya's southern Gregory Rift Valley, through Magadi and Olorgesailie to Mount Suswa, and south into Tanzania past Lake Natron to

Red Spitting Cobra, *Naja pallida*. **Left and centre:** Kenya (Stephen Spawls). **Right:** Kenya (Sebastian Kirchof).

Olduvai Gorge. It has also been recorded from Mjohoroni, just south of Mount Kilimanjaro. It is probably widespread in extreme south-east South Sudan, as it is known from near Torit and just across the border on the lower Omo River in Ethiopia; likewise, it is probably in north-east Uganda, but there are few records from those areas. A small yellowish cobra, possibly this species, was photographed in eastern Central African Republic (not shown on the map).

Natural History: Terrestrial, but it may climb into bushes and low trees. It is quick-moving and alert. Adults are mostly nocturnal, hiding in holes (particularly in termite hills), brush piles, old logs or under ground cover during the day, but juveniles are often active during the day. When disturbed, it rears up relatively high and spreads a long, narrow hood, and may spit twin jets of venom to distances of 2–3m. It lays 6–15 eggs, about 50 × 25mm. A juvenile was taken on the eastern Tharaka Plain, Kenya, in October and at Olorgesailie in July. Amphibians are its favourite food; when the storms arrive in the dry country where this snake lives, the frogs all come out and the snake gorges itself. A specimen in Tsavo was found at night up in a thorn tree, eating foam-nest tree-frogs. It also takes rodents and birds, is known to raid chicken runs, and probably eats other snakes. It will come around houses in dry country looking for water, and for frogs that live in or near water tanks.

Red Spitting Cobra, *Naja pallida*. **Left:** Kenya (Johan Kloppers). **Right:** Kenya (Konrad Mebert).

Medical Significance: Often common in parts of its range, active on the ground at night and liable to be trodden on, so probably a significant hazard, but there seem to be very few documented bites from this snake. The venom is cytotoxic, causing swelling and local necrosis. The venom is probably produced in large quantities. As with other spitting cobras, a bite should be treated in a well-equipped hospital. Antivenom is available. A surprising number of bites from this snake result from the snake entering buildings at night; it is feasible that, as with the Mozambique Spitting Cobra (*Naja mossambica*), a sleeping human gives off some signal (warmth or smell) that the snake associates with prey. A child treated in Garissa in Kenya for a bite on the elbow had no pain, but after 18 hours considerable swelling prompted incisions on the hand and elbow. There was no subsequent necrosis and recovery was complete. A bite at Voi in Kenya, on the sole of the foot, caused considerable pain; the patient received antivenom and an immediate anaphylactic reaction complicated matters; this was treated. Subsequent pain and swelling lasted four days and necrosis caused a 2cm hole around the bite puncture, which took months to heal. Many of the Somali people living around Wajir, in Kenya, had friends or relatives who had been bitten by this snake, but nobody knew of anyone who had died from a bite; the usual treatment was to eat something that caused vomiting. Venom in the eye is also a frequent result of an encounter with this snake, causing agonising pain, and, if untreated, corneal lesions and complications. The affected eye must be gently irrigated with large quantities of water. Anyone who has been spat at should visit a hospital or eye clinic for an assessment of the damage to the eye.

Nubian Spitting Cobra *Naja nubiae*

Identification: A relatively small, usually brown or grey cobra of the Nile Valley and north-east Africa. It has a smallish head and a large eye with a round pupil. The body is cylindrical, and the tail fairly long, 15–19% of total length. The scales are smooth; in 23–27 rows at midbody; ventrals 207–226; cloacal scale entire; subcaudals 58–72. Maximum size about 1.5m (possibly larger), most adults 0.7–1.2m; hatchlings about 16–20cm. The colour is variable, from warm brown to maroon, grey or olive; juveniles are usually lighter. The scales are sometimes black-edged, giving

Naja nubiae

a net-like effect. There is usually a dark band on the neck, which may cross the underside of the hood, and below this are two dark bands that always cross the throat. There is a black 'teardrop' created by the black edging of the labial scales below and behind the eye. The underside is white, cream or brown.

Habitat and Distribution: Dry savanna, semi-desert, vegetated rocky hills and riverside vegetation, from sea level to about 1,500m. This species occurs from Asyut in Egypt, on the Nile, south along the Nile Valley to Khartoum in Sudan, and possibly further south. It is also found in extreme eastern Sudan and Eritrea. Apparently isolated populations are known from Gebel Elba in south-east Egypt, on Jebel Marra in western Sudan, in the Ennedi Massif in north-east Chad and in

Nubian Spitting Cobra, *Naja nubiae*. **Left top:** hatchling (Tomáš Mazuch). **Right top:** Sudan (Abubakr Mohammed). **Left bottom and right bottom:** Egypt (Stephen Spawls).

the Air Mountains in northern Niger; this distribution suggests these snakes were more widespread when the Sahara area was wetter and have become isolated in pockets of highland with higher rainfall than the surrounding desert.

Natural History: Terrestrial, but it may climb into bushes and low trees. This snake is quick-moving and alert. Adults are mostly nocturnal, hiding in holes, rock fissures or under ground cover during day, but juveniles are often active during the day. When disturbed, it rears up and spreads a hood, and may spit twin jets of venom. It apparently swims well. Females lay eggs. The diet is probably similar to that of the Red Spitting Cobra (*Naja pallida*), although captive animals feed readily on mice. This species is often used by Egyptian 'snake charmers', despite the fact that it can spit and thus represents a hazard to all observers.

Medical Significance: Fairly common in parts of its range, active on the ground at night and liable to be trodden on, so probably a significant hazard, but there seem to be very few documented bites from this snake. The venom is cytotoxic, causing swelling and local necrosis. The venom is probably produced in large quantities. As with other spitting cobras, a bite should be treated in a well-equipped hospital. Venom in the eye is also a frequent result of an encounter with this snake, causing agonising pain, and, if untreated, corneal lesions and complications. The affected eye must be gently irrigated with large quantities of water. Anyone who has been spat at should visit a hospital or eye clinic for an assessment of the damage to the eye.

Tree cobras *Pseudohaje*

Tree cobras occur in forested tropical Africa. Two species are known. They are large, agile black or dark brown tree snakes with huge eyes. They look like a cross between a mamba (*Dendroaspis*) and a cobra; recent molecular work indicates they are the sister species to true cobras. In medical terms they are not very significant; being alert tree snakes that live in forest, they rarely come into contact with people. They have potent neurotoxic venoms, presumably causing a descending paralysis. A bite victim in DR Congo died in 12 hours. No antivenom is available, but the polyvalent antivenom produced by South African Vaccine Producers (SAVP) is reported to be effective at neutralising tree cobra venom.

Key to the genus *Pseudohaje*, tree cobras

1a Scales usually in 13 rows on the anterior body; ventrals 180–187; occurs mostly west of the Dahomey Gap......*Pseudohaje nigra*, Black Tree Cobra (p. 184)

1b Scales usually in 15–17 rows on the anterior body; ventrals 185–205; occurs mostly east of the Dahomey Gap......*Pseudohaje goldii*, Gold's Tree Cobra (p. 182)

Gold's Tree Cobra *Pseudohaje goldii*

Identification: A large, thin-bodied, shiny tree cobra with a short head. The eye is huge and dark, with a round pupil. The body is cylindrical and the tail long and thin, 22–25% of total length; it ends in a spike, which is probably a climbing aid. If held by the head this snake will drive the spike into the restraining arm. The scales are smooth and very glossy; in 15 rows (occasionally 13 or 17) on the front half of the body, reducing to 13 rows slightly past midbody; ventrals 185–205; subcaudals 76–96. Maximum size 2.7m (perhaps slightly larger); most adults average

Pseudohaje goldii

1.5–2m; hatchlings 40–50cm. The skin is very fragile, and tears or abrades easily. The body is glossy black above, and the scales on the chin, throat and side of the head are yellow edged with black. Juveniles have several yellow cross-bars or bands, narrowing towards the tail, becoming scattered yellow scales on the back half; some juveniles have upward-pointing yellow triangles low on the flanks, coincident with the cross-bars.

Habitat and Distribution: In primary and altered forest, forest islands, swamp and mangrove forest and, surprisingly, in suburbs and even wooded savanna, between sea level and about 1,700m. It occurs from the Albertine Rift, westwards across the great Central African forests to south-eastern Nigeria, south to north-west Angola, and north into scattered woodlands in Central African Republic. There are isolated records from the Kakamega area in Kenya, the shore of Lake Victoria, Mabira Forest and the Kibale area of Uganda, and from forests in southern Ghana; there is a curious record from Ivory Coast, and this species is historically reported from Bioko Island.

Gold's Tree Cobra, *Pseudohaje goldii*. **Left and right:** Kenya (Stephen Spawls).

Natural History: Very poorly known. This snake is arboreal, but will descend to the ground. It is quick and alert, climbs fast and well, and moves rapidly on the ground, often with the head raised. It is active during the day, but might also be nocturnal. In Kenya, specimens have been found in squirrel traps and in trees along rivers, specimens in eastern DR Congo are reportedly common in oil palm plantations, in West Africa this species has been found in fish traps, and a survey in Nigeria found most individuals fairly close to waterbodies. Gold's Tree Cobra is not an aggressive species, but if molested or restrained it can flatten the neck into a very slight hood (nothing like as broad as a cobra's), and may move forward with the head raised and try to bite; specimens have been seen to make rapid direct strikes. If disturbed, this snake will pause with the head up, moving it from side to side like a metronome. Male–male combat has been observed in Nigeria in January. Females lay 10–20 eggs, approximately 50 × 25mm. A hatchling was captured in September in the Ituri Forest in DR Congo. Nigerian specimens had eaten frogs, mudskippers and other fish, a small hinged tortoise and rats; specimens were taken in squirrel traps in Kakamega, suggesting it eats arboreal mammals. It has been reported as descending to the ground to feed on terrestrial amphibians, but ascending to then digest its prey. It is prized as a human food in eastern DR Congo, but in the same area it has a fearsome reputation, being said to retaliate readily if molested. Captive animals remain alert and are quick to attempt to escape.

Medical Significance: Rarely comes into contact with humans, being an alert diurnal tree snake with good eyesight; it moves rapidly away from confrontation, although a collector described it as being quick to try to bite if it thinks it is cornered. The venom is a potent neurotoxin; a snake collector in eastern DR Congo was bitten during a demonstration and died within 12 hours. South African Vaccine Producers (SAVP) polyvalent antivenom, although not prepared for this snake, is reportedly effective against the venom. A bite case would need to be transported rapidly to hospital, and respiratory support may be needed.

Black Tree Cobra *Pseudohaje nigra*

Identification: Virtually identical to Gold's Tree Cobra (*Pseudohaje goldii*). A large thin-bodied, shiny tree cobra with a short head. The eye is huge and dark, with a round pupil. The body is cylindrical and the tail long and thin, 20–22% of total length; it ends in a spike. The scales are very smooth and glossy; in 15 rows on the neck but 13 rows at midbody; ventrals 180–187; subcaudals 74–82. Maximum size 2.2m (possibly larger); most adults average 1.6–2.1m; hatchling size unknown, but probably 35–45cm. The colour is black, shiny brownish black or shiny deep brown above, and the scales on the chin, throat and side of the head are yellow edged with black.

Pseudohaje nigra

Habitat and Distribution: Forest or thick woodland, between sea level and about 800m, or possibly higher. It is often found near rivers, and might extend out of the forest proper in riverine forest; a specimen from Kintampo in Ghana was in Guinea savanna woodland with stands of tall trees and almost continuous canopy. This is essentially a West African forest snake; it occurs from Ghana west to Sierra Leone and Guinea (although records are lacking from Ivory Coast.). There are isolated records from hill forests on the Ghana–Togo border, and from Enugu in Nigeria, although recent research in Nigeria failed to find the species and suggests it might no longer occur there.

Natural History: Very poorly known; almost nothing is published, but it is probably similar to Gold's Tree Cobra. It is presumably arboreal, but will descend to the ground. It is very quick-moving in trees, climbing fast and well, and moves rapidly on the ground, often with the head raised and alert; if disturbed it will pause with the head up. Both active and resting specimens have been captured during the day, and it might be predominantly diurnal or nocturnal. This snake has been described as 'a cross between a mamba and a cobra'. It might be able to flatten the neck slightly, as can Gold's Tree Cobra. Females lay eggs. This snake probably eats amphibians, and maybe mammals.

Medical Significance: As for the previous species.

Black Tree Cobra, *Pseudohaje nigra*. **Left:** Guinea (Jean-Francois Trape). **Right:** Ghana (Barry Hughes).

Desert black snakes (black desert cobras) *Walterinnesia*

Black desert elapids; two species occur in the dry country and deserts of Egypt and the Middle East, the eastern species (*Walterinnesia morgani*) being recently elevated from the synonymy of the western form (*Walterinnesia aegyptia*). They are fairly large, quick-moving black elapid snakes with a medium-sized, broad head. The body is stocky and cylindrical, and the scales smooth. Although rare, they are alert and will bite if cornered; a few bite cases are documented. They are not aggressive, but can strike a long way. The venom is neurotoxic and has caused deaths, although no recent fatalities are documented. They are closely related to cobras but do not spread a hood, so the name 'black desert cobras', although widely used, is not appropriate as it may lead to confusion with black-coloured Egyptian Cobras (*Naja haje*).

Desert Black Snake *Walterinnesia aegyptia*

Identification: A large black snake of the Egyptian desert, usually active at night. It has a fairly broad head, slightly distinct from the neck, and a smallish, dark eye with a round pupil (which is hard to see). The body is cylindrical and the tail long and thin, 12–15% of total length, ending in a spike. The scales are smooth and glossy; in 23 rows at midbody; ventrals 186–198; subcaudals 43–53, largely single but, unusually, some are often in pairs. Maximum size about 1.3m, average 0.8–1.2m; hatchlings 20–35cm. The colour is black or dark grey, often with

Walterinnesia aegyptia

a bluish sheen, and paler whitish grey or sometimes yellow-brown beneath. This snake becomes grey before sloughing. The sides of the head and lip scales are often lighter in Egyptian animals, becoming light grey or yellow-brown on the snout.

Habitat and Distribution: Known from rocky desert, gravel or sandy plains, vegetated wadis and hilly and mountainous desert. However, it is seemingly more common near agriculture and irrigated areas, and in and around desert and desert-fringe settlements and oases (although this might be simply due to the fact that the collectors favoured these spots rather than deep desert). It is absent from extensive areas of soft sand and dunes. This species occurs between sea level and 1,200m (possibly higher; *Walterinnesia morgani* ascends to over 2,000m in mountains). In Africa, it occurs eastwards and north-eastwards from Cairo into the Sinai Peninsula (there are few records from the interior of the Sinai), then into Israel, Jordan and the north-west corner of Saudi Arabia; *Walterinnesia morgani* occurs further east. There are no credible records west of the Nile, but it might occur in Egypt's eastern desert, as people living there know of a large black desert snake.

Natural History: Secretive, terrestrial and largely nocturnal, it lives in holes, rock fissures and under ground cover. It emerges at night to hunt, and is known to be active on relatively cold nights. This is a quick-moving and alert snake; if

Desert Black Snake, *Walterinnesia aegyptia*. **Left:** Israel (Yannick Francioli). **Right:** Egypt (Stephen Spawls).

confronted, it hisses loudly and raises the body, preparing to strike, but in an angled S-shaped coil, not vertically elevated like in a cobra. If teased it may also vibrate the tail vigorously, and then hide the head and neck under its coils, defecate and release a foul-smelling fluid from the cloaca. However, it seems reluctant (at least in captivity) to actually strike, tending instead to bang the aggressor with the mouth closed. It lays 2–20 eggs; captive animals laid in May and the eggs hatched 74–77 days later, which fits with the idea of the newborns being able to put on weight before the winter. This snake feeds on a wide range of vertebrates; it is fond of toads around oases, and in the desert feeds on dabb or spiny-tailed lizards, and is often found in areas where these lizards occur. Captive juveniles are very reluctant feeders but accepted lacertid lizards and dead carpet vipers (*Echis*); most African elapids will eat other snakes. This species is sometimes used by snake charmers (including one who thought it was harmless). Animal collectors exporting these snakes from Egypt often flush them from their holes using petrol or exhaust fumes, leading to damage to, and subsequent high mortality of, captive animals.

Medical Significance: A seemingly rare snake, and apparently reluctant to bite, so not of major clinical significance. However, it is active on the ground at night and liable to be trodden on, and a few daytime bites are also recorded, where the snake was disturbed during building or agricultural work. The venom is neurotoxic, and laboratory studies also indicate some haemorrhagic activity. Antivenom is available. A series of bite cases in Israel resulted in localised pain and swelling, fever, generalised weakness, nausea and vomiting; there were no fatalities. A bite in Iraq from *Walterinnesia morgani* caused severe weakness, intense numbness at the bite site, respiratory distress and double vision; the victim was treated with antivenom and recovered completely. A known fatal case occurred in Iraq in the 1930s; the victim died several hours after the bite. As with other large elapids, a bite should be treated in a well-equipped hospital.

Shield cobras *Aspidelaps*

A genus containing two small, attractive, rather stout nocturnal elapids of southern Africa, originally known as the Coral Snake and the Shield-nose Snake. They have a large, prominently protruding rostral scale that shields the nose, hence the name. Although not true cobras, they will rear up and spread a very narrow hood. They lie within a clade that includes all true cobras, and their closest relatives (at present) appear to be the desert black snakes, *Walterinnesia*. They are not regarded as medically significant, as they are secretive and nocturnal. However, they can be locally common, especially in the rainy season, and they are active on the ground at night and willing to bite. No antivenom is available. There are conflicting reports about their venom, with reports of both cytotoxic and neurotoxic effects, and at least one is known to have caused the death of a child. They need to be treated with care.

Key to the genus *Aspidelaps*, shield cobras

1a Ventral scales 139–179; dorsal surface usually finely banded; usually found south of 27°S (except in Namibia)......*Aspidelaps lubricus*, Coral Shield Cobra (p. 187)

1b Ventral scales 108–125; dorsal surface usually blotched or speckled but not finely banded; usually found north of 27°S......*Aspidelaps scutatus*, Speckled Shield Cobra (p. 189)

Coral Shield Cobra (Coral Snake) *Aspidelaps lubricus*

Identification: A small, stocky snake, often banded, of the dry country of south-western Africa. It has a short head with a prominent rostral scale, and a huge dark eye with a round pupil. The body is cylindrical and the tail relatively short, 9–14% of total length. The scales are smooth and glossy; in 19–21 rows at midbody; ventrals 139–179; cloacal scale entire; subcaudals 17–38. Maximum size just under 80cm; most adults average 40–60cm; hatchlings 15–20cm. The colour is variable; southern animals of the subspecies *Aspidelaps lubricus lubricus* are orange

Aspidelaps lubricus

with black bands, and with a black bar on the head through the eyes, leading to the jocular name 'bandit snake'. In large adults the dorsal banding may be less prominent. The bands fade going northwards, and some northern Namibian and Angolan animals are uniform grey or grey-brown; this is the subspecies *A. l. cowlesi*. The bars extend to the underside, which is thus barred black and cream or white.

Habitat and Distribution: In shrub and heathland (fynbos), on rocky hills, and in dry, stony soil or sandy regions in dry savanna, semi-desert and desert, between sea level and 1,800m, or sometimes slightly higher. It extends westwards from the eastern Free State and Port Elizabeth in the Eastern Cape to the western coastline in South Africa, then north up through Namibia. Aus is about the northernmost

Coral Shield Cobra, *Aspidelaps lubricus*.
Left top: South Africa (Tyrone Ping). **Right top:** Namibia (Stephen Spawls).
Left bottom: Angola (Bill Branch). **Right bottom:** Namibia (Richard Boycott).

record for the subspecies *Aspidelaps lubricus lubricus*; the subspecies *A. l. cowlesi* occurs north of about 25°S, extending into south-western Angola.

Natural History: A nocturnal terrestrial snake, said to be active throughout the night, even at relatively low temperatures; a number of specimens have been captured while crossing roads at night. It shelters during the day under rocks or in holes, abandoned termitaria and rock crevices. Captive animals have been observed excavating shelters beneath rocks by digging with the nose and scooping the sand with the neck. When threatened, this snake rears up and flattens the neck like a true cobra, presenting a narrow hood. If further molested, it will produce a loud, short, explosive hiss and may lunge forward in an attempt to bite. Male–male combat is recorded. Mating occurs in the southern spring. Three to 11 eggs are laid, roughly 20 × 30mm, and hatch in around two months. The diet includes lizards, small snakes, reptile eggs and rodents; captive snakes take rodents readily.

Medical Significance: Its fairly small range, mostly in inhospitable country, lessens the danger posed by this snake. However, it is active on the ground at night, and sometimes present in numbers, which increases the risk. Its venom is neurotoxic. Early work found the venom caused respiratory distress, leading to death in mice. Lethal doses similar to that of the Black Mamba (*Dendroaspis* *polylepis*) were recorded, although the amounts produced were very small. No antivenom is available. Although most bites on humans seem to cause only pain, and no definite fatalities are reported, a snake keeper who was bitten twice was vomiting an hour after the bite; his upper body muscles became paralysed and he stopped breathing and needed respiratory support. Twelve hours later the symptoms resolved. In addition, one of these snakes was found in a dwelling in northern Namibia where two children had died from snakebite, although another snake might have been responsible. Bite victims should visit a hospital, where symptomatic treatment can be given.

Speckled Shield Cobra (Shield-nose Snake)
Aspidelaps scutatus

Identification: A small, stocky snake, often with a black neck, of the savannas and semi-desert of southern Africa. It rears up like a cobra. It has a short head with a prominent rostral scale, and a huge dark eye with a round pupil. The body is cylindrical and the tail relatively short, 12–17% of total length. The scales are smooth and glossy; in 21–23 rows at midbody; ventrals 108–125; cloacal scale entire; subcaudals 19–39. Maximum size around 75cm; most adults average 40–60cm; hatchlings 14–20cm. The colour is variable. The two main subspecies

Aspidelaps scutatus

(*Aspidelaps scutatus scutatus* and *A. s. intermedius*) are both usually various shades of brown or pinky-brown, or sometimes orange or grey, and uniform or spotted or speckled to a greater or lesser extent. In some individuals (especially of the extreme eastern subspecies, *A. s. fulafula*) the back has large black spots that almost become cross-bars. The head and neck are often black, to a greater or lesser extent, and the underside is cream, white or grey, usually with a broad, deep black neck bar.

Habitat and Distribution: In semi-desert and moist and dry savanna, between sea level and 1,400m, or sometimes slightly higher. It occurs in northern Namibia, most of Botswana except the south-west corner, the northern provinces of South Africa, southern Mozambique, and western, southern and south-east Zimbabwe. Those in eastern Mpumalanga and Limpopo are the subspecies *Aspidelaps scutatus intermedius*; those in south-eastern Zimbabwe and Mozambique are the subspecies *A. s. fulafula*; and those in the remainder of the range are the nominate subspecies, *A. s. scutatus*.

Natural History: A nocturnal and terrestrial snake, said to be active throughout the night, even at relatively low temperatures; a number of specimens have been captured while crossing roads at night. It shelters during the day under ground cover, in holes and in abandoned termitaria, and may bury itself in leaf litter or loose sand. When threatened, this snake rears up and flattens the neck like a true cobra, presenting a narrow hood. If further molested, it will produce a very loud, short, explosive hiss and may lunge forward in an attempt to bite. If repeatedly teased, it may sham (feign) death. Male–male combat is recorded.

Speckled Shield Cobra, *Aspidelaps scutatus*.
Left: South Africa (Bill Branch). **Centre:** Botswana (Stephen Spawls). **Right:** South Africa (Bill Branch).

Speckled Shield Cobra, *Aspidelaps scutatus*.
Left: South Africa (Bill Branch). **Centre:** Botswana (Stephen Spawls). **Right:** South Africa (Bill Branch).

Mating occurs in the southern spring. Four to 11 eggs are laid, roughly 20 × 30mm, and hatch in just over two months. This species seems to feed readily on most small vertebrates, including snakes. Six individuals were found in a single night in south-east Zimbabwe, waiting in ambush for gerbils; they have also been observed taking winged termites. Captive animals readily eat rodents and toads.

Medical Significance: This snake is quite widespread in southern Africa, and parts of its range are densely populated; it is also active on the ground at night, and sometimes present in numbers, which increases the risk. On one wet night in Mpumalanga five were found on the road. There are conflicting reports on the effects of its venom; it has caused both cytotoxic (swelling and local pain) and neurotoxic symptoms (facial muscle weakness and respiratory distress). The lethal dose for a mouse was about twice that of a Cape Cobra (*Naja nivea*). No antivenom is available. A single fatality is documented: a four-year-old child died 16 hours after a bite from this snake; she stopped breathing. Bite victims should visit a hospital, where symptomatic treatment can be given.

Rinkhals *Hemachatus*

A monotopic genus containing a single species, a stocky southern African endemic that looks identical to a true cobra; in defence it rears up and spreads a broad hood. If differs from true cobras in having keeled scales, giving birth to live young and having a maxillary (upper jaw) bone with fangs but without solid teeth. It can spit its venom. It lies within a clade that includes all true cobras (*Naja*), but split from them more than 30 million years ago. It is a medically significant species, as it is widespread across a wide area of well-populated South Africa, is tolerant of urbanisation and can be locally common. Antivenom is available. The venom is both cytotoxic and neurotoxic.

Rinkhals *Hemachatus haemachatus*

Identification: A medium-sized, stocky relative of the cobras, of temperate southern Africa, with a minute relict population in Zimbabwe. It has a short head and a large, dark eye with a round pupil. The body is cylindrical and the tail 15–19% of total length (usually longer in males). The dorsal scales are largely keeled; in 17–19 rows at midbody; ventrals 116–150; cloacal scale entire; subcaudals

30–47. Maximum size around 1.5m; most adults average 0.9–1.3m; hatchlings 16–22cm. The colour is variable, usually olive, brown or dull black above, with one or two (sometimes three) white or yellow cross-bars on the throat. A black form with yellow, tan or white bands also exists. The underside is usually dark brown, grey or black.

Hemachatus haemachatus

Habitat and Distribution: In grassland, including high montane areas, moist savanna, rocky outcrops, lowland forest and heathland (fynbos), between sea level and 2,500m. It occurs from sea level in the Western Cape, South Africa, all around Cape Town, then through the Cape Fold Mountains into the Eastern Cape, although it is seemingly absent from some areas. From the Eastern Cape it spreads northwards along the eastern escarpment through Kwazulu-Natal and the Lesotho and Free State grasslands, into Gauteng, eastern North West Province, Mpumalanga and western Eswatini (Swaziland). It has been reported from Kimberley, but there are no recent records. An isolated population occurs around the Nyanga National Park in the highlands of eastern Zimbabwe, mostly over 1,500m, in grassland and stunted miombo woodland. It is speculated that this population may be extinct; no specimens have been collected there for over 25 years.

Natural History: A largely terrestrial snake, but a good climber and quick to escape up trees. Mostly active by day, it basks in the morning before hunting, but it is also active on summer nights. It is quick-moving and alert, and will usually try to escape, often to a nearby refuge. If cornered, it rears up and spreads a hood, displaying the white throat bands; if approached it will spit venom, to a distance of over 3m, while hissing and lunging forward. If further threatened, it will often play dead, dropping down and twisting sideways or upside down, with the mouth open and tongue hanging out. If may remain limp if handled, or suddenly try

Rinkhals, *Hemachatus haemachatus*. **Left:** South Africa (Johan Marais). **Centre top:** banded form (Johan Marais). **Centre bottom:** shamming death (Bill Branch). **Right:** close-up of hood (Bill Branch).

to bite. It gives birth to usually 20–30 live young, but as many as 63 have been recorded. A variety of small vertebrates are eaten; it is fond of toads but also takes rodents, birds and their eggs, lizards and snakes. In South Africa it seems very tolerant of urbanisation, still being present in the Greater Johannesburg area, living near dams and grassy areas, especially those with open water or swamps (vleis); it has been found living in rockeries right next to major highways.

Medical Significance: This snake is quite widespread in South Africa and is tolerant of urbanisation, clinging on because of its secretive habits. Parts of its range are densely populated, and it is active on the ground by day and night, so it represents a significant snakebite hazard. A polyvalent antivenom is available and effective. The venom is both cytotoxic and neurotoxic, with a curiously broad symptomology that includes painful swelling of the bitten limb, nausea, dizziness and respiratory distress. However, there are no credible recent reports of any fatalities. As with other large elapids, a bite should be treated in a well-equipped hospital. Venom in the eye is also a frequent result of an encounter with this snake, causing agonising pain, and, if untreated, corneal lesions and complications can occur. The affected eye must be gently irrigated with large quantities of water. Anyone who has been spat at should visit a hospital or eye clinic for an assessment of the damage to the eye.

Subfamily Hydrophiinae
Viviparous sea snakes and Australian land elapids

A family of front-fanged, venomous snakes, living on the land and in the sea; the terrestrial species look quite different from the marine ones. Their taxonomy is still under investigation, but at present there are around 50 species in the genus *Hydrophis* (sea snakes). One species, the Yellow-bellied Sea Snake (*Hydrophis platurus*), occurs in the western Indian Ocean and occasionally comes ashore on the eastern coasts of Africa. No other sea snakes are known from the shores of mainland Africa; reports of other sea snakes on African coasts are invariably either snake eels or moray eels. These may be told from sea snakes by their pointed tails; the sea snake has an oar-like tail. In general, sea snakes are not a hazard to those who work on or near the sea as they avoid confrontation in the water and are clumsy on land. A few bites are documented.

Yellow-bellied Sea Snake *Hydrophis platurus*

Identification: A medium-sized black and yellow sea snake. The head is long and tapering, the eye is small with a round pupil, and the nostrils are on top of the head. The body is laterally compressed and the tail is short, vertically flattened and oar-like, about 10% of total length. The scales are smooth; in 49–76 rows at midbody; ventrals 264–406. Maximum size about 90cm, average 60–85cm; hatchlings 24–25cm. It has bright warning colours; the broad dorsal stripe is usually black, greeny-black or very dark brown, the belly is yellow, and the tail is spotted black and cream or yellow. The dorsal stripe may be straight-edged (most African specimens are this colour), wavy, break up into black saddles or totally disappear.

Habitat and Distribution: Widespread across the Indian and Pacific Oceans; in all tropical seas except the Atlantic. It is not recorded from the Red Sea, but is found from Djibouti southwards all along the coast to Cape Town in South Africa; a few vagrants have been recorded further north on the Atlantic coast of South Africa and Namibia (not shown on map). This species is also known from Madagascar and the Seychelles.

Hydrophis platurus

Natural History: Lives mostly in open water, not around reefs, but usually within 300km of land. It is usually washed ashore only when dying or accidentally in storms; it can only wriggle in an ungainly fashion on land. It is a superb swimmer, moving with side-to-side undulations like a snake on land, but it can swim backwards and forwards and is capable of very rapid bursts of speed. This species is often found in the vicinity of sea slicks (long, narrow lines of floating vegetation and accumulated debris on the sea surface), where it waits quietly to ambush fish; it may attack individual fish or rush into a small shoal, indiscriminately biting everything it can grab. It is afraid of humans and moves off if approached by swimmers. Three to eight young are born live; gravid females have been collected between March and October off the South African coast. Concentrations of up to 10 snakes have been seen in East African waters; a group of six was found in Turtle Bay, Watamu, Kenya. The sea snake sloughs its skin as do land snakes, but since it doesn't have convenient objects to rub against, it forms tight loops, rubbing coil against coil; in this way it also frees itself from sea animals like barnacles.

Medical Significance: This is a dangerously venomous snake, although bites are extremely rare (and often not noticed). Those at risk are fishermen who may mishandle the snake while emptying their nets; a snakebite to a swimmer, diver or snorkeller is unheard of. The venom is a powerful neurotoxin, and a bite may lead to paralysis, muscle destruction, renal failure and cardiac arrest. Early symptoms include headache, a thick feeling of the tongue, thirst, sweating and vomiting, then pain, stiffness and tenderness, followed by paralysis as in other neurotoxic poisoning; this may lead to respiratory failure. There is no swelling. No antivenom is produced in Africa, but it is available in Australia.

Yellow-bellied Sea Snake, *Hydrophis platurus.*
Below: Kenya (Royjan Taylor).

Family Lamprophiidae
African snakes (house snakes and allies)

A family of African, Madagascan and Asian snakes, with eight subfamilies and over 50 genera; over 30 are mainland African. Most are harmless or rear-fanged, but the burrowing asps (*Atractaspis*) and harlequin snakes (*Homoroselaps*) are front-fanged.

Subfamily Atractaspidinae
Burrowing asps and harlequin snakes

A subfamily of two genera and 24 species, found in Africa, with a limited penetration into the Arabian Peninsula. The two genera are easily distinguished (see the following key).

Key to the genera of the *Atractaspidinae*

1a   Fangs short and immoveable; has a yellow or red dorsal stripe; found only in South Africa......*Homoroselaps*, harlequin snakes (pp. 218–20)

1b   Fangs large and moveable; never has a yellow or red dorsal stripe; found throughout Africa......*Atractaspis*, burrowing asps (pp. 194–217)

Burrowing asps *Atractaspis*

A genus of just over 20 species of small (usually less than 70cm), dark, blunt-headed nocturnal snakes, with cylindrical bodies and a tail that ends in a short spike. They all look very similar: most are black, dark brown or grey, although just before sloughing they become silvery-grey. They have hollow, hinged front fangs, and were thus thought to be vipers, but are now recognised as specialised members of the Lamprophiidae (African snakes) within their own subfamily, the Atractaspidinae. The confusion over their status has resulted in numerous different common names being used for the group, including burrowing vipers, burrowing asps, mole vipers, stiletto snakes (or just 'stilettos'), side-stabbing snakes and back-stabbing snakes. We have opted for the name 'burrowing asps', which avoids confusion with vipers. The taxonomy of the genus is incomplete, hampered by the lack of museum specimens of these secretive, burrowing snakes. Some species are known from a handful of specimens alone; others have several subspecies, often based on minor (possibly insignificant) scalation differences. The status of some taxa is uncertain.

    Burrowing asps are widely distributed across Africa, and two (possibly three) species occur in the Middle East. They spend much time underground in holes, and feed largely on small vertebrates. Females lay eggs. These snakes can be hard to identify, as they very closely resemble purple-glossed snakes (*Amblyodipsas*), the Kwazulu-Natal Black Snake (*Macrelaps microlepidotus*) and, to a lesser extent, wolf snakes (*Lycophidion*), dark centipede-eaters (*Aparallactus*), snake-eaters (*Polemon*) and blind snakes (*Afrotyphlops*). Even expert herpetologists have been bitten by misidentified burrowing asps. Have a look at our pictures of similar species in the look-alikes section (pp. 252–78). But even when you have established you are looking at a burrowing asp, identification to species level is not easy. You will need a binocular microscope to count scales on a dead specimen (important; see the next paragraph), although the range may give some clues. Living animals often exhibit a curious and distinctive display, arching the neck and pointing the head at the ground, looking like a croquet hoop or an inverted 'U'. They may also release a distinctive-smelling chemical from the cloaca (which might serve to stop one burrowing asp eating another). If further teased, they may wind the body into tight coils, or turn the head and neck upside down, lash from side to side or jerk violently, often springing some distance; this may have a defensive purpose.

Burrowing asps cannot, under any circumstances, be held safely by hand. Their heads are too short and their fangs too long; this enables them to bite any restraining finger, no matter how tight the grip. Do not try it. The snake can slide a fang out of its mouth, without actually opening the mouth. The fang is then driven into the finger by a downwards stab. Such bites consequently usually show a single fang puncture. If held by the tail they can jerk upwards and bite the hand. The fangs are primarily used for stabbing and manipulating their prey, an adaptation that may be useful for feeding in holes, and they have been reported to enter a rodent nest and stab all the young before commencing feeding. Apart from the fangs, burrowing asps are virtually toothless, and thus struggle to ingest prey, having to flex the neck and body to swallow (other snakes manipulate prey with their solid teeth, a process called cranial kinesis).

Burrowing asps are not aggressive snakes, and move away from humans; even if teased they do not open their mouths and lunge forwards. However, they are medically significant; they cause a lot of bites, particularly in Africa's warm tropical lowlands. This is for several reasons. They have a huge range; you will find at least one species of burrowing asp virtually everywhere in sub-Saharan Africa (except true desert and some highland areas), and often more. They are often relatively common, and they emerge from their holes and move around on the ground at night, particularly during the wet season. Being slim and active, they often enter buildings and shelters, and people tread on them, roll on them in their sleep or otherwise inadvertently restrain them. The snake turns and bites immediately. In the daytime, people farming or gardening with short tools are also at risk.

Burrowing asps have very long venom glands. Their venom is largely cytotoxic, although a handful of species have been shown to have sarafotoxins (which constrict coronary blood vessels, causing angina–endothelin-like effects) in their venom. The potential effects of the sarafotoxins on humans are as yet unquantified, but some bite victims have shown elevated blood pressure, ECG changes and anaphylaxis. Bite symptoms nearly always include swelling, pain, lymphadenitis and discolouration, often with some local necrosis; symptoms usually resolve in a couple of weeks. Occasionally amputation has been needed. Fever is often present. No antivenom is available for African species (one has been produced for the Arabian Small-scaled Burrowing Asp, *Atractaspis andersonii*, but this is not commercially available at the time of writing); bites are treated symptomatically.

African burrowing asps are not generally regarded as deadly, although several recent deaths are recorded for the Arabian *Atractaspis andersonii*. In southern Africa, to date, no fatalities have been recorded from burrowing asp bites, although they cause a lot of pain and suffering. A few fatal cases (some recent, most not) have been recorded from other areas of Africa, but these largely lack enough clinical details to guarantee that the venom was solely responsible or that unusual factors were not involved. But a note of caution is in order. In virtually all documented fatal cases, systemic symptoms appeared within minutes. Acute abdominal pain, profuse salivation and comas were features. Death occurred in hours. In recent years a few fatalities have occurred, leading to speculation that, in certain circumstances, burrowing asp venom may be more dangerous than is generally thought. Professor David Warrell regards burrowing asps as the African equivalent of the Asian kraits (*Bungarus*), as both enter homes and bite people in their sleep, as well as biting people walking home after dark and heavy rain. Perhaps their danger is underplayed.

The technical key below will help identify burrowing asps to specific level, we hope. It should never be used on a living animal; they cannot be safely held. You will also need a binocular microscope and some knowledge of head and body scalation (see Figures 16–22, pp. 20–21), but even identification is not easy. Note well: some species are very hard to tell apart, but the locality and general appearance may give clues. The final identification is not important where a burrowing asp bite is concerned, however, as no antivenom is available at present and bites are therefore treated symptomatically, regardless of the species.

Key to the genus *Atractaspis*, burrowing asps

1a In our area, known only from the Sinai Peninsula......*Atractaspis engaddensis*, Palestine Burrowing Asp (p. 205)
1b Not in Sinai, but elsewhere in Africa......2

2a White markings on the head; always in the Horn of Africa......3
2b No white markings on the head......4

3a Ventrals 215–230; no white vertebral stripe......*Atractaspis scorteccii*, Somali Burrowing Asp (p. 210)
3b Ventrals 230–243; has a white vertebral stripe......*Atractaspis leucomelas*, Ogaden Burrowing Asp (p. 209)

4a Ventral counts more than 300......*Atractaspis reticulata*, Reticulate Burrowing Asp (p. 211)
4b Ventral counts fewer than 300......5

5a Found only south of 20°S; snout profile distinctly curved......*Atractaspis duerdeni*, Duerden's Burrowing Asp (p. 204)
5b Either found north of 20°S, or, if south of there, snout profile pointed and shark-like......6

6a Cloacal scale divided......7
6b Cloacal scale entire......10

7a In southern Somalia and north-east Kenya; midbody scale rows 19......*Atractaspis engdahli*, Engdahl's Burrowing Asp (p. 206)
7b Not in southern Somalia or north-east Kenya; midbody scale rows 19–27......8

8a Ventral scales more than 275......*Atractaspis branchi*, Bill Branch's Burrowing Asp (p. 200)
8b Ventral scales fewer than 265......9

9a Midbody scale rows 21–27 (usually 23–27); mental separated from chin shields; usually north of 5°S; usually in forest......*Atractaspis irregularis*, Variable Burrowing Asp (p. 207)
9b Midbody scale rows 19–23 (usually 19–21); mental in contact with the chin shields; usually south of 5°S; usually in savanna......*Atractaspis congica*, Congo Burrowing Asp (p. 201)

10a One anterior temporal scale; often slim; usually in moist savanna or forest......11
10b Two anterior temporal scales; often stocky; in savanna and semi-desert......15

11a Second pair of lower labials is absent (fused to the chin shields)......*Atractaspis corpulenta*, Fat Burrowing Asp (p. 202)
11b Second pair of lower labials is present (and distinct from the chin shields)......12

12a Midbody scale rows 29–35......*Atractaspis dahomeyensis*, Dahomey Burrowing Asp (p. 203)
12b Midbody scale rows fewer than 27......13

13a First upper labial is in contact with the posterior nasal......*Atractaspis aterrima*, Slender Burrowing Asp (p. 197)
13b First upper labial is not in contact with the posterior nasal......14

14a Mental scale separated from the chin shields by the first pair of lower labials; subcaudals single; in savanna south of 5°S......*Atractaspis bibronii*, Bibron's Burrowing Asp (p. 198)

14b Mental scale in contact with the chin shields by the first pair of lower labials; some subcaudals paired; in forest almost always north of 5°S......*Atractaspis boulengeri*, Central African Burrowing Asp (p. 200)

15a Found west of 25°E......16
15b Found east of 25°E......18

16a More than seven lower labials......*Atractaspis watsoni*, Watson's Burrowing Asp (p. 214)
16b Fewer than seven lower labials......17

17a Dorsal scale rows 29–33; subcaudals 21–26; only in Mauritania and Senegambia......*Atractaspis microlepidota*, Western Small-scaled Burrowing Asp (p. 212)

17b Dorsal scale rows 25–27; subcaudals 26–32; widespread across the Sahel...... *Atractaspis micropholis*, Sahelian Burrowing Asp (p. 213)

18a Midbody scale rows 25–27 (rarely 29); largely north of 15°N......*Atractaspis magrettii*, Magretti's Burrowing Asp (p. 215)
18b Midbody scale rows 27–37; largely south of 15°N......19

19a Ventral counts strongly sexually dimorphic, much higher in males (246–252) than females (211–238); occurs in Sudan, South Sudan and low western Ethiopia......*Atractaspis phillipsi*, Sudan Burrowing Asp (p. 215)
19b Ventral counts not strongly sexually dimorphic, with males (227–246) little different from females (229–257); occurs in South Sudan, eastern Ethiopia, Somalia, Kenya and northern Tanzania......*Atractaspis fallax*, Eastern Small-scaled Burrowing Asp (p. 216)

Slender Burrowing Asp *Atractaspis aterrima*

Identification: A small, slim, fast-moving burrowing asp. The head is blunt and the eyes tiny. The tail is short and blunt, 3–5% of total length, and ends in a pointed cone. The scales are in 19–21 (rarely 23) rows at midbody; ventrals 239–300; cloacal scale entire; subcaudals single, 17–26. Maximum size about 80cm, average 30–50cm; hatchling size unknown, but a Garamba juvenile was 20cm. The colour is usually black or blackish grey (occasionally blackish brown or blackish purple), both above and below; juveniles may be brown. Some West African specimens have a white dot near the tail tip.

Atractaspis aterrima

Slender Burrowing Asp, *Atractaspis aterrima*. **Left and right:** Tanzania (Michele Menegon).

Habitat and Distribution: Has an unusually wide choice of habitat; it is known from dry and moist savanna, woodland and forest, from sea level to 2,000m. It occurs from Senegal (south of the Gambia River) eastwards through Nigeria, across the top of Central Africa to the extreme south of South Sudan and north-west Uganda, and is also known from three isolated localities (Udzungwa and Uluguru Mountains and Mwanihana Forest Reserve) in Tanzania.

Natural History: Poorly known. This species lives and hunts in holes, but may emerge at night (especially during or after rain) and move around on the ground. If molested, it shows the distinctive responses mentioned in the generic introduction (pointing the head, releasing a chemical, and jerking and lashing). The diet is mostly rodents and smooth-bodied reptiles in the west of its range, but Tanzanian specimens have been recorded eating caecilians, and some captive specimens are reported to eat invertebrates. Females lay eggs, but the details are unknown.

Medical Significance: A widespread species that is sometimes common, is active on the ground at night and may enter buildings; it is also found in agricultural land where people farm with short tools, so is a significant snakebite hazard within parts of its range. Swelling, pain and subsequent lymphadenopathy resulted from a bite, with some slight local necrosis. No antivenom is available, and the use of polyvalent antivenom is contraindicated; treat bites symptomatically.

Bibron's Burrowing Asp *Atractaspis bibronii*

Identification: A small, fairly slim, dark, slow-moving snake with a prominent snout. The tongue is white. The body is cylindrical and the tail is short and blunt, 5–7% of total length, ending in a pointed cone. The scales are in 19–25 rows at midbody; ventrals 213–262; cloacal scale entire; subcaudals single, 18–28. Maximum size about 70cm, average 30–50cm; hatchlings 20cm. The body may be black, brown, pinkish or purplish brown (especially in Kenyan coast animals), or grey, often with a purplish sheen on the scales, and the belly may be lead grey,

Atractaspis bibronii

brownish, white or pale with a series of dark blotches. Animals with dark ventrals may have white patches on the chin and throat, or around the tail base (see the bottom picture, opposite). If the belly is pale, the pale colour may extend up to the lowest two to three scale rows on the sides and onto the upper labials.

Habitat and Distribution: In semi-desert, dry and moist savanna, woodland and coastal thicket; not in closed forest. This snake is found from sea level to about 1,800m. It occurs south from the north Kenyan coast, inland in Tanzania, then north-west to Rwanda, Burundi and across the south of DR Congo to northern and western Angola, and south to Botswana, most of Namibia and northern and eastern South Africa. There are sporadic records further south and west, along with isolated records in Kenya from Kitui, Kitobo Forest and the Maasai Mara, and from the Afgooye area (formerly Afgoi) in southern Somalia. A specimen

Bibron's Burrowing Asp, *Atractaspis bibronii*.
Left: South Africa (Wolfgang Wuster). **Right:** DR Congo (Colin Tilbury).

from Garamba National Park in north-east DR Congo has been reidentified as the Slender Burrowing Asp (*Atractaspis aterrima*).

Natural History: Nocturnal and terrestrial, spending much time in holes, in rotting logs or under ground cover, but it is active on the ground on wet nights. Like the previous species, when molested it shows a distinctive response, arching the neck, pointing the snout into the ground, jerking convulsively and releasing a distinctive chemical smell. Clutches of 3–8 eggs have been recorded, roughly 35 × 15mm. Known prey items for this species include rodents, shrews, burrowing skinks, skink eggs, snakes and worm lizards.

Medical Significance: A widespread species that is sometimes common, is active on the ground at night and may enter buildings; it is also found in agricultural land, where people farm with short tools, so is a significant snakebite hazard within parts of its range. In a survey in Kwazulu-Natal, South Africa, six out of eight people bitten by this snake trod on it; three of these were inside a building, and one rolled on the snake while sleeping. The venom manifests itself as largely cytotoxic. Swelling, pain, skin discolouration, lymphangitis, subsequent lymphadenopathy and occasionally necrosis (especially to digit tips), as well as systemic symptoms like angioedema, resulted from bites. No antivenom is available, and the use of polyvalent antivenom is contraindicated; treat bites symptomatically. In coastal South Africa and Tanzania this snake is responsible for many bites, but no fatalities are recorded.

Bibron's Burrowing Asp, *Atractaspis bibronii*,
Below: underside, Botswana (Stephen Spawls).

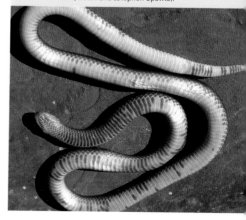

Taxonomic Notes: Recent analysis of this small, widespread snake suggests that it represents a species complex and it may be split in the future; the northern animals (originally regarded as a subspecies, *Atractaspis bibronii rostrata*) form a separate clade.

Central African Burrowing Asp *Atractaspis boulengeri*

Identification: A medium-sized burrowing asp of the Central African forest. The head is blunt and the eyes tiny. The tail is short and blunt, 6–8% of total length, ending in a pointed cone. The scales are in 21–25 rows at midbody; ventrals 192–218; cloacal scale entire; subcaudals usually paired but sometimes single, 22–27. Maximum size about 65cm, average 30–50cm; hatchling size unknown. The colour is usually black or blackish grey above, and paler below.

Atractaspis boulengeri

Habitat and Distribution: Essentially a forest snake. It occurs from southern Cameroon eastwards to the western watershed of the Albertine Rift, south to about 5°S. It extends south almost to the Angolan border; its distribution on the lower Congo River is uncertain. There are no definite records from Angola, but it might occur there. It is reported rather doubtfully from northern Cameroon.

Natural History: Poorly known. It lives and hunts in holes, but may emerge at night (especially during or after rain) and move around on the ground. If molested, it shows the distinctive responses mentioned in the generic introduction (pointing the head, releasing a chemical, and jerking and lashing). Females lay eggs, but the details are unknown.

Central African Burrowing Asp,
Atractaspis boulengeri. **Below:** Central African Republic (Jean-Francois Trape).

Medical Significance: A widespread species that is sometimes common, is active on the ground at night and may enter buildings; it is also found in agricultural land where people farm with short tools, so is a significant snakebite hazard within parts of its range. No antivenom is available, and the use of polyvalent antivenom is contraindicated; treat bites symptomatically.

Bill Branch's Burrowing Asp *Atractaspis branchi*

Identification: A small, slender burrowing asp, recently described from the West African forest. The head is blunt and the eyes tiny. The tail is fairly short and blunt, 4–6% of total length, ending in a pointed cone. The scales are in 19–20 rows at midbody; ventrals 276–288; cloacal scale divided; subcaudals single, 19–25. Described from two specimens, one 28cm, the other 72cm (a third animal was too damaged to use). The colour is pinky-grey and iridescent, the dorsal scales with a paler margin.

Atractaspis branchi

Bill Branch's Burrowing Asp, *Atractaspis branchi*. **Left and right:** Liberia (Mark-Oliver Rödel).

Habitat and Distribution: Known from only three specimens, one collected in forest in extreme north-west Liberia, and the other two from agriculture in a deforested area of southern Guinea, between 300m and 500m.

Natural History: Unknown, but presumably similar to other burrowing asps (i.e. nocturnal, terrestrial, etc.). The type was crawling along the banks of a creek at night; when picked up, it tried to hide its head under its coils, and to strike. It tried to crawl away and abruptly coiled and uncoiled, often springing distances almost equal to its body length. This species was described in 2019, and named after Dr William Roy (Bill) Branch, one of the authors.

Medical Significance: It has a small range, and its venom is probably similar to that of other burrowing asps, so it is not a major hazard. No antivenom is available, and the use of polyvalent antivenom is contraindicated; treat bites symptomatically.

Congo Burrowing Asp (Eastern Congo Burrowing Asp)
Atractaspis congica

Identification: A medium-sized burrowing asp of south-central Africa. The head is blunt and the eyes tiny. The tail is short and blunt, 5–8% of total length, ending in a pointed cone. The scales are in 19–21 (sometimes 23) rows at midbody; ventrals 190–237; cloacal scale divided; subcaudals paired, 18–25. Maximum size about 55cm, average 30–45cm; hatchling size unknown. It is quite variable in colour for a burrowing asp, being black, blue-black, steel grey or pinkish grey; the scales (especially on the forepart of the body) are light-edged.

Atractaspis congica

Habitat and Distribution: Forest and moist savanna, from sea level to 1,600m, or maybe higher. It occurs from the lower Congo River south to south-central Angola, east to the western shore of Lake Tanganyika, and also in north-west and northern Zambia. An isolated population occurs in the Caprivi Strip, Namibia, and in southern Cameroon; this species seems to be absent from Gabon.

Congo Burrowing Asp, *Atractaspis congica*.
Left top: captive (Matthijs Kuijpers). **Right top:** Zambia (Philipp Wagner).
Left bottom: Angola (Warren Klein). **Right bottom:** DR Congo (Konrad Mebert).

Natural History: Poorly known. This snake lives and hunts in holes, but may emerge at night (especially during or after rain) and move around on the ground. If molested, it shows the distinctive responses mentioned in the generic introduction (pointing the head, releasing a chemical, and jerking and lashing). It lays 3–6 eggs. The diet includes rodents and other small vertebrates.

Medical Significance: A widespread species that is sometimes common, is active on the ground at night and may enter buildings; it is also found in agricultural land where people farm with short tools, so is a significant snakebite hazard within parts of its range. No antivenom is available, and the use of polyvalent antivenom is contraindicated; treat bites symptomatically. A bite victim in Zambia suffered pain (initially intense), swelling, lymphadenitis and blistering.

Fat Burrowing Asp *Atractaspis corpulenta*

Identification: A medium-sized, stocky, slow-moving snake. The scales are in 23–29 rows at midbody; ventrals 178–208; cloacal scale entire; subcaudals single, 22–28. Maximum size about 70cm, average 30–50cm; hatchling size unknown. The colour is dark brown, slaty grey or black; occasional specimens from parts of its West African range (Cote D'Ivoire and Ghana) may have a white tip to the tail. Grey individuals may have a bluish sheen to the scales. The body is paler below.

Atractaspis corpulenta

Fat Burrowing Asp, *Atractaspis corpulenta.*
Left: Cameroon (Jean-Francois Trape). **Centre:** captive (Matthijs Kuijpers). **Right:** Gabon (Tomáš Mazuch).

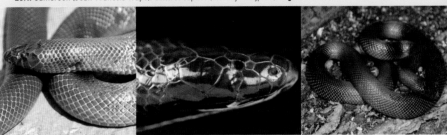

Habitat and Distribution: Its range indicates it is a forest snake. It occurs in two populations separated by the Dahomey Gap: in West Africa from Ghana westwards to Liberia and south-east Sierra Leone; and from Nigeria eastwards to north-eastern DR Congo, south on the coast of Central Africa into Gabon. It is found below 1,000m.

Natural History: Poorly known, but presumably similar to other burrowing asps.

Medical Significance: A widespread species that is sometimes common, is active on the ground at night and may enter buildings; it is also found in agricultural land where people farm with short tools, so is a significant snakebite hazard within parts of its range. No antivenom is available, and the use of polyvalent antivenom is contraindicated; treat bites symptomatically. Most reported cases involved only pain, local swelling, paraesthesia, feverishness and lymphadenopathy; one case showed haematuria. However, a recent fatality was reported for this snake in Republic of the Congo. The victim had initial severe pain, swelling, shortness of breath, vomiting and diarrhoea. He received antivenom, but two and a half hours after the bite, he went into cardiorespiratory arrest. The actual cause of death was not certain; it was possibly due to anaphylaxis from the venom, but might have been caused by the effect of the venom on the myocardium.

Dahomey Burrowing Asp *Atractaspis dahomeyensis*

Identification: A medium-sized snake; juveniles are fairly thin but large adults can become very stout. The tail is 5–9% of total length. The scales are in 29–35 rows at midbody; ventrals 210–250; cloacal scale entire; some subcaudals single, some paired, 22–30. Maximum size about 60cm, average 30–50cm; hatchling size unknown, but probably 14–16cm. The body is dark brown, grey, pinky-grey or black, and paler below. An albino specimen has been recorded from Ghana.

Atractaspis dahomeyensis

Habitat and Distribution: In woodland and moist and dry savanna, between sea level and 1,400m, or sometimes higher. There are a few records from within forest. This species occurs from the Gambia and Senegal eastwards to north-east Nigeria, mostly north of the forest, reaching the coast in the Dahomey Gap. There

Dahomey Burrowing Asp, *Atractaspis dahomeyensis*, **Left and right:** Sierra Leone (Bill Branch).

are few records for Cameroon, except the extreme south-west, but it is known from south-western Chad. It is doubtfully listed from southern Niger. Specimens reported from the northern Central African Republic are Watson's Burrowing Asp (*Atractaspis watsoni*).

Natural History: Terrestrial and nocturnal; it lives underground but emerges at night in the rainy season. It is a rather slow-moving snake, but quick to bite if restrained. Females lay eggs. The diet consists of small vertebrates.

Medical Significance: A widespread species that is sometimes common, is active on the ground at night and may enter buildings; it is also found in agricultural land where many people farm with short tools, so is a significant snakebite hazard within parts of its range. No antivenom is available, and the use of polyvalent antivenom is contraindicated; treat bites symptomatically. In a series of cases from Nigeria, victims experienced pain at the bite site and developed local swelling, which reached its maximum within 24 hours and usually resolved within five days. Some patients had enlarged and tender lymph nodes and mild fever. No necrosis was observed.

Duerden's Burrowing Asp *Atractaspis duerdeni*

Identification: A medium-sized, moderately thick snake with a distinctly rounded snout, in disjunct populations in southern Africa. It has a short, deep head. The scales are in 21–25 (usually 23) rows at midbody; ventrals 193–225; cloacal scale entire; subcaudals single, 19–27. Maximum size about 55cm, average 30–45cm; hatchling size unknown. The body is black, greyish black or grey, and white, cream or pinkish below, the paler colour extending up onto the lower scale rows on the flanks and the lips.

Atractaspis duerdeni

Habitat and Distribution: In sandy soils in dry savanna and grassland, at 1,100–1,500m; this species is absent from true Kalahari sands. It is found in five discrete

Duerden's Burrowing Asp, *Atractaspis duerdeni*.
Left: Botswana (Stephen Spawls). **Right:** head profiles: *A. bibronii*, above; *A. duerdeni*, below (Stephen Spawls).

populations on opposite sides of the Kalahari, one in north-central Namibia, two in Botswana (around Serowe, and between Otse and Gaborone) and two in South Africa (western Limpopo and Gauteng, and north of the Northern Cape); it is speculated that it lived on the shores of a massive ancient lake in the central Kalahari, which has now shrunk.

Natural History: Terrestrial and nocturnal; it lives underground but emerges at night in the rainy season, and is active mostly during the summer (October to April). It is a rather slow-moving snake, and quite good-natured for a burrowing asp, being surprisingly reluctant to bite. It lays eggs. The diet consists of small vertebrates; known prey items include lacertid lizards and snakes. It is less common than Bibron's Burrowing Asp (*Atractaspis bibronii*) within its range.

Medical Significance: It has a small range, and its venom is probably similar to that of other burrowing asps, so it is not a major hazard. No antivenom is available, and the use of polyvalent antivenom is contraindicated; treat bites symptomatically.

Palestine Burrowing Asp *Atractaspis engaddensis*

Identification: A relatively large burrowing asp, and the only one in Egypt. The body is cylindrical and the tail short, 7% of total length. The scales are in 25–29 rows at midbody; ventrals 263–284 (usually 275–282 in Egypt); cloacal scale entire; subcaudals single, 32–40 (usually 34–36 in Egypt). Maximum size over 80cm (the largest Egyptian snake measured 68.5cm), average 40–60cm; hatchling size unknown, but probably about 15cm. The colour is shiny black, above and below.

Atractaspis engaddensis

Habitat and Distribution: Semi-desert, often in large wadis and around oases. It is not found on mainland Africa, but has been recorded from Wadi Feiran and the Mount Sinai area in the Sinai Peninsula, at around 200–300m. It also occurs in western Jordan, southern and eastern Israel, and western Saudi Arabia.

Palestine Burrowing Asp, *Atractaspis engaddensis*. **Left and right:** Israel (Yannick Francioli).

Natural History: Terrestrial and nocturnal, and active by night in the warmer months of the year; it hides under ground cover and in holes during the day. If threatened, it shows the typical nose-pointing behaviour, then hides its head under its coils and may elevate and wave the tail. Three to four eggs are laid between September and November; incubation time is three months. This species feeds mostly on smooth-bodied lizards and snakes, but will take rodents.

Medical Significance: It has a small range in inhospitable country, so relatively few bites occur. However, some severe cases are documented, including fatalities, and may be connected with the sarafotoxins present; this is probably the most dangerous African species. The usual burrowing asp symptomology of immediate pain, swelling, paraesthesia and lymphadenitis is often present, but more alarming effects sometimes seen include violent autonomic symptoms (nausea, vomiting, abdominal pain, diarrhoea, muscle weakness, breathing difficulties, sweating, profuse salivation and sudden elevation of blood pressure) within minutes of the bite. One patient developed severe dyspnoea with acute respiratory failure, one had weakness, impaired consciousness and transient hypertension, and in three there were electrocardiographic changes. A child bitten by this snake in Saudi Arabia died following an atrioventricular block. At present, no antivenom is available, and the use of polyvalent antivenom is contraindicated; treat bites symptomatically.

Engdahl's Burrowing Asp *Atractaspis engdahli*

Identification: A small burrowing asp of the dry country of north-eastern Kenya and southern Somalia. The snout appears rounded from above but shark-like in profile. The body is cylindrical and the tail short, 5–6% of total length. The scales are in 19 rows at midbody; ventrals 219–232; cloacal scale divided; subcaudals paired, 19–23. Maximum size about 45cm, average 25–40cm; hatchling size unknown, but probably about 15cm. The body is black, blue-black or red-brown (juveniles can be quite light brown), and paler below.

Atractaspis engdahli

Engdahl's Burrowing Asp, *Atractaspis engdahli*. **Left and right:** preserved specimen (Tomáš Mazuch).

Habitat and Distribution: Endemic to the Horn of Africa. It occurs in coastal plain, thicket, grassland, dry savanna and semi-desert, from sea level to 250m. In Kenya, it is known only from a single specimen from Wajir Bor, 50km east of Wajir in the north-east; it is also known from the middle and lower Juba River and Kismayu in Somalia. Its distribution is probably continuous, but there are few specimens.

Natural History: Poorly known, but it is presumably similar to other *Atractaspis*, living underground and eating largely smooth-bodied reptiles. Specimens have been captured in termitaria, in holes and prowling in semi-desert at night; the Wajir Bor specimen was in a dry well.

Medical Significance: It has a small range, and its venom is probably similar to that of other burrowing asps, so it is not a major hazard. No antivenom is available, and the use of polyvalent antivenom is contraindicated; treat bites symptomatically.

Variable Burrowing Asp *Atractaspis irregularis*

Identification: A small burrowing asp of the great Central and West African forest. The body is cylindrical and the tail short, 5–8% of total length, ending in a pointed cone. The scales are smooth; in 23–27 rows at midbody (occasionally 21 but usually 23 on the eastern side, and 25–27 on the western side); ventrals 213–263; cloacal scale divided; subcaudals paired, 20–32 (higher counts in males). Maximum size about 66cm (possibly larger; see the taxonomic notes on p. 208), average 30–50cm; hatchling size unknown. The body is usually shiny black or blackish grey, and the scales are iridescent. The ventrals are dark grey, black or rufous, sometimes with white edging.

Atractaspis irregularis

Habitat and Distribution: Moist savanna, woodland, forest–savanna mosaic and forest, from sea level to 1,800m. It occurs from southern Ghana westwards to Liberia and southern Guinea, and from south-western Nigeria east to South

Variable Burrowing Asp, *Atractaspis irregularis*. **Left:** Liberia (Bill Branch). **Right:** Uganda (Jelmer Groen).

Sudan, parts of central Uganda, Rwanda and Burundi, just reaching north-west Tanzania, and south to north-west Angola. There are isolated populations in southern Togo, Jos in Nigeria, parts of the western and central highlands of Kenya, Mount Bizen in Eritrea, and the Chercher Mountains and the forest of south-west Ethiopia. Oddly, there seem to be virtually no records from Gabon, although this snake occurs in Cameroon right up to the border, and there are few records from central DR Congo.

Natural History: Nocturnal and terrestrial, spending much time in holes or under ground cover, but it is active on the ground on wet nights. It shows the nose-pointing response if molested. A female from the Mabira Forest in Uganda contained six eggs; mating was observed there in September. This species is known to eat rodents as well as small reptiles.

Medical Significance: A widespread species that is sometimes common, is active on the ground at night and may enter buildings; it is also found in agricultural land where many people farm with short tools, so is a significant snakebite hazard within parts of its range. It is said to be a major cause of snakebite in DR Congo. No antivenom is available, and the use of polyvalent antivenom is contraindicated; treat bites symptomatically. In most documented cases, the usual symptoms of pain, swelling and lymphadenopathy are present. More extensive general muscle pain may be experienced, and elevated heart rates have been noted, which may be due to the sarafotoxins. There have been two recorded fatalities, both of which involved exceptional circumstances and an incomplete clinical picture. In the first, a victim in Liberia was bitten a number of times after rolling on the snake in his sleep; he died quickly without treatment. In similar circumstances, a 10-month-old baby was bitten while asleep and died within minutes.

Taxonomic Notes: In 1959, Gaston de Witte described *Atractaspis battersbyi* from Bolobo, on the lower Congo River. The two specimens were 82.5cm and 52.5cm; midbody scale count 23; ventral scales 341 and 348; subcaudals 22 and 23. Most authorities regard these specimens as aberrant examples of *Atractaspis irregularis*, despite the large size of one animal and high ventral counts. Likewise, in 1960 Jean-Luc Perret described *Atractaspis coalescens* from south-west Cameroon, on the basis of head scalation that differed very slightly from *A. irregularis*; no more specimens have been found, and it seems *A. coalescens* is probably an aberrant example of *A. irregularis*.

Ogaden Burrowing Asp *Atractaspis leucomelas*

Atractaspis leucomelas

Identification: A small, slim burrowing asp of the northern Horn of Africa, with a white head and white vertebral stripe. The scales are in 23 rows at midbody; ventrals 230–243; cloacal scale entire; subcaudals single, 27–29. Maximum size uncertain, but probably around 50cm, average 20–30cm. It is distinctly marked, with a white or pale head and a dark crown; the back is dark with a fine yellow-white vertebral stripe. The four lowest scale rows and the ventrals are white, with a broad black throat band. This is one of two distinctly marked burrowing asps from the Horn of Africa; this species and the Somali Burrowing Asp (*Atractaspis scorteccii*) are the only burrowing asps to have any sort of large-scale markings, all the rest being uniform in colour.

Habitat and Distribution: Known from two specimens alone, from dry savanna and grassland: one from the Ogaden in eastern Ethiopia (although the exact locality is unknown); the second on the Somali–Djibouti border, in a region of sandy soil with grass and stunted thornbush at an altitude of 300m.

Natural History: Nothing is known of its biology, and most of its range lies within a militarily unsafe area. One specimen is reported as having been captured during the day, but it is not stated if the snake was active or not; burrowing asps are usually nocturnal. Its habits are presumably similar to those of other burrowing asps.

Medical Significance: A small, rare snake with a small range, so it is unlikely to deliver much venom in a bite; the venom is probably similar to that of other burrowing asps, so it is not a major hazard. No antivenom is available, and the use of polyvalent antivenom is contraindicated; treat bites symptomatically.

Ogaden Burrowing Asp, *Atractaspis leucomelas*. **Left and right:** preserved specimen (Tomáš Mazuch).

Somali Burrowing Asp *Atractaspis scorteccii*

Identification: A large, stout burrowing asp with a white head, from the dry country of far north-east Africa. The scales are in 23–25 rows at midbody; ventrals 215–230; cloacal scale entire; subcaudals single, 30–34. Maximum size about 85cm, average 50–75cm; hatchling size unknown. The colour is usually black or dark brown, with a broad white ring right around the neck. The top of the head is mottled dark brown. The lips, chin and throat are white, and there is often a dark patch behind and slightly below the eye. The belly is white, mottled with dark brown, and the outer edges of the belly scales are dark. This is one of two distinctly marked burrowing asps from the Horn of Africa; this species and the Ogaden Burrowing Asp (*Atractaspis leucomelas*) are the only burrowing asps to have any sort of large-scale markings, all the rest being uniform in colour.

Atractaspis scorteccii

Habitat and Distribution: Known from only a handful of specimens from sandy and stony country in the dry grassland and semi-desert of eastern Ethiopia and northern Somalia, between 600m and 1,100m.

Natural History: First described in 1949, but virtually nothing is known of its biology, and most of its range lies within a militarily unsafe area. The significance of the white neck band is unknown, but another African snake with a near-identical pattern, the Pale-collared Snake-eater (*Polemon graueri*), found in the forested lake regions of western Uganda and DR Congo, feeds on snakes, suggesting that the Somali Burrowing Asp may do so too.

Somali Burrowing Asp, *Atractaspis scorteccii*. **Below:** preserved specimen (Stephen Spawls).

Medical Significance: A rare snake with a small range; the venom is probably similar to that of other burrowing asps, so it is not a major hazard. No antivenom is available, and the use of polyvalent antivenom is contraindicated; treat bites symptomatically.

Reticulate Burrowing Asp *Atractaspis reticulata*

Identification: A huge, broad-headed burrowing asp of Central Africa. The head is short and broad. The scales are in 21–23 (occasionally 19) rows at midbody; ventrals 308–344; cloacal scale divided; subcaudals paired, 21–26. Maximum size about 1.14m, average 60–90cm; hatchling size unknown. The body is usually grey, and the dorsal scales are pale-edged, giving the snake its reticulate or 'net-like' appearance.

Atractaspis reticulata

Habitat and Distribution: Forests and moist savanna, between sea level and 1,400m. It occurs from southern Cameroon eastwards, just reaching south-west Central African Republic, across DR Congo to the southern Albertine Rift, and southwards to north-west Angola. There are isolated records from Mbeya in Tanzania and Bauchi in Nigeria, and a possible record from Ghana.

Natural History: Poorly known, despite its huge range. It is terrestrial and nocturnal, and lives underground but emerges at night in the rainy season. It lays eggs. The diet consists of small vertebrates.

Medical Significance: A widespread species that is sometimes common, is active on the ground at night and may enter buildings; it is also found in agricultural land where many people farm with short tools, so is a significant snakebite hazard within parts of its range. The symptomology is probably similar to that of other burrowing asps, with pain, swelling and lymphadenopathy. No antivenom is available and the use of polyvalent antivenom is contraindicated; treat bites symptomatically.

Reticulate Burrowing Asp, *Atractaspis reticulata*.
Top: Cameroon (Jean-Francois Trape).
Bottom: Tanzania (Ignas Safari).

The small-scaled burrowing asp (*Atractaspis microlepidota*) complex

This is a complex of six species of relatively large, stocky, broad-headed burrowing asps. They occur eastwards from Senegal across the Sahel and savanna of West and Central Africa, encircle the Ethiopian highlands and extend south to northern Tanzania. They were all regarded as subspecies of *Atractaspis microlepidota* for a long time (or even just differing examples of a single, rather variable species). But recent systematic and molecular work suggests that at least six taxa are involved. However, exactly where one taxon ends and another begins is not certain. Many specimens in the low country surrounding high central Ethiopia are problematic, where the various taxa are separated by slight differences in lepidosis. Where the identity of specimens is equivocal, we have assigned them zoogeographically, which may differ slightly from some published data. Historically, members of this complex have caused some deaths, although no recent, clinically well-documented cases that indicate the venom was directly responsible for the fatality are known.

Western Small-scaled Burrowing Asp
Atractaspis microlepidota

Identification: A fairly large black burrowing asp of the savanna of western West Africa. It is a stocky snake with a short, broad head. The body is cylindrical and the tail short, about 7% of total length, ending in a spike. The scales are smooth; in 29–33 rows at midbody; ventrals 198–222; cloacal scale entire; subcaudals single, 21–26. Maximum length 67cm, average 40–60cm; hatchling size unknown. The dorsal colour is glossy uniform black, and the underside is usually lighter, often grey.

Atractaspis microlepidota

Habitat and Distribution: In the Sudan savanna and Sahel of West Africa, at low altitude, between sea level and 200m. This species is known only from Senegal, the Gambia, and southern and south-western Mauritania.

Western Small-scaled Burrowing Asp, *Atractaspis microlepidota*. Below: Senegal (Jean-Francois Trape).

Natural History: Poorly known. It is nocturnal and terrestrial, usually underground, but active at night on the ground in the rainy season. If molested, it shows the usual distinctive burrowing asp response, arching the neck and pointing the head at the ground, looking like an inverted 'U', then writhing and hiding the head. It may also release a distinctive-smelling chemical from the cloaca. It presumably lays eggs. The diet is toads, lizards and small snakes, and it might take rodents.

Medical Significance: A snake with a small range, and whose venom is probably similar to that of other burrowing asps (i.e. causes pain, swelling, etc.), so it is probably not a major hazard. However, it can grow fairly large, and a similar species has caused fatalities, so a bite needs to be monitored and, if necessary, treated at a competent hospital. No antivenom is available, and the use of polyvalent antivenom is contraindicated; treat bites symptomatically.

Sahelian Burrowing Asp *Atractaspis micropholis*

Identification: A fairly large brown burrowing asp of the Sahel. It is a stocky snake with a short, broad head. The body is cylindrical and the tail short, less than 10% of total length, ending in a spike. The scales are smooth; in 25–27 rows at midbody; ventrals 211–230; cloacal scale entire; subcaudals single, 26–32. Maximum length 90cm, average 40–75cm; hatchling size unknown. The dorsal colour is shiny brown, and the back and lower edges of the dorsal scales are pale, especially on the flanks, giving the snake a speckled appearance. The underside is usually light, clear brown.

Atractaspis micropholis

Habitat and Distribution: In the Sudan savanna and Sahel of West Africa, at low altitude, between sea level and 800m. It occurs eastwards from Senegal and the Gambia through Mali, Burkina Faso, northern Nigeria and Niger to western Chad, with an apparently isolated population in eastern Chad; it is nearly always between 11°N and 17°N. A curious record from south-western Cameroon is not shown on the map.

Natural History: Poorly known. It is nocturnal and terrestrial, usually underground, but active at night on the surface in the rainy season. If molested, it shows the usual distinctive burrowing asp response, arching the neck and pointing the head at the ground, looking like an inverted 'U', then writhing and hiding the head. It may also release a distinctive-smelling chemical from the cloaca. It presumably lays eggs. The diet comprises mostly lizards, but it might take other small vertebrates. It has been reported to be more common in Senegal than elsewhere.

Sahelian Burrowing Asp, *Atractaspis micropholis*.
Below: Senegal (Jean-Francois Trape).

Medical Significance: A snake with a small range, and whose venom is probably similar to that of other burrowing asps (i.e. causes pain, swelling, etc.), so it is probably not a major hazard. However, it can grow fairly large; a bite needs to be monitored and, if necessary, treated at a competent hospital. No antivenom is available, and the use of polyvalent antivenom is contraindicated; treat bites symptomatically. A French soldier bitten by a large example of this snake in Senegal died in under an hour.

Watson's Burrowing Asp *Atractaspis watsoni*

Identification: A fairly large black burrowing asp of the savanna and Sahel of West and Central Africa. It is a stocky snake with a short, broad head. The body is cylindrical and the tail short, less than 10% of total length, ending in a spike. The scales are in 29–31 (occasionally 27) rows at midbody; ventrals 213–242; cloacal scale entire; subcaudals single, 21–30. Maximum recorded length 72cm, average 30–65cm; hatchling size unknown. The colour is black or brownish black above, and slightly lighter beneath.

Atractaspis watsoni

Habitat and Distribution: The Sudan savanna and Sahel of West Africa, at low altitude, between sea level and 800m. It occurs eastwards from the Senegal–Mauritania border country through Mali, Burkina Faso, southern Niger and northern Nigeria to northern Cameroon, southern Chad and the extreme northern Central African Republic. A British Museum specimen from Juba, South Sudan, is probably another species.

Natural History: Poorly known. It is nocturnal and terrestrial, usually underground, but active at night on the ground. If molested, it shows the usual distinctive burrowing asp response, arching the neck and pointing the head at the ground, looking like an inverted 'U'. It may also release a distinctive-smelling chemical from the cloaca. It presumably lays eggs. The diet comprises mostly reptiles, but it might take other small vertebrates.

Watson's Burrowing Asp, *Atractaspis watsoni*.
Below: Chad (Jean-Francois Trape).

Medical Significance: A large burrowing asp with a huge range, including populated areas, so it may well be a snakebite hazard. The venom is probably similar to that of other burrowing asps (i.e. causes pain, swelling, etc.), and is reported to cause local necrosis. Such a large snake may deliver a lot of venom (it may have large venom glands), so bites need to be monitored and, if necessary, treated at a competent hospital. No antivenom is available, and the use of polyvalent antivenom is contraindicated; treat bites symptomatically.

Magretti's Burrowing Asp *Atractaspis magrettii*

Identification: A rare burrowing asp from Eritrea and nearby Sudan, and possibly Somalia and Ethiopia. It is a stocky snake with a short, broad head. The body is cylindrical and the tail short, less than 10% of total length, ending in a spike. The scales are in 25–29 rows at midbody; ventrals 228–248; cloacal scale entire; subcaudals single, 31–33. Maximum recorded length 65cm, average 30–60cm; hatchling size unknown. The colour is uniformly black above, and slightly lighter beneath.

Atractaspis magrettii

Habitat and Distribution: Known from semi-desert, dry savanna and semi-arid high plateau, between 400m and 2,400m. It has been recorded from a handful of localities in Eritrea and adjacent Sudan at Kassala. The type came from Mandafena, near Adi Keyh, at 2,400m in the mountains of central Eritrea, a curiously high-altitude location for a large burrowing asp. A handful of specimens from the border area between the northern Ogaden in Ethiopia and Somalia are tentatively also assigned to this species.

Natural History: Essentially unknown, but presumably similar to other burrowing asps.

Medical Significance: As for other burrowing asps with restricted ranges.

Sudan Burrowing Asp *Atractaspis phillipsi*

Identification: A fairly large steel-grey or black burrowing asp of central Sudan, South Sudan and western Ethiopia. It is a stocky snake with a short, broad head. The body is cylindrical and the tail short, less than 10% of total length, ending in a spike. The scales are in 29–33 rows at midbody; ventrals 222–252 (possibly as low as 211); cloacal scale entire; subcaudals single, 23–33. Maximum recorded length about 90cm, average 50–70cm; hatchling size unknown. The colour is uniform black, steel grey or grey-black above and below.

Atractaspis phillipsi

Habitat and Distribution: Dry savanna, grassland and swamp, between 400m and 600m on the plains of the Sudanese and South Sudanese Nile. It occurs south from Khartoum, up the Blue Nile to Sinjah, up the White Nile to Juba, and east to Gambela and western Benishangul-Gumuz in Ethiopia.

Sudan Burrowing Asp, *Atractaspis phillipsi.*
Left: Sudan (Sebastian Kirchof). **Right:** Ethiopia (Tomáš Mazuch).

Natural History: Essentially unknown, but presumably similar to other burrowing asps.

Medical Significance: As for other burrowing asps with restricted ranges. Two victims who were bitten by the same 25cm snake received injections of antivenom. One recovered uneventfully, while the other lost the nail and tip of the finger to necrosis.

Taxonomic Notes: This species was re-elevated to species status from the synonymy of *Atractaspis microlepidota* by Don Broadley in 1994, following an examination of a snake collection from Al Jazirah State, Sudan.

Eastern Small-scaled Burrowing Asp *Atractaspis fallax*

Identification: A really large burrowing asp of the Horn of Africa and north-east East Africa. It is often stout, with a broad, blunt head. The body is cylindrical and the tail short, 7–9% of total length. The scales are smooth; in 27–37 rows at midbody; ventrals 227–257; cloacal entire; subcaudals single (occasionally paired, or a mixture of both), 20–39. One of the largest burrowing asps, reaching 1.1m in eastern Kenya; there are anecdotal reports of larger specimens, up to 1.3m, on the Kenyan coast. Average size 60–90cm; hatchling size unknown. The

Atractaspis fallax

colour is quite variable, being black, grey, brown or often distinctly purple-brown; the scales may be iridescent. Coastal specimens are often light purple-brown, darkening towards the head, with the head and the neck black. Juveniles are often light brown. The belly is usually dark.

Habitat and Distribution: Coastal bush and thicket, dry and moist savanna, grassland and semi-desert, from sea level to about 1,800m, but most common in dry savanna below 1,000m. It occurs from northern Somalia and the Ogaden in Ethiopia, southwards through Somalia into northern and eastern Kenya and north-east Tanzania. There are seemingly isolated records along the lower Awash River in Ethiopia, the Maasai Mara and Serengeti in Kenya and Tanzania, and

Eastern Small-scaled Burrowing Asp, *Atractaspis fallax*.
Left: Ethiopia (Stephen Spawls). **Right:** Tanzania (Michele Menegon).

parts of north-west Kenya. South Sudanese records of this species are probably best assigned to the Sudan Burrowing Asp (*Atractaspis phillipsi*).

Natural History: Nocturnal and terrestrial; it is usually underground, but active at night on the surface in the rainy season. It is sometimes active by day. If molested, it shows the usual distinctive burrowing asp response, arching the neck and pointing the head at the ground, looking like an inverted 'U', then writhing and hiding the head. It may also release a distinctive-smelling chemical from the cloaca. Captive specimens have been photographed elevating the head and flaring the neck into a moderate hood, like a cobra. A Kenyan female laid eight eggs. The diet is snakes, smooth-bodied lizards and occasionally rodents. This species is said to be relatively common in parts of its range, especially the north Kenyan coast, from Kilifi to Malindi. In Somali areas it is known as 'Jilbris', 'the snake of seven steps' (meaning that if you are bitten, you take seven steps and die), or 'father of 10 minutes' (for obvious reasons). It is a very difficult snake to catch; large adults are strong, active and bite furiously if restrained.

Medical Significance: Widespread in the low and coastal regions of Kenya and Somalia, sometimes common, active on the ground at night and may enter buildings; it is also found in agricultural land, where people farm with short tools, so it is a significant snakebite hazard within parts of its range. The venom manifests itself as largely cytotoxic; swelling, pain, skin discolouration, subsequent lymphadenopathy and occasionally necrosis (especially to digit tips) resulted from bites. Occasional bites in Kenya have caused necrosis requiring excision or amputation. No antivenom is available, and the use of polyvalent antivenom is contraindicated; treat bites symptomatically. The large size of some adults means that bites may need effective symptomatic treatment.

Eastern Small-scaled Burrowing Asp, *Atractaspis fallax*.
Below: Tanzania (Michele Menegon).

Harlequin snakes *Homoroselaps*

Two species of colourful, small, slender front-fanged snakes, both endemic to southern Africa. They have been classified as elapids (their original common name was 'dwarf garter snake'), but molecular work now indicates they belong in the subfamily Atractaspidinae. Due to their small size, restricted distribution and secretive way of life, they are of minor medical importance. No antivenom is available and no fatalities are recorded; a bite can be treated symptomatically.

Key to the genus *Homoroselaps*, harlequin snakes

1a Ventral scales 160–215; markings variable, red and/or yellow on a black background......*Homoroselaps lacteus*, Spotted Harlequin Snake (p. 218)

1b Ventral scales 215–240; always with a well-defined single yellow vertebral stripe......*Homoroselaps dorsalis*, Striped Harlequin Snake (p. 219)

Spotted Harlequin Snake *Homoroselaps lacteus*

Identification: A small, slim, vividly marked black, yellow and often red snake of southern and eastern South Africa. The head is small and bullet-shaped, and the eyes are small and dark. The tail is 9–16% of total length (longer in males). The scales are in 15 rows at midbody; ventrals 160–209; cloacal shield divided; subcaudals 23–43. Maximum size about 65cm, average 30–40cm; hatchlings 13–15cm. The colour is quite variable, usually a mix of yellow, black and often red. The ground colour is usually dark, heavily speckled with yellow or cream,

Homoroselaps lacteus

Spotted Harlequin Snake, *Homoroselaps lacteus*. **Below:** Eswatini/Swaziland (Richard Boycott).

sometimes with a black collar and a vertebral stripe of either speckled yellow or orangey-red. In some specimens the vertebral stripe is broken by black bars; others are barred black and yellow. The underside is often yellow or cream, usually with a darker midventral line, but may be uniform black.

Spotted Harlequin Snake, *Homoroselaps lacteus*. **Below:** South Africa (Bill Branch).

Habitat and Distribution: In grassland, semi-desert, dry and moist savanna, thicket and heathland (fynbos), between sea level and 1,800m. This species is endemic to southern Africa. In South Africa it is found in most of the Western Cape and Eastern Cape, the southernmost part of the Northern Cape, Kwazulu-Natal, the eastern side of the Free State, Gauteng, Mpumalanga and central Limpopo; there is an isolated population on the north-western coast in the Port Nolloth/Kleinsee area, and some historical records from around Kimberley. It is also found in western Eswatini (Swaziland) and on the western border of Lesotho.

Natural History: A ground-dwelling snake, usually in holes and rarely on the surface, except after rainstorms. It shelters in termite mounds, under rocks and in holes. If molested, it usually does not try to strike, but wriggles violently to try to get away. If handled, it will try to inflict slow, deliberate bites. It lays 6–16 eggs in the summer (December or January); these take about 50 days to hatch. The diet is mostly small snakes and lizards, including worm lizards.

Medical Significance: A rare, secretive snake with a relatively small range. No antivenom is available. No fatalities are reported and the use of polyvalent antivenom is contraindicated; treat bites symptomatically. A few bite cases were characterised by local pain, sometimes severe, swelling that was sometimes slow to resolve and lymphadenitis; one victim had a persistent headache, while another had local numbness. The venom does not seem to have been analysed, but the symptomology is similar to that from burrowing asp bites – perhaps not unexpectedly, as they are in the same clade.

Striped Harlequin Snake *Homoroselaps dorsalis*

Identification: A small, slim, vividly striped yellow-on-black snake of northern and eastern South Africa. The head is small and bullet-shaped, and the eyes are small and dark. The tail is 8–11% of total length (longer in males). The scales are in 15 rows at midbody; ventrals 210–239; cloacal shield divided; subcaudals 22–33. Maximum size about 32cm, average 20–25cm; hatchling size unknown. The colour is black above, with a broad yellow stripe extending from the tip of the nose to the tip of the tail. The lips, underside and lower scale rows on the

Homoroselaps dorsalis

Striped Harlequin Snake, *Homoroselaps dorsalis*. **Below:** South Africa (Johan Marais).

flanks are cream to pale yellow, becoming more vivid chrome yellow posteriorly.

Habitat and Distribution: In grassland, dry and moist savanna and coastal thicket, between sea level and 1,800m. This species is endemic to the eastern side of southern Africa. In South Africa it is found in the central Free State, Kwazulu-Natal, Gauteng, central Limpopo and north-eastern Mpumalanga, and it also occurs in western Eswatini (Swaziland).

Natural History: A ground-dwelling snake, usually under cover and rarely on the surface, except after rainstorms. It shelters in termite mounds, under rocks and in holes. If molested, it does not try to bite, but wriggles violently to try to get away. It lays 2–4 eggs in the summer (December or January). It eats worm snakes of the family Leptotyphlopidae.

Medical Significance: A rare, secretive snake with a relatively small range. No antivenom is available. No bites or fatalities are reported and the use of polyvalent antivenom is contraindicated; treat bites symptomatically.

Dangerous rear-fanged and fangless snakes

Boomslang,
Dispholidus
typus, juvenile
(Bill Branch).

This section covers some 17 species of African snake that are dangerous/potentially dangerous but do not have front fangs. They do not form any natural taxonomic group, but are a mixed bag of snakes that are (or may be) dangerous to humans under certain circumstances. It includes the two large African pythons, which have no venom but are strong enough to kill people; some rear-fanged snakes that have a deadly venom and have killed a few people (mostly snake handlers); some rear-fanged snakes with potentially dangerous venoms; and, perhaps surprisingly, a handful of fangless snakes that have bitten people (usually their keepers) and caused some alarming symptoms on account of their toxic saliva. We have not attempted to create a key to this disparate group of snakes (although keys to the vine snakes, *Thelotornis*, and broad-headed tree snakes, *Toxicodryas*, are provided); instead, the generic introductions provide pointers to identify the snakes described here.

Apart from the Boomslang (*Dispholidus typus*) and the vine snakes, the other species here are not of concern to medical professionals. The likelihood of an accidental and significant bite from one is just about non-existent, no relevant antivenom exists, and a snake keeper presenting with a bite from one of them will hopefully know exactly what bit them and can be treated symptomatically. Note: the vine snakes can be identified with near certainty by their slim grey bodies and keyhole-shaped pupils; no other African snakes have such pupils. The Boomslang, if seen clearly, can usually be identified by a combination of large size, keeled scales, short head, and large eye with a subtly teardrop-shaped pupil.

Family Pythonidae Pythons

A family of just over 40 species, found in Australasia, Asia and Africa, that includes some of the world's largest snakes.

Pythons *Python*

A genus of 10 relatively large, constricting Old World snakes; six occur in Asia and four in Africa. The two large African species of rock python do not have fangs, only solid teeth. They may be considered dangerous in terms of their size; their strength means they are capable of constricting humans, and they can also deliver a severe bite that may need suturing (stitching). Some authorities regard the rock python as a single variable species.

Southern African Rock Python *Python natalensis*

Identification: Unmistakable; a huge, thickset snake (rock pythons are the largest snakes in Africa) of southern and south-eastern Africa, with a subtriangular head. The snout is rounded, the eye is fairly large and the pupil is vertical (though this is hard to see). The top of the head is covered with small to medium smooth scales. There are two heat-sensitive pits at the front of the upper labials and three smaller such pits on the lower labials. The stout body is cylindrical in juveniles and slightly depressed in large adults. The tail is fairly

Python natalensis

short, 9–10% of total length in females and 12–16% in males. The scales are smooth; in 78–99 rows at midbody; 260–291 narrow ventrals; subcaudals 63–84. There are small claw-like spurs on either side of the anal scale (vestigial legs), which are larger in males; legend in East Africa says these are used to block the nostrils of the prey being constricted. Maximum size about 5.5m (anecdotal reports of larger animals exist), average 2.8–4m (in general, females are larger than males); hatchlings 45–60cm. The colour is a mixture of brown, tan, yellow and grey blotches. The tail sometimes has a light central stripe. The underside is white, with irregular dark speckling. Large adults may look almost black. Freshly sloughed specimens are iridescent.

Habitat and Distribution: Coastal thicket, grassland, moist savanna and woodland, often in the vicinity of water or rocky hills; it is known from semi-desert in part of its range, and also thick woodland and forest. Found from sea level to 2,200m, but it is rare above 1,800m. This species is most abundant near low-altitude rivers, lakes and swamps; it likes to drink, but may be in quite dry country. This snake is widespread through southern and eastern Africa. It occurs from central Namibia to southern Angola, and from eastern and northern South Africa, through Mozambique, Zambia and Malawi to south-eastern DR Congo, Tanzania and Kenya; it intergrades with the Central African Rock Python (*Python sebae*) in

Southern African Rock Python, *Python natalensis*. **Left:** Kenya (Stephen Spawls). **Right:** Tanzania (Michele Menegon).

parts of Tanzania and southern and central Kenya. There are sporadic records in northern Angola and the Eastern Cape in South Africa.

Natural History: Usually nocturnal, but it will bask and hunt opportunistically during the day. It is mostly terrestrial, but juveniles will climb trees. This snake is often aquatic, and adults may spend a lot of time in water, hunting and feeding there (they are excellent swimmers), but emerging to bask. When inactive, it shelters in holes (especially warthog and porcupine burrows), thickets, reedbeds, up a tree, in a rock fissure or underwater. It often curls up in a heap of coils, with the head resting on top. Juveniles are active hunters, climbing trees to check nests and holes, prowling about and swimming around looking for prey; large adults tend to hunt from ambush, waiting quietly in a coiled strike position beside a game trail or under a bush until prey passes. Medium-sized juveniles (1.5–2.5m) in Kenya were observed to become active near dusk. Pythons will try to escape if confronted, but if cornered they will strike; they have very sharp teeth, causing deep wounds and cuts that require stitching. Many captive specimens remain permanently bad-tempered and are unsafe to handle (which has, perhaps fortunately, made them unpopular in the pet trade). From 16 to 100 tennis-ball-sized eggs are laid in a thick bush, rock fissure or deep, moist hole. The female coils around the eggs to protect them until just before they hatch, leaving only to drink; during this time she elevates her body temperature through a mixture of basking and an unexplained mechanism. Incubation varies from two to three months. There is some evidence that the neonates wait underground until all have hatched, then all emerge at once. Growth rates can be rapid, with juveniles reaching 1.4m in a year. Captive specimens have lived over 27 years and fasted over two and a half years. They eat a range of prey, from small mammals, birds and frogs up to medium-sized antelope and domestic stock; other recorded prey items include crocodiles, fish, lizards, goose eggs, domestic dogs and poultry. In rural East Africa these snakes are often found after they have entered a chicken run, eaten something large and been unable to get out. Small pythons may be attacked by snake-eating snakes, birds and mammalian carnivores, but large adults have few enemies except Leopards (*Panthera pardus*), Lions (*Panthera leo*), hyenas, crocodiles and humans; in southern Africa some leopards become specialised python-eaters. Pythons may be killed for their skin or their perceived danger to stock, but they do a lot of good,

Southern African Rock Python, *Python natalensis*.
Below: Tanzania (David Moyer).

eating rodents. In some areas they are venerated; a common belief is that if they are killed, rain will not fall.

Medical Significance: Very small. However, a large rock python is capable of killing an adult, and children in Africa are obviously more at risk. Large constrictors are dangerous; in the United States between 1978 and 2012, 17 people were killed by captive constricting snakes, including a three-year-old who was killed by a 2.3m African python. In early 2018 a British pet keeper was killed by his captive 2.4m African python. In South Africa in 1979 a 13-year-old boy was killed by a python, and in 1961 a man who had been squeezed by a large African python later died of a ruptured spleen. In addition, a bite from a python, with its many large, needle-sharp, bacteria-encrusted teeth, can cause severe lacerations. Pastoralists have sometimes been badly bitten while attempting to protect or rescue their stock from pythons. However, substantiated attacks by these large snakes are extremely rare. They usually avoid confrontation and do not seem to consider humans as potential prey, although an unaccompanied child in python country might be at risk. Do not allow small children to wander alone in python country, and don't leave small children unattended. If you are a pastoralist protecting your stock, discourage the snake with a large stick. Don't go close. Anyone who has been squeezed by a large constrictor should be monitored and X-rayed to check for internal damage; a bite should be cleaned, the bleeding stopped, debris and teeth removed, the lacerations stitched if necessary, and the wound treated with antibiotics until it is fully healed.

Taxonomic Notes: In 1999, the African Rock Python was split into two full species, the Central African Rock Python (*Python sebae*) and the Southern African Rock Python (*Python natalensis*), on the basis of some physical differences (including head scales and head patterns). This has been generally accepted, but intermediates exist; molecular evidence may clarify the situation.

Central African Rock Python *Python sebae*

Identification: Unmistakable; a huge, thickset snake of West and Central Africa. Most details are as for the Southern African Rock Python (*Python natalensis*), but it differs as follows: the top of the head is covered with medium to large scales; scales in 76–99 rows at midbody; 265–283 narrow ventrals; subcaudals 67–76. Maximum size about 6.5m (although there are anecdotal reports of larger specimens up to 9.8m). This species is more brightly coloured than the Southern African Rock Python, usually black above with large, yellow-edged light

Python sebae

Central African Rock Python, *Python sebae*.
Left: Kenya (Bio-Ken Archive). **Right:** juvenile, Kenya (Anthony Childs).

brown blotches on the back and flanks. There is a dark arrowhead on the crown, a yellow stripe runs through the eye, and in front of and behind the eye is a broad dark patch (which is narrow in the Southern African Rock Python). The underside of the tail is white, and this is sometimes curled up in the air as a distraction technique, attracting predators such as Leopards away from the head.

Habitat and Distribution: Similar to the Southern African Rock Python; coastal thicket, grassland, dry and moist savanna, woodland, forest, and along rivers and around waterbodies in semi-desert. It is found from sea level to 2,200m, although it is most common below 1,500m. It occurs from southern Mauritania and Senegambia eastwards across Africa, largely south of 15°N, as far as western Ethiopia and the Ethiopian Rift Valley, south throughout the Congo basin to northern Angola, and in the Lake Victoria basin. It also occurs from central Somalia southwards along the East African coast to Dar es Salaam, Tanzania, and inland up the Tana River, and there are isolated populations on the upper northern Uaso Nyiro River in Kenya, and around Lake Manyara and the Kilombero Valley in Tanzania; here it overlaps with the Southern African Rock Python, although the exact area of the overlap is unclear, and there are known 'hybrid' specimens.

Natural History: As for the Southern African Rock Python. In general, large pythons are incompatible with humans, as they will take domestic stock, and children are at risk; for this reason it is important to protect areas of wilderness. However, some societies in Africa do venerate pythons. A common belief in parts of East Africa is that if they are killed, rain will not fall; in places in Rwanda, the souls of deceased kings were said to live on in pythons. The Luo people in the Lake Victoria basin respect rock pythons; the snakes are called 'Omieri' (supposedly after a man on whose land the original Omieri lived), and if one is seen near the beginning of the rainy season then general good luck, bountiful rains and a decent harvest will follow. In such areas, pythons are tolerated, often living in sugar cane plantations.

Central African Rock Python, *Python sebae*.
Below: Ghana (Stephen Spawls).

Medical Significance: As for the previous species.

Family **Colubridae** Colubrid snakes

An intercontinental snake family with around 1,900 species, found on every continent except Antarctica (although there are very few in Australasia) The colubrids might be loosely described as 'ordinary' or 'typical' snakes. None are very large or very small, most being 0.5–2m long. They do not have front fangs; most have obvious eyes, broad belly scales and tapering tails. There remains vigorous debate about what taxa should be placed in the family, but a recent analysis suggests there are eight subfamilies.

Subfamily Colubrinae

A subfamily of around 100 genera and nearly 750 species. Most are harmless, but the group includes some significantly dangerous species. These include the African Boomslang (*Dispholidus typus*), a rear-fanged snake that has killed people, and the four vine or twig snakes of the genus *Thelotornis*, two species of which have caused fatalities. We have also described the Dagger-tooth Vine Snake (*Xyelodontophis uluguruensis*), which has large rear fangs and will probably turn out to be a type of vine snake, two large tree snakes with known neurotoxic venoms but no reported deaths (*Toxicodryas*, the broad-headed tree snakes), and one species of black tree snake (*Thrasops*), which has no fangs but appears to have fairly toxic saliva.

Boomslang *Dispholidus*

A large, widespread, rear-fanged tree snake. There is a single species in the genus, found virtually throughout sub-Saharan Africa; it will probably turn out to represent a species complex (see the taxonomic notes). Unlike most rear-fanged snakes, it has a deadly venom. However, it poses little risk to the ordinary person, as it is arboreal and unaggressive. A specific monovalent antivenom is available from South African Vaccine Producers (SAVP).

Boomslang *Dispholidus typus*

Identification: A large, highly venomous, rear-fanged tree snake that is usually green, grey or brown, but may be several other colours. It has a huge eye, short egg-shaped head and keeled dorsal scales. The pupil is almost round, but elongate at the front (sometimes dumbbell-shaped). The iris is yellow with black veins in adults, and it is light green in juveniles. The body is laterally compressed and the tail is long, 25–30% of total length in males and 22–28% in females. The dorsal scales are strongly keeled; in 17–21 (usually 19) rows at midbody; ventrals

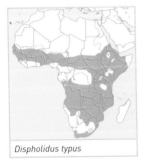

Dispholidus typus

161–201, with double keels; cloacal scale divided; subcaudals 91–142. Maximum size about 2.1m (anecdotal reports of larger ones exist), average 1.2–1.5m; hatchlings 30–40cm. The colour is very variable; males are usually shades of

Boomslang, *Dispholidus typus.*
Left top: threat display, Botswana (Stephen Spawls). **Centre top:** South Africa (Luke Verburgt).
Left bottom: Nigeria (Gerald Dunger). **Centre bottom:** Kenya (Stephen Spawls).
Right: juvenile Boomslang eating sunbird (Tyler Davis).

green, and females usually brown or grey. Other colour morphs include green or yellow with black-edged scales, uniform black, black or brown with yellow or cream spotting, brown with a rufous head, striped black and grey-white, light yellow-brown with white spots, and blue-grey. Juveniles have a different colour and pattern to adults, with a blue-spotted grey or brown vertebral stripe; black-speckled light grey or pinkish flanks; brown, olive or grey top to the head; a light chin and throat, often with a large yellow blotch on the side of the neck; and a vivid apple-green eye. This pattern gradually changes to adult colouration between lengths of 60–90cm; occasional adults retain the vertebral stripe. Large adults sometimes darken; captive green Boomslangs have gradually become black.

Habitat and Distribution: Lives in coastal thicket, woodland, moist and dry savanna and semi-desert, around forest edges and clearings, but not usually within dense forest or high grassland. It occurs from sea level to 2,200m. This is a snake with a huge range, occurring virtually throughout the savannas and woodland of sub-Saharan Africa, and known from every country there apart from Lesotho, Gabon, mainland Equatorial Guinea and Liberia. It is absent from some high-altitude areas (central Kenya and Ethiopia) and drier, more grassy parts of Namibia and South Africa.

Natural History: A diurnal snake, almost totally arboreal, although it will descend to cross open areas, seize prey and lay eggs. It is fast-moving, alert and graceful, climbing effortlessly. When prey is spotted it pauses, while curious lateral waves run down its body; it then darts forward to seize the victim. It appears to have binocular vision and, unusually for a snake, can spot prey animals before they move. It is non-aggressive, and if approached will slip quietly away. However, if threatened it has a remarkable display, inflating the neck and forepart of the body

Boomslang, *Dispholidus typus.*
Below: Zambia (Bill Branch).

to reveal the skin between the scales; it also flicks the tongue up and down in a deliberate manner. If further molested or restrained it will strike. Boomslangs lay 8–27 eggs, roughly 20 × 40cm, in deep, moist holes, damp tree hollows or vegetation heaps. Mating has been observed in March, April, May and July in southern Tanzania, and in July in Kisumu, Kenya. A southern Tanzanian female laid 10 eggs in September; a female from Voi, Kenya, laid 11 eggs in August. Hatchlings have been captured in Kakamega, Kenya, in September, and in January in southern Tanzania. Incubation time is 3–4 months in East Africa. Prey items include arboreal lizards such as chameleons, agamas and large geckoes, birds, fledglings and eggs, frogs and rodents. It often raids nests; a study in Namibia found that 80% of Sociable Weaver (*Philetairus socius*) communal nests had a resident Boomslang; a captive female ate a dead bird that had been disgorged by another freshly caught Boomslang. This species is often mobbed by birds. Boomslangs are said to sometimes congregate in huge numbers, the purpose of which is unknown.

Medical Significance: Although this snake has a huge range, a deadly venom and is relatively common in some areas, the threat it poses to the ordinary person is negligible. This is because it is diurnal, unaggressive, active and alert, lives in trees and has good eyesight; it moves quickly away from confrontation. There is an incorrect belief that, being a rear-fanged snake, it has to chew to engage the fangs, and thus if one bites you have time to pull it off. This is incorrect; Boomslangs have an enormous gape and huge fangs, and can embed these fangs on first contact. However, most documented bites are to snake handlers who tried to catch, handle or restrain the snake. A few accidental bites are known: a child put his hand into a weaverbird's nest where a snake was hiding and was bitten; a victim trod on a snake as it crossed her path; and a toddler on the Kenyan coast kicked a Boomslang that slid between her legs. Although the venom is usually slow-acting, often taking 24–48 hours or more to kill, it contains enzymes that activate prothrombin and factor X, leading to a consumptive coagulopathy, severe hypofibrinogenaemia and fatal bleeding if untreated. Local effects are usually trivial, although local swelling has occasionally been observed. Symptoms include nausea, vomiting, colicky abdominal pain and headache. A general bleeding tendency develops, from old and recent wounds, in the gums and under the skin, and there may be internal haemorrhaging. Bleeding in the eye causes the victim to see in shades of yellow. Any victims should be hospitalised for 48 hours, while being monitored for clotting abnormalities. Death is frequently due to renal failure. A specific monovalent antivenom is available from South African Vaccine Producers (SAVP). Although it would be rather pointless for a regional hospital to stock the antivenom, any institute keeping and maintaining Boomslangs might be advised to stock the antivenom; likewise, a poisons centre in any African capital city might be advised to maintain a supply. In a surprising recent bite case, a young South African died from cardiac arrest eight hours after a Boomslang bite; this is unusually quick, but he had been bitten twice, presumably while handling the snake, and may have had an intracranial bleed.

Taxonomic Notes: It is suggested that this snake, with its huge range, is a species complex, with cryptic species concealed within the group evolving separately. Several species and subspecies have been described, none of which have found much acceptance. A recent analysis suggests there are at least two separate lineages in southern Africa, and those on Pemba Island, Tanzania, seem to vary significantly from mainland animals, with a different colour pattern and very long tails. For the time being, however, the Boomslang remains a single species.

Vine or twig snakes *Thelotornis*

An African genus of medium-sized, largely diurnal tree snakes with a curious dumbbell- or keyhole-shaped eye pupil. They are slim, with a long tail, and are rear-fanged, gentle and non-aggressive. Four species are known, but there is some overlap of diagnostic characters (especially between the savanna species and subspecies), and exactly where one taxon ends and another begins is uncertain; intergrades exist. The slim body and pointed head have led to the legend in some parts of Africa that they can shoot through humans like an arrow. In Swahili, vine snakes are sometime called 'Mfunga Kuni' ('tie up the firewood'), the suggestion being that when people are gathering firewood, they mistake the snake for a vine that they use to tie the wood bundles.

Although they have a deadly venom that causes non-clotting blood, as with the Boomslang (*Dispholidus typus*), vine snakes pose a negligible threat to humans; they are diurnal, secretive, unaggressive, active and alert, with good eyesight, and they live in trees. They either remain motionless, or move quickly away from confrontation. A number of bites, some fatal, are documented, but all were inflicted on snake handlers who were bitten by the snake they were handling or trying to catch. As with the Boomslang, the venom causes a consumptive coagulopathy, severe hypofibrinogenaemia and fatal bleeding if untreated. No antivenom is produced; a bite victim would need screening for clotting abnormalities over 48 hours, and should be treated symptomatically, possibly by replacement therapy.

Key to the genus *Thelotornis*, vine or twig snakes

1a Top of head and temporals brown, grey or green, but both the same colour, and usually immaculate, unspeckled......2

1b Top of head usually green or blue-green, speckled black, and temporals grey or brown, differing in colour from the top of the head, often margined with black or brown......3

2a Top of head always green or grey; ventrals 162–189; never a black triangle or spotting on the sixth labial; inhabits forest west of 33°E......*Thelotornis kirtlandii*, Forest Vine Snake (p. 230)

2b Top of head green or brown; ventrals 156–169; nearly always a black triangle or diagonal or black spotting on the sixth labial; inhabits forest east of 36°E......*Thelotornis usambaricus*, Usambara Vine Snake (p. 234)

3a Top of head bright green to pale brown, uniform or lightly speckled with black; temporals brown, speckled with black......*Thelotornis mossambicanus*, Eastern Vine Snake (p. 231)

3b Top of head blue-green, with black speckling forming a 'Y' or 'T' marking or speckling covering the entire head......*Thelotornis capensis*, Southern Vine Snake (p. 233)

Forest Vine Snake (Forest Twig Snake) *Thelotornis kirtlandii*

Identification: A long, very thin tree snake with a green head, from the Central and West African forests. The head is long and arrow-shaped, and the large, pale eye has a horizontal keyhole-shaped pupil; this pupil is diagnostic of vine snakes, with no other African snake having a pupil of this shape. The iris is green and yellow. A narrow groove runs from each eye to the tip of the snout, giving the snake binocular vision. The tongue is bright red with a black tip. The body is cylindrical and the tail is very long and thin, 33–42% of total length. The scales are feebly keeled; in 19 rows at midbody (very occasionally 17); ventrals 162–189; subcaudals 132–172. Maximum size about 1.6m, average 0.9–1.4m; hatchlings 25–32cm. The body is grey-brown, heavily speckled with darker grey or black, and the underside is also distinctly speckled. From a distance this snake appears silvery-grey. The head is green above and the lips and chin are white, sometimes with black spotting. Specimens from Rwanda have deep grey-green heads, and the lips are extensively mottled with green or grey; the females have orange throats.

Thelotornis kirtlandii

Forest Vine Snake, *Thelotornis kirtlandii.*
Top: Angola (Warren Klein).
Centre: DR Congo (Konrad Mebert).
Bottom: Ivory Coast (Mark-Oliver Rödel).

Habitat and Distribution: In thicket, forest, forest patches and woodland, from sea level to about 2,200m. Within the forest itself, it seems to be most common around natural glades. It is often found in farmland within forests, and in and around parks and gardens in forest towns. It occurs westwards from the Ugandan Albertine Rift right across DR Congo, to western Nigeria, and southwards to northern Angola and extreme north-west Zambia. It reappears west of the Dahomey Gap, from south-western Ghana west to Guinea-Bissau. Isolated populations occur in the Imatong Mountains of South Sudan, the Mahale Peninsula in Tanzania, and hill forests of southern Benin and Togo.

Natural History: Diurnal and arboreal, in trees, bushes, thickets and reedbeds. This species can climb quickly and elegantly, but rarely seems to go very high, preferring lower branches. It will descend to the ground to pursue prey or cross to other trees, and it moves quickly on the ground and in trees. It relies on camouflage for defence, and spends much time sitting totally motionless in trees, sometimes with the front part of the body sticking out and swaying back and forth, resembling a grey branch with a green leaf at the end moving in the wind. It may remain motionless even if closely approached or poked. However, if disturbed it will often flick the tongue up and down in a deliberate and highly visible manner, and if sufficiently molested and unable to escape it will inflate the neck to a considerable size, as described for the Eastern Vine Snake (*Thelotornis mossambicanus*), and make a determined strike. Its fangs are set well back, but it has a wide gape, and should it be determined to bite it may seize and chew the victim. Females lay clutches of 4–12 eggs, 35 × 15mm. An Ivory Coast snake laid two pairs of eggs in an arboreal ant nest, which may provide protection from predators; the eggs hatched in January. This species eat mostly lizards, especially arboreal ones such as chameleons, agamas and geckoes, but will also ambush ground-dwelling species from a perch, or drop down and pursue them. It is also known to take other snakes and amphibians, and is reputed to take eggs. Another name for this snake (although little used and inaccurate) is 'Bird Snake', and there is a curious legend that it hypnotises birds with its flickering red and black tongue. This snake has binocular vision and is one of the few species that can detect prey by sight before it moves. In some areas of its forest range it is regarded as totally harmless, which it is unless severely molested. In other areas it is greatly feared, and is believed to be able to fly through humans like an arrow. It may be abundant in areas without being seen, due to its excellent camouflage; in a collection from Ivory Coast it was the second most common species.

Medical Significance: Although this snake has a huge range, a deadly venom and is relatively common in some areas, the threat it poses to the ordinary person is negligible. This is because it is diurnal, unaggressive, active and alert, with good eyesight, and lives in trees; it moves quickly away from confrontation. Most documented bites are to snake handlers who tried to catch, handle or restrain the snake. No antivenom is available; bites will need to be treated symptomatically, probably by replacement therapy while screening for clotting abnormalities.

Eastern Vine Snake *Thelotornis mossambicanus*

Identification: A very thin, medium-sized grey tree snake of south-eastern Africa. The head is long and the snout pointed. The eye is large, the pupil is keyhole-shaped and the iris is yellow. The tongue is red with a black tip. The body is thin and the tail long, 35–40% of total length. The body scales are keeled; in 19 rows at midbody (occasionally 17, 21 or 23); ventrals 144–172; cloacal scale divided; subcaudals paired, 123–167. Maximum size about 1.4m, average 0.9–1.3m; hatchlings 23–30cm. From a distance, this

Thelotornis mossambicanus

snake appears silvery-grey or grey-brown; close up it is pale grey or whitish with intense dark speckling. The temporal region behind the eye is brown, heavily speckled with black. The belly is usually paler than the back, but also speckled with black.

Eastern Vine Snake, *Thelotornis mossambicanus*.
Left top: Malawi (Gary Brown). **Left bottom and right:** Tanzania (Michele Menegon).

Habitat and Distribution: Savanna, woodland and coastal thicket, between sea level and 1,700m. It occurs in the country around Mount Kilimanjaro, including Kitobo Forest, the North Pare Mountains and Arusha National Park, between Lakes Eyasi and Manyara, and throughout most of south-eastern Tanzania, extending up the east side of Lake Tanganyika almost to the Burundi border, and south into northern Malawi, most of central Mozambique, northern and southern Malawi (but oddly not central Malawi) and extreme eastern Zimbabwe. Isolated populations have been reported from parts of northern and southern Mozambique and the south Somali coastline.

Natural History: Diurnal, arboreal and secretive, it waits motionless in the branches for prey, often with the forepart of the body sticking out like a twig. It will descend to the ground to cross open areas and to seize prey, and it may also strike down at passing terrestrial animals. It is gentle and non-aggressive, moving away if threatened, but if cornered or seized it inflates the neck and front half of the body, exposing vivid black, pale grey and pale blue cross-bars, and flicks its bright tongue up and down slowly. If further molested, it may strike; sometimes it bites, and sometimes it is deliberately off target, or may bang the aggressor with its snout, with the mouth closed. In the breeding season, male combat has been observed, with rivals trying to force their opponent's head down. Mating was observed in August in southern Tanzania. Females lay 4–16 eggs, roughly 15 × 35mm. Hatchlings were collected in February in southern Tanzania. The diet is mostly lizards, including chameleons (taken in trees), lacertids and agamas (often ambushed from above); when prey is seized from a tree, the snake hangs down and swallows the animal upwards, if feasible. Odd prey items include other snakes, frogs and fledgling birds. This snake is often mobbed by birds; one attacked in southern Tanzania by a Black-crowned Tchagra (*Tchagra senegalus*) fell from the tree and died shortly afterwards.

Medical Significance: As for the previous species. The eminent Zimbabwean herpetologist Don Broadley kept three of these snakes; when handled, one always tried to bite, one never tried to bite, and the other very occasionally tried to bite.

Southern Vine Snake (Cape Vine Snake/Twig Snake)
Thelotornis capensis

Identification: A very thin, medium-sized grey tree snake of southern Africa. The head is long and the snout pointed. The eye is large, the pupil is keyhole-shaped and the iris is greenish yellow. The tongue is red with a black tip. The body is thin and the tail is long, 35–40% of total length. The body scales are keeled; in 19 rows at midbody (occasionally 15–17); ventrals 144–177; cloacal scale divided; subcaudals paired, 126–173. There are two subspecies: the Cape Vine Snake (*Thelotornis capensis capensis*) and Oates' Vine Snake (*T. c. oatesi*). Maximum size of

Thelotornis capensis

Oates' Vine Snake is about 1.7m, average 1–1.5m. Maximum size of the Cape Vine Snake is 1.4m, average 0.9–1.3m. Hatchlings of both forms are 23–33cm. From a distance, these snakes appear silvery-grey or grey-brown, with darker and lighter blotches; close up they are pale grey or whitish with intense dark speckling. The temporal region behind the eye is brown, heavily speckled with black. The top of the head is greenish or grey, with speckling, although Oates' Vine Snake usually lacks the random speckling but has a black-speckled brown Y-shape between the eyes. The belly is slightly paler, but also speckled with black.

Habitat and Distribution: Semi-desert, dry and moist savanna, woodland and coastal thicket, between sea level and 1,700m. Oates' Vine Snake occurs from south-eastern DR Congo and north-central Angola south-east through Zambia, north-west Namibia, most of Malawi (except the south), northern Botswana,

Southern Vine Snake, *Thelotornis capensis*.
Left top: South Africa (Wolfgang Wuster). **Right top:** South Africa (Johan Marais).
Left bottom: Botswana (Stephen Spawls). **Right bottom:** subspecies *oatesi*, Zambia (Bill Branch).

northern Zimbabwe and western Mozambique. The Cape Vine Snake occurs in southern Zimbabwe, eastern Botswana, northern and eastern South Africa and Eswatini (Swaziland).

Natural History: Diurnal, arboreal and secretive, this snake waits motionless in the branches for prey, often with the forepart of the body sticking out like a twig. It will descend to the ground to cross open areas and to seize prey, and may also strike down at passing terrestrial animals. It is gentle and non-aggressive, moving away if threatened, but if cornered or seized it inflates the neck and front half of the body, exposing vivid black, pale grey and pale blue cross-bars, and flicks its bright tongue up and down slowly. If further molested, it may strike; sometimes it bites, and sometimes it is deliberately off target, or may bang the aggressor with its snout, with the mouth closed. In the breeding season, male combat has been observed, with rivals trying to force their opponent's head down. Mating occurs in spring. Females lay 4–18 eggs in the summer. Hatchlings have been collected in March and April. The diet comprises mostly lizards, including chameleons (taken in trees), lacertids and agamas (often ambushed from above); when prey is seized from a tree, the snake hangs down and swallows the animal upwards, if feasible. Odd prey items include other snakes, frogs and fledgling birds; it has also been observed eating amphibians at night.

Medical Significance: As for other vine snakes.

Usambara Vine Snake *Thelotornis usambaricus*

Identification: A very thin, medium-sized grey tree snake. The head is long and the snout pointed. The eye is large, the pupil is keyhole-shaped and the iris is yellow centrally, edged with green and sometimes white. The tongue is red with a black tip. The body is thin and the tail long, roughly 37–40% of total length. The body scales are keeled; in 19 rows at midbody; ventrals 145–169; cloacal scale divided; subcaudals paired, 143–175. Maximum size about 1.4m (reports of larger specimens exist), average 0.9–1.3m; hatchling size unknown, but probably

Thelotornis usambaricus

around 25cm. From a distance, this snake appears silvery-grey or grey-brown; close up it is pale grey or whitish with intense dark speckling. The top of the head can be any shade of green, brown or rufous, and some captive snakes have undergone a change of head colour. The lips are white, and the sixth labial is nearly always speckled with black, often in the form of a triangle or diagonal. The underside is slightly paler, but also speckled with black.

Habitat and Distribution: Savanna, woodland and coastal thicket, between sea level and 2,000m. It occurs along the length of the Kenyan coast, and into north-eastern Tanzania (but slightly away from the coastal strip), including the Usambara, Nguu and Nguru Mountains, south to Dar es Salaam. Inland in Kenya it is found to the Taita Hills and Mount Kasigau, and across the border to the southern side of Kilimanjaro, although some of the southern Kenyan animals seem intermediate

Usambara Vine Snake, *Thelotornis usambaricus*. **Left, centre and right:** Kenya (Stephen Spawls).

between this species and the Eastern Vine Snake (*Thelotornis mossambicanus*). There is a curious record (not on the map) of a female with a yellow head from Vamizi Island, off the north Mozambique coast.

Natural History: Similar to the previous species: diurnal, arboreal and secretive. It waits motionless in the branches for prey, often with the forepart of the body sticking out like a twig. It will descend to the ground to cross open areas and to seize prey, and it may also strike down at passing terrestrial animals. It is gentle and non-aggressive, moving away if threatened, but if cornered or seized it inflates the neck and front half of the body. The diet is as for other members of the genus.

Medical Significance: As for other vine snakes.

Dagger-tooth vine snake *Xyelodontophis*

A large, rear-fanged tree snake, described in 2002. There is a single species in the genus, found in the hills of eastern and south-eastern Tanzania. Nothing is known of its venom, but the species has proved to be nested within the vine snake (*Thelotornis*) clade, so presumably it has highly toxic venom similar to that of the vine snakes (and will shortly be renamed).

Dagger-tooth Vine Snake *Xyelodontophis uluguruensis*

Identification: A medium to large, rather thin arboreal snake of the Eastern Arc Mountains in Tanzania. It has an elongate head and large eyes with horizontal pear-shaped pupils. In general, it resembless *Thelotornis* species, but it lacks the distinctive keyhole-shaped pupil typical of that genus and differs in the development of strongly curved, rear maxillary teeth, which are blade-like, with sharp edges. The dorsal scales are elongated and moderately to feebly keeled, and in 19 rows at midbody, with a single large apical pit; ventrals

Xyelodontophis uluguruensis

Dagger-tooth Vine Snake, *Xyelodontophis uluguruensis*.
Left: Tanzania (Bill Branch). **Right:** Tanzania (Lorenzo Vinciguerra).

166–169; subcaudals 132–159. Maximum recorded length is over 1.9m, with other specimens being over 1m; hatchlings 37cm. The head is clearly bicoloured bronze, brown or green above, with immaculate white to yellow labials, throat and chin. The body is brown to blue-grey or black, speckled with light brown, green and yellow. When the snake inflates the neck for defensive purposes, the neck appears bright yellow with three large black patches on both sides; the body appears barred yellow and brown.

Habitat and Distribution: A montane and submontane forest species. It was originally known from the Uluguru Mountains only, but recent investigations revealed its presence in the forests of the Nguru, Udzungwa and Mbarika (Mahenge) Mountains, from 1,000m to about 2,000m. There is a curious report of a dead specimen from Manda Island, on the north Kenyan coast (not shown on the map).

Natural History: Diurnal, and individuals were found both in trees and moving on the ground. When disturbed, it inflates its neck in the manner of a Boomslang (*Dispholidus typus*), enhancing the vivid barring on its throat. It bites readily. A female of 1.47m in length laid 10 eggs. Little is known about its diet, but it is probably similar to *Thelotornis*. One of the type specimens had eaten an Uluguru Pygmy Chameleon (*Rhampholeon uluguruensis*).

Medical Significance: Nothing is known about the toxicity or composition of the venom, but it is presumably similar (and thus equally toxic) to that of the species in the genus *Thelotornis*, so bites must be avoided. No antivenom is available; any suspected bite victim will need to be monitored for clotting abnormalities for 48 hours or so and treated symptomatically.

Taxonomic Notes: Recent molecular investigation shows that records of the Forest Vine Snake (*Thelotornis kirtlandii*) from the Udzungwa Mountains of Tanzania are probably assignable to this species, and that the genus *Xyelodontophis* is nested within *Thelotornis* and will probably be put in synonymy with the latter.

Broad-headed tree snakes *Toxicodryas* (formerly *Boiga*)

An African genus containing two species of broad-headed, large-eyed, rear-fanged tree snakes that are widespread across forested Africa. These were originally in the genus *Boiga* (a diverse Asian genus, with 30-plus species), and some authorities still classify the African species as *Boiga*. Fascinatingly, molecular analysis indicates that the egg-eaters (*Dasypeltis*) are their closest relatives. Laboratory studies have shown that their venom contains neurotoxins and is produced in relatively large quantities, so they might represent a danger to humans; however, bites to humans by tree snakes are very rare. No antivenom is available.

Key to the genus *Toxicodryas*, broad-headed tree snakes

1a Midbody scale rows 19; small, never more than 1.25m in length......*Toxicodryas pulverulenta*, Powdered Tree Snake (p. 239)

1b Midbody scale rows 21–25; large, up to 2.8m in length......*Toxicodryas blandingii*, Blanding's Tree Snake (p. 237)

Blanding's Tree Snake *Toxicodryas blandingii*

Identification: A very large, stocky tree snake with a thin neck; a short, broad, flattened head; and prominent eyes set well forward, with vertical pupils. The nostrils are large. The body is triangular in section and rubbery in texture, and the head is also rubbery and flexible; this is particularly noticeable when the snake is held by the head. The tail is long and thin, 20–26% of total length. The scales are large and velvety; in 21–25 (21–23 in East Africa) rows at midbody; ventrals 240–289; subcaudals 120–141. Maximum size about 2.8m,

Toxicodryas blandingii

average 1.4–2m; hatchlings 28–35cm. There are two main colour morphs. One is glossy black above and yellow underneath; the yellow may extend up the sides or be confined to a narrow stripe in the middle of the belly. The lips are yellow, bordered with black, and the eye is dark. Black specimens are usually male. The second colour morph is brown, grey or yellow-brown, and yellow-brown below, with faint or clear darker cross-bars or diamonds on the flanks. The skin between the scales is bluish grey, and this is obvious when the snake inflates the body. The head is brown, the lips may be yellow-brown, and the eye is yellowish or brown. Brown specimens are usually female. Juveniles are all brown (both sexes), with clear, irregular black bars that are roughly diamond-shaped, sometimes with pale edging and/or lighter centres. Black adults are startlingly similar in colour to Gold's Tree Cobra (*Pseudohaje goldii*), but have more velvety skin.

Habitat and Distribution: Forest, woodland, forest–savanna mosaic and riverine woodland, from sea level to about 2,200m. This species occurs from the Ugandan Albertine Rift westwards across DR Congo to Togo, and extends south to the northern tip of Zambia and north-east Angola. It reappears in Ghana beyond

Blanding's Tree Snake, *Toxicodryas blandingii*.
Left top: DR Congo (Konrad Mebert). **Right top:** Cameroon (Yannick Francioli).
Left bottom: Gabon (Olivier Pauwels). **Centre bottom:** Uganda (Mike McLaren). **Right bottom:** captive (Darrell Raw).

the Dahomey Gap, westwards to Guinea-Bissau. There are isolated records from the Imatong Mountains in South Sudan, the Kakamega area in Kenya, the forest of the northern shore of Lake Victoria and the Ssese Islands in Uganda, and Rubondo Island and the Mahale Peninsula in Tanzania.

Natural History: Arboreal, it climbs quickly but ponderously, and its great length enables it to go very high, up to 30m or more in large forest trees. It will descend to the ground to cross open spaces and roads. It is active at night (one of Africa's few nocturnal tree snakes). When inactive (mostly during the day), it rests in leaf clumps, hollows in trees, etc. When threatened, this snake responds with a distinctive display, inflating the body to a huge extent, flattening the head and lifting the forepart of the body off the ground into wide C-shaped coils. If further molested, it will open the mouth enormously wide, exposing the pink lining, and make a huge, lunging strike. Active specimens prowl about trees, crawling along branches and investigating nests, recesses and hollows. They seem able to smell sleeping birds, especially in nests, and when one is detected the snake will make a very slow, careful approach and try to catch the bird in the nest. This species quite often occurs around human habitation, and in parks and gardens in cities, where it has been found in hedges and even quite small ornamental trees. It is known to enter buildings, often in search of bats, and is also known to frequent trees outside caves and buildings where bats roost, where it can catch the mammals as they emerge at dusk. Females lay 3–14 eggs, approximately 20 × 40mm; Nigerian animals had clutches of 4–6 large eggs. This snake eats a range of prey: birds, bird eggs, arboreal lizards such as chameleons and agamas, frogs, arboreal rodents and bats; juveniles eat largely lizards, while adults eat mostly birds and mammals.

Medical Significance: This snake has a huge range, and is relatively common in places, but the threat it poses to the ordinary person is negligible; it is unaggressive, active and alert, with good eyesight, lives in trees, is active by night and moves quickly away from confrontation. However, experimentally the venom contains a powerful neurotoxin, and although there are no documented human fatalities, the species should be treated with care and bites avoided. Bite victims have experienced local pain and swelling, dizziness, chest and muscle pain and abdominal cramps, hyperaesthesia and lymphadenopathy. No antivenom is available; bites should be treated symptomatically.

Powdered Tree Snake *Toxicodryas pulverulenta*

Identification: A fairly large, broad-headed, rear-fanged tree snake that is usually pink or red-brown. The head is broad, the eye is large and prominent, and the pupil is vertical. The body is laterally compressed, with a vertebral ridge. The tail is 20–25% of total length. The scales are smooth; in 19 rows at midbody; ventrals 236–276; subcaudals 96–132. Maximum size about 1.25m, average 0.8–1.2m; hatchling size unknown. The colour is pinkish to brown, red-brown or pinky-grey, with darker cross-bars that are alternately narrow and uniform or broadening to subtriangular,

Toxicodryas pulverulenta

enclosing a pale spot. The back is usually very finely dusted with brown or black specks (hence the common name), as is the belly, which is pale pink with dashed dark lines on the side of each ventral scale. The tongue is pink with a white tip.

Habitat and Distribution: Forest, woodland and forest–savanna mosaic, between sea level and 2,000m, or possibly higher. This species occurs from the Albertine Rift in Uganda westwards across DR Congo to Nigeria, Togo and Benin, and south to parts of Gabon and central DR Congo. It is also in forest west of the Dahomey Gap, from Ghana west to western Guinea. There are isolated records from the Kakamega area in Kenya, forests around Entebbe, Kampala and the Kibale area

Powdered Tree Snake, *Toxicodryas pulverulenta*.
Left and right top: DR Congo (Konrad Mebert). **Right bottom:** Gabon (Bill Branch).

in Uganda, the Talanga Forest in South Sudan, and parts of north-west Kwanza Norte and Kwanza Sul in north-west Angola.

Natural History: Arboreal, it is an elegant, careful climber, and nocturnal, which is unusual for a large tree snake. During the day it shelters in holes and cracks in trees, creeper and epiphyte tangles, disused plant pods and bird nests, and sleeps curled up; it will also sleep in a fork or on a branch. When threatened, it elevates and inflates the forepart of its body, flickers its red tongue, hisses loudly and makes lunging strikes. It lays eggs; a Ghanaian specimen contained three developing eggs, roughly 30 × 10mm, in September (the end of the rainy season), a specimen from DR Congo contained eggs in June, and a Nigerian specimen contained two eggs in January. The diet apparently comprises rodents and arboreal lizards.

Medical Significance: A snake with a huge range, but being nocturnal, arboreal and rear-fanged, it is unlikely to be a threat to the ordinary person. However, its venom is unstudied and a near relative, Blanding's Tree Snake (*Toxicodryas blandingii*), has a toxic venom, so this snake should be treated with care.

Black tree snakes *Thrasops*

A tropical African genus of large, harmless black forest-dwelling tree snakes with keeled or smooth scales. They inflate the neck as a threat display. Four species are known. Jackson's Tree Snake (*Thrasops jacksonii*) looks like a black Boomslang (*Dispholidus typus*), and recent molecular analysis reveals the two species are actually very close; see our note below about care needed to distinguish the two. Black tree snakes have no fangs, but severe symptoms (albeit under unusual circumstances) are reported following the bite of a Jackson's Tree Snake.

Jackson's Tree Snake *Thrasops jacksonii*

Identification: A large black tree snake with a large, dark eye and a short head. It smells strongly of liquorice, especially if freshly sloughed. The pupil is round but the iris is black, so the whole eye just looks black. The body is laterally compressed and the tail is long, a third of the total length. The body scales are usually unkeeled; in 19 rows at midbody (sometimes 17 or 21); ventrals 187–214; cloacal scale divided; subcaudals 129–155. Maximum size about 2.3m, average 1.4–1.8m; hatchlings 32–35cm. Adults are uniform glossy black above and below, though

Thrasops jacksonii

the throat may be grey or white (a dull green specimen is reported from Zambia). Juveniles are very variable depending on the locality; they are checked black and yellow, often with a scattering of blue, green or orange scales, above and below; the tail is spotted yellow and the head and neck are yellow, olive-green or brown. The juvenile colour changes when the snake is 40–60cm long; some young adults may retain traces of the juvenile pattern. Note well: the adult is virtually indistinguishable from the black colour form of the Boomslang (*Dispholidus*

Jackson's Tree Snake, *Thrasops jacksonii*. **Left:** Angola (Warren Klein). **Right:** juvenile (Matthijs Kuijpers).

typus), so treat with caution. Looking in the mouth doesn't always help; Jackson's Tree Snakes have enlarged back teeth that resemble fangs.

Habitat and Distribution: Forest, forest islands, woodland and riverine forest, from sea level to over 2,400m. From western Kenya it extends across Uganda (records are lacking from the centre), west across the northern half of DR Congo and the southern Central African Republic, to Cameroon, then south to northern Angola. It is also found in western and southern Rwanda (Nyungwe and Cyamudongo Forests) and north-western Burundi. In Tanzania, it is confined to the extreme north-west (Kabare and Rubondo Island), and there are isolated records from Talanga Forest in the Imatong Mountains of South Sudan and Ikelenge in North-western Province, Zambia. There are very few records from inside the lower curve of the Congo River.

Natural History: Diurnal and arboreal, it is a superb climber, ascending high into trees, to 30m or more. If approached or threatened in a tree and unable to slide away, it will launch itself into the air, and its long, light body enables it to move sideways while falling; when it hits the ground it quickly slides away. Kenyan specimens were seen to jump from trees into the water. If cornered, it will inflate its neck and anterior body like a Boomslang, move sideways and make huge lunging strikes, looking very large and threatening. It lays 7–12 eggs, roughly 15 × 30mm. It is a generalist feeder; known prey items include arboreal lizards (especially chameleons) and mammals, including bats. It raids nests for eggs, nestlings and adult birds, and has been seen to drop out of a tree to catch a frog.

Medical Significance: Although widespread, this secretive, harmless tree snake is not a risk to humans in Africa. However, a snake keeper who allowed a captive Jackson's Tree Snake to chew on him for over two minutes (he didn't want to damage the snake by removing it forcibly) suffered alarming symptoms. The wound bled for several days and the victim experienced immediate nausea and dizziness, intense pain, lymphadenitis, and swelling extending to the elbow; the hand became twice its normal size, and the finger turned black and remained stiff for many days. Another snake keeper bitten by a captive snake of the same genus (the Yellow-throated Tree Snake, *Thrasops flavigularis*) suffered pain and similar swelling, with the hand also becoming twice its normal size. Most people do not allow a potentially dangerous snake to chew on them, but this case serves as a warning to those who keep these snakes; any species may have toxic saliva.

Family Lamprophiidae African snakes

A family of African, Madagascan and Asian snakes, with eight subfamilies and more than 50 genera; over 30 occur in mainland Africa. Some African rear-fanged snakes in this family have been known to cause significant symptoms following a bite, in particular the fast-moving diurnal snakes in the subfamily Psammophiinae, in the genera *Psammophis* and *Malpolon*.

Subfamily Psammophiinae Sand snakes and allies

A subfamily of rear-fanged snakes, with just over 50 species and eight genera. They are mostly African, with a limited penetration into Europe and Asia and a single Madagascan species. Most are longitudinally striped, relatively large, fast-moving and diurnal. Four species in the subfamily have venom that has caused serious, largely local effects, but there are no recorded fatalities.

Sand snakes *Psammophis*

A genus of fast-moving, rear-fanged diurnal snakes. They have long heads and large eyes with round pupils, slim, elongate bodies and long tails, and have evolved to move fast in lightly vegetated or grassy habitats. Most are striped above and below. About 30 species occur in Africa. Sand snakes have relatively large, grooved fangs and some humans have suffered local symptoms of envenomation, such as swelling, haemorrhage, intense itching, local pain and nausea, following the bites of some species. One large species, the Olive Sand Snake (*Psammophis mossambicus/sibilans/rukwae*), has occasionally caused severe (but never life-threatening) symptoms.

Olive Sand Snake (Hissing Sand Snake)
Psammophis mossambicus/sibilans/rukwae

Identification: A large brown snake, common almost throughout the moist savannas of Africa. It has a rounded snout, overhanging supraocular and preocular scales (giving a 'penetrating expression'), and a large eye with a round pupil; the iris may be shades of yellow or brown. The body is cylindrical and muscular, and the tail is long, 27–33% of total length. The scales are smooth; in 17 rows at midbody; ventrals 151–185; cloacal scale divided; subcaudals paired, 82–120. Maximum size about 1.8m (possibly larger), average 1–1.5m; hatchlings 28–33cm. This snake is quite variable in colour; it

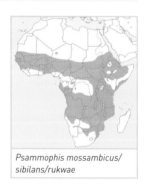

Psammophis mossambicus/ sibilans/rukwae

is usually olive-brown above, but may be any shade of brown or grey. All the dorsal scales or just the central ones may be black-edged, giving a finely striped appearance; some specimens have faint or low-contrast (but never distinct) longitudinal striping. The labial scales, chin and sides of the neck may be yellow, orange or dull red; the lips may be black-speckled, as may the lower flanks. The

Olive Sand Snake, *Psammophis mossambicus/sibilans/rukwae.*
Left: Kenya (Stephen Spawls). **Right:** Nigeria (Gerald Dunger).

belly is usually yellow, cream or white (sometimes white anteriorly, shading to yellow posteriorly), sometimes with a faint dark line or a series of dashes on each side, but never with a pair of clear, continuous black lines.

Habitat and Distribution: Widespread in a range of savanna habitats, between sea level and 2,500m. It is often found in waterside vegetation or associated with water sources, and will follow these into drier areas. This species occurs from southern Mauritania and Senegambia east to Ethiopia and parts of Somalia, and south to South Africa. The map shows the distribution of all members of the complex.

Natural History: Diurnal and partially arboreal. This snake is fast-moving and alert; if threatened, it dashes away and hides. It is often mistaken for more dangerous species due to its size and colour. It emerges in the mid-morning to bask and then hunt. If restrained, it will bite vigorously. From eight to 30 eggs, roughly 15 × 30mm, are laid in a suitable moist hole or damp tree crack, or in leaf litter. In South Africa these take about two months to hatch. Hatchlings were collected in November and December in southern Tanzania, and eggs were laid in September there. A wide range of prey has been recorded, including snakes (one specimen ate a young Black Mamba, *Dendroaspis polylepis*), lizards, rodents, frogs and even birds; juveniles eat mostly lizards. This snake is often eaten by birds of prey, caught as it basks on top of bushes.

Medical Significance: Despite this snake's huge range and relative abundance, it poses little threat to the average person, as it is alert, very fast-moving and readily avoids confrontation; it must also chew to engage its rear fangs. However, occasional accidental bites do occur, where the snake was inadvertently cornered. In Ghana a farmer opened a chicken coop and was bitten on the hand by a snake inside; in Kenya a snake that was trodden on delivered a bite. Many snake handlers have

Olive Sand Snake, *Psammophis mossambicus/sibilans/rukwae.* **Below:** Kenya (Stephen Spawls).

been bitten when handling these snakes. Symptoms of envenomation recorded following bites include pain, which may be severe, swelling, discolouration, nausea, abdominal cramps and intense local itching. No antivenom is available and it is not necessary; treat symptomatically.

Taxonomic Notes: Names used for the large brown savanna snakes of this complex include *sibilans, mossambicus, phillipsi, rukwae, brevirostris* and *afroccidentalis,* and members of a closely related group go under the names *Psammophis sudanensis, P. orientalis* and *P. subtaeniatus*. There is impassioned debate about the status of the various forms, which names have priority and so on, but this is only of academic interest as far as snakebite is concerned.

Montpellier and Moila snakes *Malpolon*

A genus of three species of medium to large, fast-moving diurnal snakes, associated with North Africa and the Sahara. They have large rear fangs, long heads and large eyes with round pupils, slim, elongate bodies and long tails, and have evolved to move fast in lightly vegetated or grassy habitats. They have relatively large, grooved rear fangs. Some human bite victims have suffered systemic effects as well as local symptoms of envenomation. One species is able to spread a cobra-like hood.

Key to the African members of the genus *Malpolon*, Montpellier and Moila snakes

1a One loreal scale; snout profile convex; widespread across the Sahara......*Malpolon moilensis*, Moila Snake (p. 247)

1b Two loreals; dorsal snout profile concave or flattened; mostly near North African coastline......2

2a Usually has 19 dorsal scale rows at midbody; males have a dark 'saddle' on the foreparts; Western Sahara to central north Algeria......*Malpolon monspessulanus*, Western Montpellier Snake (p. 244)

2b Usually has 19 (rarely 17) dorsal scale rows at midbody; no 'saddle' is present in males, and colouring often includes narrow, longitudinal pale stripes; eastern Algeria (possibly Morocco) and Tunisia eastwards along the North African coast to Egypt, and from there further east......*Malpolon insignitus*, Eastern Montpellier Snake (p. 246)

Western Montpellier Snake *Malpolon monspessulanus*

Identification: A large, fast-moving brown or olive snake of the western Maghreb. It has a rounded snout, overhanging supraocular and preocular scales (it is said to have a 'penetrating expression'), and a large eye with a round pupil; the iris may be shades of yellow or brown. The body is cylindrical and muscular, and the tail is long, 25–28% of total length. The scales are smooth; in 19 rows at midbody; ventrals 146–210; cloacal scale divided; subcaudals paired, 62–102. Maximum size about 2.5m (although African animals are rarely larger

Malpolon monspessulanus

Western Montpellier Snake, *Malpolon monspessulanus*.
Left: Morocco (Stephen Spawls). **Right:** Morocco (Konrad Mebert).

than 2m), average 1–1.5m; hatchlings 20–36cm. This snake is quite variable in colour. The juveniles are usually brown or grey, with heavy, dark dorsal speckling often forming vague cross-bars; the underside is white with extensive light brown mottling. Adults usually become uniform brown, olive or grey, but the lower flank scales are often black-edged, and sometimes blue-grey; males may have a darker 'saddle', and in parts of Morocco and Western Sahara almost the entire dorsal surface may be dark, but with a light-coloured head and neck. Some adults retain the juvenile colouration.

Habitat and Distribution: Widespread in a range of near-coastal habitats, provided there is some vegetation, including semi-desert, open woodland, rocky hills and valleys, Mediterranean scrubland (maquis), salt marshes, orchards and gardens; in desert it occurs in oases. It is found between sea level and 2,300m (or possibly higher; a skin was found above 3,000m in the Atlas Mountains). This species occurs north-west from the Dakhla Peninsula in Western Sahara through Morocco, north of the desert, into north-west Algeria. It also occurs in Spain, Portugal and southern France.

Natural History: Usually diurnal, but it is sometimes crepuscular during the summer. It is terrestrial, but will climb into vegetation. This species is fast-moving and alert, and hunts its prey by sight; if threatened it dashes away and hides, but will also hiss loudly and flatten the neck. If restrained, it will bite vigorously. Males take part in combat, neck-wrestling each other. The glands on the raised ridges between the eye and the nostril secrete a liquid that the snake rubs over its body after sloughing; this probably serves to spread pheromones. It lays 4–20 eggs, 40–50 × 20mm, in a suitable refuge, often a deep hole; communal nesting has been reported. A wide range of prey has been recorded; most small vertebrates are taken, including snakes, tortoises and even small rabbits. Juveniles eat mostly lizards and may take arthropods.

Western Montpellier Snake, *Malpolon monspessulanus*.
Below: Morocco (Stephen Spawls).

Medical Significance: Despite the snake's huge range and relative abundance, it poses little threat to most people, as it is alert, very fast-moving and readily avoids confrontation. A series of documented bites in Spain were all inflicted on snake handlers. However, it has large, grooved rear fangs, and although most bites merely result in short-lived bleeding from the fang punctures, discolouration and local swelling, a handful of cases reported more significant symptoms such as major swelling, nausea and muscle spasms. In some cases symptoms such as stiffness, numbness and drowsiness that might be due to neurotoxicity were reported. In one case, the victim (who had managed to insert his finger into the back of the snake's mouth), after initial swelling, began to exhibit weakness and signs of cranial nerve involvement, including blurred vision, drooping eyelids and lack of eye control (including inability to focus); these took six days to resolve. No antivenom is available.

Eastern Montpellier Snake *Malpolon insignitus*

Identification: A large, fast-moving brown or olive snake of the eastern Maghreb east to Egypt. It has a rounded snout, overhanging supraocular and preocular scales (it is said to have a 'penetrating expression'), and a large eye with a round pupil; the iris may be shades of yellow or brown. The body is cylindrical and muscular, and the tail is long, 25–28% of total length. The scales are smooth; in 19 rows at midbody; ventrals 166–180; cloacal scale divided; subcaudals paired, 80–102. Maximum size about 2.1m (although the largest Egyptian snake

Malpolon insignitus

was 1.8m), average 1–1.5m; hatchlings 20–36cm. The colour is as for the previous species: the juveniles are usually olive, brown or grey, the heavy, dark dorsal speckling often forming vague cross-bars; the underside is white with extensive light brown mottling. Adults usually become uniform brown, olive or grey.

Habitat and Distribution: Widespread in Mediterranean scrub and semi-desert, as well as in cultivated areas, grasslands, wetlands, salt marshes and oases, between sea level and 2,000m. There is some debate about where the distribution of this species starts and the Western Montpellier Snake (*Malpolon monspessulanus*) ends. They were originally regarded as subspecies; there may be some overlap in eastern Morocco and Algeria, and a female specimen from Tamanrasset in southern Algeria could not be assigned to either species. At present, this species is considered to occur along the North African coastline, from eastern Algeria through Tunisia to Libya and Egypt, but it does not

Eastern Montpellier Snake, *Malpolon insignitus*.
Below: Syria (Konrad Mebert).

extend south into the desert, although there is a record from Siwa Oasis in Egypt. There is a curious isolated record from Kufrah in south-eastern Libya. From Egypt it extends east to Iran, and north to Turkey; another subspecies is widespread in eastern Europe. An unsubstantiated report from Sudan is not shown on the map.

Natural History: Similar to the previous species; that is, largely terrestrial, diurnal, hunts by sight, lays up to 20 eggs, and eats a wide range of vertebrate prey.

Medical Significance: As for the previous species; not a threat to the average person, although there is a strange anecdotal report of a fatality from this species in Libya.

Moila Snake (False Cobra) *Malpolon moilensis*

Identification: A slim, attractive, spotted, large-eyed snake of the Sahara that spreads a hood. It has a fairly pointed snout and a large eye with a round pupil; the iris is orange or rufous. The body is cylindrical and muscular, and the tail is long, 19–27% of total length. The scales are smooth; in 17 rows at midbody; ventrals 139–188; cloacal scale divided; subcaudals paired, 48–73. Maximum size much debated, with two 1.9m specimens reported from Libya and a 1.5m snake from Saudi Arabia, but elsewhere the maximum is about 1.3m

Malpolon moilensis

and specimens over 1m are unusual. The average size is 60–80cm; hatchlings 17–18cm. The colour is tan, brown or grey-brown, with darker dorsal blotches forming cross-bars with matching amorphous dark lateral blotches; there is often higher contrast in males. There are up to three dark lateral bars behind the eye and sometimes one though the eye; these are very prominent in juveniles and may disguise the position of the eye (and possibly thus camouflage the head). The underside is cream or white, uniform or speckled red-brown.

Moila Snake, *Malpolon moilensis*. **Below:** spreading hood (Mark O'Shea).

Moila Snake, *Malpolon moilensis*, **Below:** Egypt (Tomáš Mazuch).

Habitat and Distribution: A snake of dry savanna, Sahel, semi-desert and vegetated areas in true desert, although it is usually absent from extensive sand dune areas and is rare on rock pavement. It is fond of vegetated wadis. This species occurs from sea level to 1,500m. It is found right across North Africa, from Mauritania and Western Sahara eastwards to Sudan and northern Eritrea, south in Niger virtually to the northern Nigerian border, and north to the Maghreb and Egypt, although it is absent from north of the High Atlas and Sahara Atlas. It extends east out of Africa to Iraq and possibly Iran, and north to Syria.

Natural History: A terrestrial snake, it is active by day, even during the hottest hours; one was found crossing a road in Morocco at 11 a.m. when the air temperature was 40°C. However, it is also crepuscular and occasionally nocturnal in summer. It hunts actively, inspecting vegetation tufts and probing down holes. If threatened and unable to escape, it elevates the front of the body at 45° and flattens the neck, looking like a cobra, and may advance towards the aggressor, but it rarely stands vertical as a cobra does. It shelters under ground cover and in holes. This species is often killed on the roads by vehicles. Mating occurs in April to June, and eggs laid between June and August hatch two to three months later; 4–18 eggs are laid. The diet includes a wide range of vertebrates, including reptiles (especially lizards, but also snakes), small rodents and birds; juveniles have been reported to eat large arthropods, and captive juveniles freely took rodents.

Medical Significance: As for the previous species; not a threat to the average person, and it must chew to engage its rear fangs. Oddly, there are historical anecdotal reports of fatalities, but this snake's resemblance to a cobra when spreading its hood has probably clouded the issue of reported bite cases. Recent bite cases to snake handlers are characterised by local pain, sometimes immediate and intense, and swelling; the swelling from one that was allowed to chew on a snake handler took a week to subside. No antivenom is produced.

Subfamily Aparallactinae African burrowing snakes

A subfamily of small snakes, with around 46 species in nine genera. They are mostly African; a couple of species reach the Middle East. Most are small, dark-coloured burrowing snakes, with small, short heads, small eyes, no loreal scale and grooved rear fangs (one centipede-eater has no fangs), but they are not dangerous to humans. However, a bite from one South African species caused some worrying symptoms.

Kwazulu-Natal black snake *Macrelaps*

A genus with a single species, a fairly large, rear-fanged black snake that looks like a burrowing asp (*Atractaspis*) and is confined to eastern South Africa. It is rare and secretive, and bites are unlikely. However, a couple of anecdotal reports indicate it may have a toxic venom. No antivenom is available.

Kwazulu-Natal Black Snake *Macrelaps microlepidotus*

Identification: A rear-fanged, blunt-headed black burrowing snake of the eastern side of South Africa; it is very similar in appearance to a burrowing asp (*Atractaspis*). It has a broad, rounded head and a tiny black eye. The body is cylindrical, stocky and muscular, and the tail is not short (distinguishing it from a burrowing asp), 12–19% of total length. The scales are smooth; in 25–27 (occasionally 23) rows at midbody; ventrals 158–172; cloacal scale entire; subcaudals paired, 35–50. Maximum size about 1.1m, average 60–90cm; hatchlings 20–30cm. The

Macrelaps microlepidotus

colour is uniform shiny jet black or grey-black, with the underside sometimes slightly paler; it becomes silvery-grey before sloughing.

Kwazulu-Natal Black Snake, *Macrelaps microlepidotus*. **Below:** South Africa (Bill Branch).

Habitat and Distribution: Savanna, grassland, woodland and coastal thicket, and suburbia, from sea level to about 1,300m. This species is endemic to South Africa, occurring from north-east Kwazulu-Natal southwards to East London in the Eastern Cape, inland to the Estcourt area in Kwazulu-Natal and the Amatole Mountains in the Eastern Cape. It might reach extreme southern Mozambique.

Natural History: Terrestrial, usually in burrows, in leaf litter or under ground cover. It moves on the surface on warm nights but has been observed to be active on warm, overcast days. It pushes through leaf litter in search of food. This snake is often found in damp localities near water sources and is said to be a good swimmer. It sometimes shams (feigns) death, turning its head sideways and opening its mouth. It lays 3–10 eggs in the summer. The diet is frogs, legless lizards, other snakes and small rodents.

Medical Significance: Being a secretive, nocturnal, inoffensive snake with a small range, it does not pose a major snakebite risk. However, two bites to snake handlers caused collapse due to loss of consciousness, which persisted for up to 30 minutes, indicating that this unstudied snake's venom might well be toxic. No antivenom is available; a bite must be treated symptomatically.

Subfamily Pseudoxyrhophiinae Gemsnakes

A subfamily of about 90 species and 20 genera, mostly from Madagascar, but a few from other African offshore islands. Two genera occur on mainland Africa. One species, the Many-spotted Snake (*Amplorhinus multimaculatus*), might have toxic venom.

Many-spotted snake *Amplorhinus*

A genus with a single species, a small rear-fanged snake that is confined to southern and eastern South Africa, Zimbabwe and western Mozambique. It is rare and secretive, and bites are unlikely. However, anecdotal reports indicate it may have a toxic venom. No antivenom is available. This snake and the harmless slug-eaters (*Duberria*) are the only mainland African examples of the subfamily Pseudoxyrhophiinae; the rest are virtually all Madagascan.

Many-spotted Snake *Amplorhinus multimaculatus*

Identification: A striped (occasionally green), rear-fanged, blunt-headed snake of southern and eastern South Africa and the highlands of eastern Zimbabwe. It has a rounded snout and a large eye. The body is cylindrical and slim, and the tail is long, 21–36% of total length. The scales are smooth; in 17 rows at midbody; ventrals 133–154; cloacal scale entire; subcaudals paired, 56–91. Maximum size about 63cm, average 35–55cm; hatchlings 12–20cm. The colour is olive or olive-brown, with dark blotches down the back and sometimes with a pair

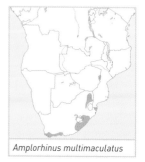

Amplorhinus multimaculatus

Many-spotted Snake, *Amplorhinus multimaculatus*. **Left and right:** South Africa (Richard Boycott).

of light dorsolateral stripes; some individuals are uniform green with black spots. The underside is bluish, olive or dull green.

Habitat and Distribution: Tends to live in wetlands, reedbeds and waterside vegetation in heathland (fynbos), moist savanna and montane forest, between sea level and nearly 2,600m. In South Africa, isolated populations occur in the Western Cape and Eastern Cape, Kwazulu-Natal, Mpumalanga and south-eastern Limpopo; this species is also in Lesotho, and in montane grassland in the eastern highlands of Zimbabwe and extreme western Mozambique.

Natural History: A slow-moving diurnal terrestrial snake that forages during the day in damp and marshy areas. If disturbed, it will coil up tightly and then strike freely. It gives birth to 4–8 live young (occasionally more, up to 13). The diet includes lizards, frogs and small rodents.

Medical Significance: Being a slow-moving, inoffensive snake with a small range, it does not pose a major snakebite risk. However, a bite to a snake handler, who was chewed, caused bleeding, immediate burning pain, inflammation and slight swelling. This snake's venom is unstudied but may well be toxic. No antivenom is available; a bite must be treated symptomatically.

Look-alikes and common species

White Lip, *Crotaphopeltis hotamboeia* (Stephen Spawls).

In this section, we provide pictures and short descriptions of three sorts of African snakes: those that resemble (to a greater or lesser extent) dangerous species and might be confused with them; those that are relatively common or widespread; and those that look dangerous, but are not. We give some pointers to help identification. Browse through it, look at the pictures and become familiar with the animals. The descriptions are in general taxonomic order. Most of the species here have not, to date, caused any sort of documented adverse reaction following a bite, although it must always be remembered that any fangless snake might prove to have toxic saliva and any rear-fanged snake has both fangs and venom. It goes without saying that, unless you are an expert, you should not take any risks. You may be fairly certain, using this guide or another source, that you have identified the green snake in your living room as a Spotted Bush Snake (*Philothamnus semivariegatus*), not a green mamba (*Dendroaspis angusticeps* or *D. viridis*), but we still advise chivvying the snake out with a broom, not picking it up. Never handle a snake unless you are certain that it has been correctly identified and is harmless. Always avoid being bitten.

We have suggested dangerous species that these snakes might be confused with, where relevant. Some of these suggestions might seem a little surprising, but they are based on our experiences. We have seen some startling mistakes. The identification of snakes is a subtle skill, and sometimes not even herpetologists have it. Differences that are obvious to some are not obvious to all, and many people do not like and have no interest in snakes; they observe them only when forced to. Imagine trying to explain the difference between a Land Rover and a Toyota Land Cruiser to someone who has no interest in cars. Some brief stories may be of interest. A Kenyan professional hunter who was emigrating to South Africa in the 1970s donated a pet python he had kept for years to Nairobi Snake Park. When the snake was tipped out of the bag it proved to be a huge Puff

Adder (*Bitis arietans*). A herpetologist colleague was examining cobras in a large European museum collection identified a forest cobra by sight, and was startled when the curator, who had vast African experience, asked him how he could tell it apart from a Black-necked Spitting Cobra (*Naja nigricollis*) without counting the scales. Some years ago, we were sent two pictures by a herpetologist working in the Serengeti; he suggested one was a Common Egg-eater (*Dasypeltis scabra*) and the other Battersby's Green Snake (*Philothamnus battersbyi*). Both were night adders: a Rhombic Night Adder (*Causus rhombeatus*) and a Velvety-green Night Adder (*Causus resimus*).

A worrying feature of our digital age is that people are also often disturbingly quick to attempt to identify snakes without any real knowledge, particularly on the internet. The Digital Farmers Kenya Facebook group has 300,000 members, who debate all things farming. Recently, a farmer posted a picture of a dead snake, asking for identification. It was clearly a harmless Brown House Snake (*Boaedon fuliginosus*). Within an hour, however, dozens of people had confidently identified it as a Black Mamba (*Dendroaspis polylepis*). Bizarrely, some attempted to justify their identification by posting pictures from the internet of real Black Mambas, which obviously looked quite unlike the house snake. Those who tried to inject some sense into the debate were abused by other posters. Even on those internet forums concerned with snake identification, mistakes are common. So, when making an identification, use your judgement. Practice will help; visit a snake park, study a good book (regional guides are listed in the Appendix, pp. 310–11) or search online for African snake pictures. It is vitally important to get it right. Misidentifying a bird will not have any unfortunate consequences; misidentifying a snake can.

Royal Python *Python regius*

A harmless nocturnal, largely terrestrial snake of the West and Central African savanna and Sahel, from Senegal to Uganda and Sudan. Average size 0.7–1.3m. Might be mistaken for a Puff Adder (*Bitis arietans*), p. 57. Note the distinctive pattern and pear-shaped head.

Royal Python, *Python regius*. **Right:** Sierra Leone (Bill Branch).

Calabar Ground Python *Calabaria reinhardtii*

A harmless burrowing python of the great African forests, from Sierra Leone and Guinea east to eastern DR Congo. Average size 60–90cm. Might be mistaken for a burrowing asp (*Atractaspis*), pp. 194–217. Note the blunt head and tail.

Calabar Ground Python, *Calabaria reinhardtii*.
Left: Cameroon (Paul Freed). **Right:** Nigeria (Gerald Dunger).

Kenya Sand Boa *Eryx colubrinus*

Kenya Sand Boa, *Eryx colubrinus*. **Left:** Kenya (Stephen Spawls). **Right:** Kenya (Stephen Spawls).
Right below: Tanzania (Stephen Spawls).

A harmless burrowing boa of the dry country of East and north-east Africa, east from Niger and Chad to Sudan and Ethiopia, and south through Somalia to Kenya and eastern Tanzania. Average size 40–90cm. Might be mistaken for a Puff Adder (*Bitis arietans*), p. 57, or carpet viper (*Echis*), pp. 83–95. Note the many tiny scales on top of the head.

Müller's Sand Boa *Eryx muelleri*

A harmless burrowing boa of the dry savanna and Sahel of West and Central Africa, east from Senegal to Sudan. Average size 30–60cm. Might be mistaken for a Puff Adder (*Bitis arietans*), p. 57, or carpet viper (*Echis*), pp. 83–95. Note the many tiny scales on top of the head.

Müller's Sand Boa, *Eryx muelleri*.
Right: Nigeria (Gerald Dunger).

Javelin Sand Boa *Eryx jaculus*

A harmless burrowing boa of the North African coast and hinterland, from eastern Morocco to Egypt. Average size 30–60cm. Might be mistaken for a horned viper (*Cerastes*), pp. 96–100, or carpet viper (*Echis*), pp. 83–95. Note the many tiny scales on top of the head.

Javelin Sand Boa, *Eryx jaculus*.
Right: captive (Paul Freed).

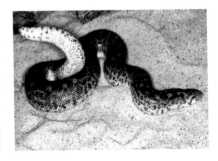

Lineolate Blind Snake *Afrotyphlops lineolatus*

A harmless burrowing blind snake that occurs across the savannas of West, Central and East Africa, from Senegal to Somalia, south to Tanzania. Average size 30–50cm. Might be mistaken for a burrowing asp (*Atractaspis*), pp. 194–217. Note the blunt, rounded head and tail, the non-tapering body and the tiny scales all the way around the body. A number of very similar-looking species occur widely throughout sub-Saharan Africa.

Lineolate Blind Snake, *Afrotyphlops lineolatus*.
Above: Tanzania (Bill Branch).

Schlegel's Beaked Blind Snake *Afrotyphlops schlegelii*

A harmless burrowing blind snake that occurs in semi-desert, dry and moist savanna and coastal thicket of southern Africa, from Botswana and South Africa to Mozambique and Zimbabwe. Average size 60–90cm. Might be mistaken for a burrowing asp (*Atractaspis*), pp. 194–217. Note the huge size, variation in colour and blunt, rounded head and tail.

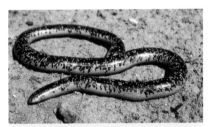

Schlegel's Beaked Blind Snake, *Afrotyphlops schlegelii*.
Above: Botswana (Stephen Spawls).

Spotted Blind Snake *Afrotyphlops punctatus*

A harmless burrowing blind snake, widespread in forest and moist savanna from Uganda west to Senegal. Average size 30–60cm. Might be mistaken for a burrowing asp (*Atractaspis*), pp. 194–217. Note the relatively large size, two colour morphs and blunt, rounded head and tail.

Spotted Blind Snake, *Afrotyphlops punctatus*.
Left: striped form, Nigeria (Gerald Dunger).
Above: blotched form, Nigeria (Gerald Dunger).

Delalande's Beaked Blind Snake *Rhinotyphlops lalandei*

A harmless burrowing blind snake that occurs in savannas, grassland and semi-desert of southern Africa, from South Africa to Mozambique and Zimbabwe. Average size 20–30cm. Might be mistaken for a burrowing asp (*Atractaspis*), pp. 194–217. Note the variation in colour and the blunt, rounded head and tail.

Delalande's Beaked Blind Snake, *Rhinotyphlops lalandei*. **Above:** Botswana (Stephen Spawls).

Ethiopian Gracile Blind Snake *Letheobia somalica*

A harmless pink burrowing blind snake that occurs in grassland and forest in Ethiopia. About 25 species of *Letheobia* occur in Africa; all look fairly similar. They range from 20cm to 60cm. All are pink. Unlikely to be mistaken for any dangerous snake. Note the colour, tiny scales all the way around the body, and blunt, rounded head and tail.

Ethiopian Gracile Blind Snake, *Letheobia somalica*. **Above:** Ethiopia (Stephen Spawls).

Merker's Worm Snake *Leptotyphlops merkeri*

A harmless, tiny worm snake that is black or grey. In savannas of East Africa, possibly further. Average size 12–20cm. Might be mistaken for a burrowing asp (*Atractaspis*), pp. 194–217. Note the minute size, blunt head and tail, and non-tapering body. Similar species occur throughout sub-Saharan Africa.

Merker's Worm Snake, *Leptotyphlops merkeri*. **Above:** Kenya (Stephen Spawls).

Hook-snouted Worm Snake *Myriopholis macrorhyncha*

A harmless, tiny pink worm snake. In dry savannas of East and north-east Africa, possibly further. Average size 12–20cm. Unlikely to be mistaken for any dangerous snake. Note the minute size, blunt head and tail, and non-tapering body. Similar species occur throughout sub-Saharan Africa.

Hook-snouted Worm Snake, *Myriopholis macrorhyncha*. **Above:** Ethiopia (Stephen Spawls).

Red and Black Striped Snake *Bothrophthalmus lineatus*

A harmless nocturnal terrestrial snake. Occurs almost throughout the forests of Central and West Africa. Average size 0.4–1m. Unlikely to be mistaken for any dangerous snake, although it looks dangerous with its vivid colouration. Note the pale head and vivid black and red striping.

Red and Black Striped Snake, *Bothrophthalmus lineatus*. **Above:** DR Congo (Konrad Mebert).

Olive House Snake *Boaedon olivaceus*

A harmless nocturnal terrestrial snake. In forests and woodlands of Central and West Africa, from Kenya west to Guinea, and south to northern Angola. Average size 40–80cm. Might be mistaken for a burrowing asp (*Atractaspis*), pp. 194–217, although the head shape is very different. Note the lack of markings, red eyes and many small scales around the body.

Olive House Snake, *Boaedon olivaceus*. **Above:** DR Congo (Kate Jackson). **Right:** DR Congo (Konrad Mebert).

Brown House Snake *Boaedon fuliginosus*

Brown House Snake, *Boaedon fuliginosus*. **Above left:** Tanzania (Michele Menegon). **Above right:** Kenya (Stephen Spawls). **Below:** Ghana (Stephen Spawls).

A harmless nocturnal terrestrial snake. Occurs almost throughout the savannas, semi-desert and woodland of West, Central and East Africa. Average size 20–80cm. Probably Africa's most common snake, especially in suburbia. Unlikely to be mistaken for any dangerous snake if seen clearly,

but often identified as a Black Mamba (*Dendroaspis polylepis*), p. 126, and killed. Note the white stripes on the python-shaped head. Various forms of this species occur widely in Africa. Their status is debated; we describe some below.

Brown House Snake, *Boaedon fuliginosus*.
Right: Tanzania (Stephen Spawls).

Cape House Snake *Boaedon capensis*

Cape House Snake, *Boaedon capensis*. **Left and right:** Botswana (Stephen Spawls).

A harmless nocturnal terrestrial snake. Occurs almost throughout the forests, woodlands, savannas and semi-desert of southern Africa. Average size 20–80cm. Probably southern Africa's most common snake, especially in suburbia. Unlikely to be mistaken for any dangerous snake if seen clearly, but often identified as a Black Mamba (*Dendroaspis polylepis*), p. 126, and killed. Note the white stripes on the python-shaped head.

Striped House Snake *Boaedon lineatus*

A harmless nocturnal terrestrial snake. In savanna and woodland of East, Central and West Africa. Average size 20–80cm. Often common, especially in suburbia. Unlikely to be mistaken for any dangerous snake if seen clearly. Note the elongate white stripes on the python-shaped head.

Striped House Snake, *Boaedon lineatus*.
Right: Ghana (Stephen Spawls).

West African Forest House Snake *Boaedon virgatus*

A harmless nocturnal terrestrial snake. In forest and woodland of West Africa, from Guinea to western DR Congo. Average size 30–80cm. Unlikely to be mistaken for any dangerous snake if seen clearly. Note the elongate white stripes on the python-shaped head.

West African Forest House Snake, *Boaedon virgatus.*
Right: Nigeria (Gerald Dunger).

Yellow House Snake *Boaedon subflavus*

A harmless nocturnal terrestrial snake. In savanna and semi-desert. Occurs from Cameroon east to Kenya and northern Tanzania. Average size 30–80cm. Might be confused with a juvenile Black Mamba (*Dendroaspis polylepis*), p. 126. Note the distinctive yellowish colour and absence of head markings.

Yellow House Snake, *Boaedon subflavus.*
Right: Tanzania (Stephen Spawls).

Aurora House Snake *Lamprophis aurora*

A harmless nocturnal (sometimes diurnal) terrestrial snake. In savanna and woodland of eastern South Africa. Average size 30–80cm. Might be mistaken for the striped morph of the Spotted Harlequin Snake (*Homoroselaps lacteus*), p. 218. Note the yellow stripe on an olive body.

Aurora House Snake, *Lamprophis aurora.*
Right: South Africa (Johan Marais).

Striped Ethiopian Mountain Snake *Pseudoboodon lemniscatus*

A harmless nocturnal terrestrial snake. In savanna and grassland in highland Ethiopia and Eritrea. Average size 40–90cm. Doesn't really resemble any dangerous species, but it is the most common snake in highland Ethiopia. Note the distinctive striping and the pit in the upper labials below the eye.

Striped Ethiopian Mountain Snake, *Pseudoboodon lemniscatus.*
Right: Ethiopia (Stephen Spawls).

Cape Wolf Snake *Lycophidion capense*

Cape Wolf Snake, *Lycophidion capense*. **Left:** Kenya (Sonya Varma). **Right:** Kenya (Stephen Spawls).

A harmless nocturnal terrestrial snake. In savannas of East Africa, from Ethiopia and Sudan south to South Africa. Average size 15–50cm. Surprisingly frequently mistaken for a burrowing asp (*Atractaspis*), pp. 194–217, although the head shape is very different. Note the flat, wedge-shaped head and the white-edged scales.

Flat-snouted Wolf Snake *Lycophidion depressirostre*

A harmless nocturnal terrestrial snake. Largely in dry savannas of East Africa, from Ethiopia and Sudan south to northern Mozambique. Average size 20–35cm. Surprisingly frequently mistaken for a burrowing asp (*Atractaspis*), pp. 194–217, although the head shape is very different. Note the flat, wedge-shaped head.

Flat-snouted Wolf Snake, *Lycophidion depressirostre*. **Above:** Tanzania (Bill Branch).

Half-banded Wolf Snake *Lycophidion semicinctum*

A harmless nocturnal terrestrial snake. In savannas of West and Central Africa, from Guinea east to Chad, and possibly Sudan. Average size 30–60cm. Sometimes mistaken for a garter snake (*Elapsoidea*), pp. 129–42, on account of its white bands, or a burrowing asp (*Atractaspis*), pp. 194–217, being small and dark. Note the flat, wedge-shaped head and the white bands.

Half-banded Wolf Snake, *Lycophidion semicinctum*. **Above:** Nigeria (Gerald Dunger).

Banded False Wolf Snake *Dendrolycus elapoides*

A harmless nocturnal terrestrial and arboreal snake. In forests of West and Central Africa, from Cameroon east to eastern DR Congo. Average size 30–50cm. Doesn't closely resemble any dangerous snake, but looks venomous. Note the flat head and fine black bands.

Banded False Wolf Snake, *Dendrolycus elapoides*.
Right: DR Congo (Konrad Mebert).

Cape File Snake *Limaformosa capensis* (formerly *Mehelya capensis*, *Gonionotophis capensis*)

A large, harmless nocturnal terrestrial snake; very slow-moving. In savannas and woodland of south-eastern Africa, from Tanzania south to South Africa and Angola. Average size 0.6–1.4m. Looks dangerous, but unlikely to be mistaken for any dangerous snake if seen clearly. Often feared for superstitious reasons. Note the distinctive triangular body shape and long tail. Several very similar-looking snakes of this genus occur widely in Africa.

Cape File Snake, *Limaformosa capensis*.
Above: Tanzania (Michele Menegon).

Mole Snake *Pseudaspis cana*

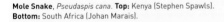

A large, harmless, diurnal terrestrial snake. In savannas of eastern, south-eastern and southern Africa, from Kenya to South Africa and Angola. Average size 0.6–1.6m. Large adults are often mistaken for cobras, and the vividly marked juveniles for night adders (*Causus*), pp. 101–11, and other small vipers. Note the large size, short head and stout body.

Mole Snake, *Pseudaspis cana*. **Top:** Kenya (Stephen Spawls).
Bottom: South Africa (Johan Marais).

Western Keeled Snake *Pythonodipsas carinata*

A small, rear-fanged, nocturnal terrestrial snake. In rocky desert and semi-desert. Occurs along the western side of Namibia, into extreme south-west Angola. Average size 30–70cm. Might be mistaken for a Horned Adder (*Bitis caudalis*), p. 67. Note the flat head and fragmented head scales.

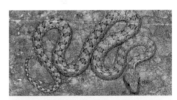

Western Keeled Snake, *Pythonodipsas carinata*. **Above:** Namibia (Johan Marais).

East African Shovel-snout *Prosymna stuhlmanni*

A small, harmless nocturnal burrowing snake. In savanna, from Kenya south to South Africa. Average size 20–25cm. Might be mistaken for a burrowing asp (*Atractaspis*), pp. 194–217. Note the yellowish snout and relatively large eyes.

East African Shovel-snout, *Prosymna stuhlmanni*. **Above:** Kenya (Stephen Spawls).

Speckled Shovel-snout *Prosymna meleagris*

A small, harmless nocturnal burrowing snake. In savanna, from Senegal east to South Sudan. Average size 20–30cm. Similar forms (*Prosymna greigerti* and *P. collaris*) also occur in the West and Central African savannas. Might be mistaken for a burrowing asp (*Atractaspis*), pp. 194–217. Note the angular snout and the speckled body.

Speckled Shovel-snout, *Prosymna meleagris*.
Above: Nigeria (Gerald Dunger).

Schouteden's Mud Snake *Helophis schoutedeni*

A small, harmless snake. Associated with watercourses and swamps in central DR Congo. Average size 30–70cm. Might be mistaken for a small cobra. Note the long tail and light and dark banding.

Schouteden's Mud Snake, *Helophis schoutedeni*.
Above: DR Congo (Eli Greenbaum).

Kenyan Bark Snake *Hemirhagerrhis hildebrandtii*

A small, rear-fanged, diurnal, largely arboreal snake. In savanna and semi-desert. Various species of this genus (all fairly similar) occur across the Sahel, from Burkina Faso to Sudan, south to South Africa. Average size 25–45cm. Not really similar to any dangerous snake, but the distinctive pattern leads people to think it is a viper. Note the grey colour, vertebral stripe and small, pointed head.

Kenyan Bark Snake, *Hemirhagerrhis hildebrandtii*. **Above:** Kenya (Stephen Spawls).

Northern Stripe-bellied Sand Snake
Psammophis sudanensis

A fairly large, rear-fanged, diurnal, largely arboreal snake; very fast-moving. In savanna and semi-desert. Occurs from Senegal east to Sudan, and south to Tanzania; a near-identical species occurs in southern Africa. Average size 0.5–1.1m. Not really similar to any dangerous snake, but relatively widespread and common. Note the distinctive striping.

Northern Stripe-bellied Sand Snake,
Psammophis sudanensis.
Above: Kenya (Stephen Spawls).

Speckled Sand Snake *Psammophis punctulatus*

A large, rear-fanged, diurnal, largely arboreal snake; very fast-moving. In savanna and semi-desert. Occurs from Egypt south to Tanzania. Average size 0.7–1.4m. Not really similar to any dangerous snake, but relatively widespread and common. Note the distinctive striping and rufous head.

Speckled Sand Snake, *Psammophis punctulatus*. **Above:** Kenya (Bio-Ken Archive).

Elegant Sand Snake *Psammophis elegans*

A large, rear-fanged, diurnal, largely arboreal snake; very fast-moving. In dry and moist savanna and semi-desert. Occurs from Senegal east to Chad. Average size 0.7–1.5m. Very slight resemblance to a vine snake (*Thelotornis*), pp. 229–35. Note the distinctive striping, round pupil and speckled head.

Elegant Sand Snake, *Psammophis elegans*. **Above:** Ghana (Stephen Spawls).

Link-marked Sand Snake *Psammophis biseriatus*

A fairly small, rear-fanged, diurnal, largely arboreal snake. In savanna and semi-desert. Occurs from Somalia south through eastern Ethiopia to Kenya and northern Tanzania. Average size 40–80cm. Might be mistaken for a vine snake (*Thelotornis*), pp. 229–35, which has a similar-shaped body and head. Note the distinctive pattern of regular dark blotches and the chain-like marks down the back. A prolonged bite from a Kenyan example of this species caused local haemorrhaging, swelling and lymphadenitis.

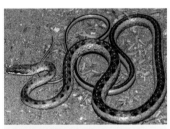

Link-marked Sand Snake, *Psammophis biseriatus*. **Above:** Kenya (Florian Fink).

Dwarf Beaked Snake *Dipsina multimaculata*

A small, rear-fanged, diurnal terrestrial snake. In semi-desert, dry savanna and rocky desert. Found from the western side of South Africa north to northern Namibia. Average size 20–45cm. Might be mistaken for a Horned Adder (*Bitis caudalis*), p. 67. Note the hooked snout and C-shaped coils, which mimic a Horned Adder.

Dwarf Beaked Snake, *Dipsina multimaculata*. **Above:** Namibia (Paul Freed).

Southern Striped Skaapsteker *Psammophylax tritaeniatus*

A medium-sized, rear-fanged, diurnal terrestrial snake. In grassland and savanna. Occurs from Tanzania and DR Congo south to South Africa and Namibia; other similar species occur in Kenya and Ethiopia. Average size 40–80cm. Not really similar to any dangerous snake. Note the distinctive striping. Said to have a toxic venom, but no adverse symptoms ever recorded from a bite.

Southern Striped Skaapsteker, *Psammophylax tritaeniatus.* **Above:** Namibia (Stephen Spawls).

Spotted Skaapsteker *Psammophylax rhombeatus*

A medium-sized, rear-fanged, diurnal terrestrial snake. In grassland and savanna. Occurs widely in eastern South Africa, north to south-west Angola. Average size 0.5–1m, but does reach 1.5m. Might be mistaken for a small adder. Note the distinctive spotting and striping. Has a potent neurotoxic venom, but no adverse symptoms ever recorded from a bite.

Spotted Skaapsteker, *Psammophylax rhombeatus.* **Above:** Eswatini (Swaziland) (Richard Boycott).

Western Beaked Snake *Rhamphiophis oxyrhynchus*

A fairly large, rear-fanged, diurnal snake; terrestrial but occasionally climbs into bushes. In dry and moist savanna. Occurs from Sudan, Ethiopia and Somalia south-west to Uganda, and west to Senegal. Average size 0.6–1.3m. Might be mistaken for a cobra or a mamba (*Dendroaspis*), pp. 120–29, if not seen clearly. Note the distinctive pointed snout.

Western Beaked Snake, *Rhamphiophis oxyrhynchus.* **Above:** Nigeria (Gerald Dunger).

Rufous Beaked Snake *Rhamphiophis rostratus*

A fairly large, rear-fanged, diurnal snake; terrestrial but occasionally climbs into bushes. In semi-desert and dry and moist savanna. Occurs from Sudan and Ethiopia southwards to South Africa. Average size 0.6–1.3m. Spreads a hood. Might be mistaken for a cobra or a Boomslang (*Dispholidus typus*), p. 226, if not seen clearly. Note the distinctive pointed snout and dark line through the eyes.

Rufous Beaked Snake, *Rhamphiophis rostratus.* **Above:** Kenya (Stephen Spawls).

Red-spotted Beaked Snake *Rhamphiophis rubropunctatus*

A huge, rear-fanged, slim diurnal snake; terrestrial but occasionally climbs into bushes. In semi-desert and dry and moist savanna. Occurs from Chad, Sudan and Ethiopia southwards to northern Tanzania. Average size 1–2m, but reaches 2.5m. If not seen clearly, might be mistaken for a Black Mamba (*Dendroaspis polylepis*), p. 126, or a red-headed morph of the Boomslang (*Dispholidus typus*), p. 226. Spreads a hood. Note the vivid orange head. Only young individuals are spotted.

Red-spotted Beaked Snake, *Rhamphiophis rubropunctatus*. **Above:** Kenya (Florian Fink).

Cape Centipede-eater *Aparallactus capensis*

A small, rear-fanged burrowing snake. In savanna and woodland, from Kenya south to South Africa. A very similar-looking species, Jackson's Centipede-eater (*Aparallactus jacksonii*), occurs in East Africa and South Sudan. Average size 20–35cm. Doesn't really look like any dangerous snakes, although some juvenile cobras have black heads. Note the tiny eyes and black head and neck.

Cape Centipede-eater, *Aparallactus capensis*. **Above:** Tanzania (Stephen Spawls).

Plumbeous Centipede-eater *Aparallactus lunulatus*

A small, rear-fanged burrowing snake. In savanna and woodland, from Ethiopia south to South Africa. Several colour morphs. Average size 20–40cm. The grey colour morph might be mistaken for a burrowing asp (*Atractaspis*), pp. 194–217. Note the tiny eyes and the long tail, which distinguishes it from a burrowing asp.

Plumbeous Centipede-eater, *Aparallactus lunulatus*. **Left:** Kenya (Tomáš Mazuch). **Right:** Kenya (Anthony Childs).

Western Forest Centipede-eater *Aparallactus modestus*

A small, slim, harmless burrowing snake. In forest and forest–savanna mosaic, from Uganda westwards to Sierra Leone, and south to southern DR Congo. Average size 30–60cm. The grey colour morph might be mistaken for a burrowing asp (*Atractaspis*), pp. 194–217. Note the tiny eyes and the long tail, which distinguishes it from a burrowing asp.

Western Forest Centipede-eater, *Aparallactus modestus*. **Above:** DR Congo (Kate Jackson).

Two-coloured Snake *Micrelaps bicoloratus*

A small, slim, rear-fanged burrowing snake. In savanna and semi-desert of eastern Kenya and north-east Tanzania. Average size 20–30cm. In several colour morphs; the dark form might be mistaken for a burrowing asp (*Atractaspis*), pp. 194–217. Note the short tail.

Two-coloured Snake, *Micrelaps bicoloratus*. **Right:** Kenya (Anthony Childs).

Guinea-fowl Snake *Micrelaps vaillanti*

A small, slim, rear-fanged burrowing snake. In savanna and semi-desert, from northern Tanzania north to Sudan and Somalia. Average size 20–35cm. Might be mistaken for a burrowing asp (*Atractaspis*), pp. 194–217, but identified by the speckling. Note the short tail.

Guinea-fowl Snake, *Micrelaps vaillanti*. **Above:** Tanzania (Bill Branch).

Black Snake-eater *Polemon ater*

A small, dark, rear-fanged burrowing snake. In forest and well-wooded savanna. Found from western Kenya south to Zambia, and west into eastern DR Congo. Average size 30–70cm. Might be mistaken for a burrowing asp (*Atractaspis*), pp. 194–217; it looks very similar. Note the very shiny body.

Black Snake-eater, *Polemon ater*. **Right:** DR Congo (Colin Tilbury).

Fawn-headed Snake-eater *Polemon collaris*

A small, dark, rear-fanged burrowing snake, usually with a yellow collar but not always. In forest and forest remnants. Found from western Uganda west to Cameroon, and south to southern DR Congo. Average size 40–80cm. Might be mistaken for a burrowing asp (*Atractaspis*), pp. 194–217; it looks very similar. Note the very shiny body.

Fawn-headed Snake Eater, *Polemon collaris*.
Above: Uganda (Jelmer Groen).

Common Purple-glossed Snake *Amblyodipsas polylepis*

Common Purple-glossed Snake, *Amblyodipsas polylepis*.
Left: South Africa (Bill Branch). **Right:** head pointing, Kenya (Stephen Spawls).

A fairly large, dark, rear-fanged burrowing snake. In coastal woodland and thicket, savanna and evergreen hill forest. Occurs from Kenya south to South Africa, and south-west to Angola. Average size 30–80cm; the southern African subspecies is larger than the eastern one. Might be mistaken for a burrowing asp (*Atractaspis*), pp. 194–217; it looks very similar, and may point its head in a similar way. Note the short tail and absence of a preocular scale (burrowing asps have preoculars).

Katanga Purple-glossed Snake *Amblyodipsas katangensis*

A small, dark, rear-fanged burrowing snake. In dry and moist savanna, woodland and forest. Occurs from Tanzania and Mozambique west to Zambia and DR Congo. Average size 40–90cm. Might be mistaken for a burrowing asp (*Atractaspis*), pp. 194–217; it looks very similar, and may point its head in a similar way. Note the short tail and absence of a preocular scale (burrowing asps have preoculars).

Katanga Purple-glossed Snake, *Amblyodipsas katangensis*. **Above:** Mozambique (Luke Verburgt).

Western Purple-glossed Snake *Amblyodipsas unicolor*

A small, dark, rear-fanged burrowing snake. In dry and moist savanna and woodland. Occurs from Kenya west to Senegal. Average size 0.5–1m. Might be mistaken for a burrowing asp (*Atractaspis*), pp. 194–217; it looks very similar, and may point its head in a similar way. Note the short tail and absence of a preocular scale (burrowing asps have preoculars).

Western Purple-glossed Snake, *Amblyodipsas unicolor*. **Above:** Ghana (Stephen Spawls).

Aboubakeur's False Smooth Snake *Macroprotodon abubakeri*

A small, rear-fanged terrestrial snake; largely nocturnal. In Mediterranean scrubland, woodland and lightly vegetated areas. In northern Morocco and Algeria. Four species in the genus, all very similar, occur right across the top of North Africa, from Western Sahara to Egypt, but don't penetrate into the desert. Average size 25–50cm. Might be mistaken for a small viper. Note the dark stripe through the eye and the round pupil.

Aboubakeur's False Smooth Snake, *Macroprotodon abubakeri*. **Above:** Algeria (Stephen Spawls).

Algerian Whip Snake *Hemorrhois algirus*

A harmless diurnal terrestrial snake; fast-moving. In semi-desert and scrubland, hills, wadis and grasslands, from northern Mauritania right across North Africa to Egypt, but nearly always north of the desert, near the coast. Average size 0.4–1m. Unlikely to be mistaken for any dangerous species if seen clearly. Note the slim body and dark cross-bars.

Algerian Whip Snake, *Hemorrhois algirus*. **Above:** Libya (Tomáš Mazuch).

Horseshoe Whip Snake *Hemorrhois hippocrepis*

A harmless diurnal terrestrial snake; fast-moving. In semi-desert and scrubland, hills and grasslands, from Morocco east to northern Tunisia, but north of the desert, near the coast. Average size 0.6–1.3m. Unlikely to be mistaken for any dangerous species if seen clearly, although the dorsal markings might lead to confusion with vipers. Note the slim body and circular back markings.

Horseshoe Whip Snake, *Hemorrhois hippocrepis*, **Above:** Morocco (Stephen Spawls).

Awl-headed Snake (Diademed Sand Snake)
Lytorhynchus diadema

Awl-headed Snake, *Lytorhynchus diadema*. **Left and right:** Egypt (Stephen Spawls).

A small, harmless burrowing snake. In dry savanna, semi-desert and desert. Occurs from Mauritania right across the Sahara to Egypt, then east to Iran. Average size 20–40cm. Might be mistaken for a small desert viper (*Cerastes*), pp. 96–100. Note the sandy colour and distinctive enlarged scale on the snout.

Diadem Snake *Spalerosophis diadema*

A fairly large, slender, harmless nocturnal or crepuscular snake; fast-moving. In dry savanna, semi-desert and vegetated or rocky areas in desert. The subspecies *Spalerosophis diadema cliffordi* occurs from Mauritania east to Sudan, north right across the Sahara (not north of the Atlas), then east from Egypt to Iran. Average size 0.6–1.5m. Unlikely to be confused with any dangerous snake, but a widespread and common species. Note the elongate, pear-shaped head and curiously fractured head scales.

Diadem Snake, *Spalerosophis diadema*. **Top:** Egypt (Stephen Spawls). **Bottom:** juvenile, Egypt (Stephen Spawls).

Flowered Racer *Platyceps florulentus*

A fairly small, slim, harmless, fast-moving diurnal snake. In semi-desert and dry and moist savanna. Occurs from northern Tanzania north to Egypt, and west to Cameroon. Average size 50–90cm. Might be mistaken for a juvenile cobra. Note the cross-bars and very large eye.

Flowered Racer, *Platyceps florulentus*. **Right:** Ethiopia (Stephen Spawls).

Smith's Racer *Platyceps brevis smithi*

A fairly small, slim, harmless, fast-moving diurnal snake. In semi-desert and dry savanna. Occurs from northern Tanzania north to Somalia. Average size 50–90cm. Might be mistaken for a juvenile Red Spitting Cobra (*Naja pallida*), p. 178. Note the cross-bars and very large eye.

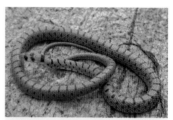

Smith's Racer, *Platyceps brevis smithi*.
Above: Kenya (Anthony Childs).

Cliff Racer *Platyceps rhodorachis*

A fairly small, slim, harmless, fast-moving diurnal snake. In semi-desert and dry savanna. Occurs from central Somalia through eastern Ethiopia to Eritrea; also widespread in the Middle East. Average size in Africa 50–90cm. Might be mistaken for a juvenile cobra. Note the cross-bars and very large eye.

Cliff Racer, *Platyceps rhodorachis*.
Above: Ethiopia (Malcolm Largen).

Western Crowned Snake *Meizodon coronatus*

A small, slim, harmless, diurnal terrestrial snake. In moist and dry savanna. Occurs from Senegal east to Chad. Average size 30–60cm. Might be mistaken for a juvenile Black Mamba (*Dendroaspis polylepis*), p. 126. Note the faint banding (more vivid in juveniles).

Western Crowned Snake, *Meizodon coronatus*.
Above: Nigeria (Gerald Dunger).

Semi-ornate Snake *Meizodon semiornatus*

A small, slim, harmless, diurnal terrestrial snake. In moist and dry savanna. Occurs from Sudan to South Africa. Average size 30–60cm. Might be mistaken for a juvenile Black Mamba (*Dendroaspis polylepis*), p. 126. Note the faint banding (more vivid in juveniles).

Semi-ornate Snake, *Meizodon semiornatus*.
Above: Tanzania (Michele Menegon).

Green snakes *Philothamnus*

About 20 species of harmless green snake occur throughout sub-Saharan Africa; we have described the seven most common or widespread. They mostly look fairly similar, being slim and green, with large eyes. They are liable to be confused with the green-coloured mambas (*Dendroaspis angusticeps, D. jamesoni* and *D. viridis*), pp. 121–26, and adult Boomslangs (*Dispholidus typus*), p. 226, from which they may be distinguished with care; green Boomslangs are hardly ever less than 80cm (smaller ones have the juvenile colouration), and have keeled scales and a teardrop-shaped pupil; green mambas have a long, coffin-shaped head and a relatively smaller eye.

Western Green Snake *Philothamnus irregularis*

A medium-sized, slim, harmless, diurnal arboreal snake. In moist and dry savanna, and occasionally forest. Occurs from Senegal east to Chad. Average size 0.6–1.1m. Might be mistaken for a green Boomslang (*Dispholidus typus*), p. 226, or a West African Green Mamba (*Dendroaspis viridis*), p. 125. Note the short, often blue head, large eye and smooth scales (never edged black).

Western Green Snake, *Philothamnus irregularis*. **Above:** Ghana (Stephen Spawls).

Angola Green Snake *Philothamnus angolensis*

A medium-sized, slim, harmless, diurnal arboreal snake. In woodland and moist and dry savanna. Occurs from Sudan and Cameroon south to Botswana. Average size 0.5–1.1m. Might be mistaken for a green Boomslang (*Dispholidus typus*), p. 226, Jameson's Mamba (*Dendroaspis jamesoni*), p. 123, or Eastern Green Mamba (*Dendroaspis angusticeps*), p. 121. Note the short head, large eye and smooth dorsal scales, with white or blue-white margins (never edged black).

Angola Green Snake, *Philothamnus angolensis*. **Above:** Tanzania (Michele Menegon).

Battersby's Green Snake *Philothamnus battersbyi*

A medium-sized, slim, harmless, diurnal arboreal snake. In woodland and moist and dry savanna. Occurs in highland Ethiopia and Kenya, southern Somalia, eastern Uganda and northern Tanzania. Average size 40–80cm. Might be mistaken for a green Boomslang (*Dispholidus typus*), p. 226, Jameson's Mamba (*Dendroaspis jamesoni*), p. 123, or Eastern Green Mamba (*Dendroaspis angusticeps*), p. 121. Note the short head, large eye, large pupil and smooth dorsal scales, with white or blue-white margins (never edged black).

Battersby's Green Snake, *Philothamnus battersbyi*. **Above:** Kenya (Stephen Spawls).

South-eastern Green Snake (Green Water Snake)
Philothamnus hoplogaster

A medium-sized, slim, harmless, diurnal arboreal snake. In woodland and moist and dry savanna. Occurs from Cameroon and northern DR Congo south to South Africa. Average size 50–90cm. Might be mistaken for a green Boomslang (*Dispholidus typus*), p. 226, Jameson's Mamba (*Dendroaspis jamesoni*), p. 123, or Eastern Green Mamba (*Dendroaspis angusticeps*), p. 121. Note the short head, large eye, yellow snout and smooth scales. There are often rufous or black dorsal blotches.

South-eastern Green Snake, *Philothamnus hoplogaster*. **Above:** Kenya (Stephen Spawls).

Thirteen-scaled Green Snake   *Philothamnus carinatus*

Thirteen-scaled Green Snake, *Philothamnus carinatus*.
Left: DR Congo (Eli Greenbaum). **Right:** Rwanda (Harald Hinkel).

A medium-sized, slim, harmless, diurnal arboreal snake. In forest and woodland from Nigeria east to Kenya. Average size 40–80cm. Might be mistaken for a green Boomslang (*Dispholidus typus*), p. 226, or Jameson's Mamba (*Dendroaspis jamesoni*), p. 123. Note the short head, large eye and black barring on the body.

Speckled Green Snake   *Philothamnus punctatus*

A fairly large, slim, harmless, diurnal arboreal snake. In semi-desert, dry and moist savanna and coastal woodland, not usually very far inland, from Somalia south to Mozambique. Average size 0.6–1.2m. Might be mistaken for a green Boomslang (*Dispholidus typus*), p. 226, or a green-coloured mamba (*Dendroaspis angusticeps, D. jamesoni* or *D. viridis*), pp. 121–26. Note the short head, large eye and raised 'eyebrow', often with black spotting. Very similar to the following species.

Speckled Green Snake, *Philothamnus punctatus*, **Above:** Kenya (Stephen Spawls).

Spotted Bush Snake *Philothamnus semivariegatus*

A fairly large, slim, harmless, diurnal arboreal snake. In semi-desert, dry and moist savanna and woodland, virtually throughout Africa south of 15°N. Average size 0.6–1.2m. Might be mistaken for a green Boomslang (*Dispholidus typus*), p. 226, or a green-coloured mamba (*Dendroaspis angusticeps, D. jamesoni* or *D. viridis*), pp. 121–26. Note the short head, large eye and raised 'eyebrow', often with black bars or spotting.

Spotted Bush Snake, *Philothamnus semivariegatus*. **Above:** South Africa (Tyrone Ping).

Black-lined Green Snake *Hapsidophrys lineatus*

A medium-sized, slim, harmless, diurnal arboreal snake. In forest and woodland from Sierra Leone east to Kenya. Average size 0.5–1m. Might be mistaken for a green Boomslang (*Dispholidus typus*), p. 226, or Jameson's Mamba (*Dendroaspis jamesoni*), p. 123. Note the short head, huge eye, heavily keeled scales and distinctive black lines along the body.

Black-lined Green Snake, *Hapsidophrys lineatus*. **Above:** Ivory Coast (Bill Branch).

Emerald Snake *Hapsidophrys smaragdinus*

A medium-sized, slim, harmless, diurnal arboreal snake. In forest and woodland from Guinea-Bissau east to Uganda, and south to Angola. Average size 0.5–1m. Might be mistaken for a green Boomslang (*Dispholidus typus*), p. 226, or Jameson's Mamba (*Dendroaspis jamesoni*), p. 123. Note the short head, huge eye, heavily keeled scales, distinctive black line through the eye and yellow underside.

Emerald Snake, *Hapsidophrys smaragdinus*. **Above:** Liberia (Bill Branch).

Large-eyed Green Tree Snake *Rhamnophis aethiopissa*

A large, slim, harmless, diurnal arboreal snake. In forest and woodland from Guinea east to Kenya, and south to southern DR Congo. Average size 0.6–1.3m. Very similar to the green and black colour form of the Boomslang (*Dispholidus typus*), p. 226, and a slight resemblance to Jameson's Mamba (*Dendroaspis jamesoni*), p. 123. Note the short head, huge eye, heavily keeled scales and black and yellow barring.

Large-eyed Green Tree Snake, *Rhamnophis aethiopissa*. **Above:** Rwanda (Fabio Pupin).

Spotted Tree Snake *Rhamnophis batesii*

Spotted Tree Snake, *Rhamnophis batesii*. **Above:** Gabon (Niels Rahola).

A large, slim, harmless, diurnal arboreal snake. In forest and woodland from Cameroon east to the Central African Republic, and south to the mouth of the Congo River. Average size 0.7–1.5m. Might be mistaken for a Boomslang (*Dispholidus typus*), p. 226, and a slight resemblance to Jameson's Mamba (*Dendroaspis jamesoni*), p. 123. Note the short head, huge eye, heavily keeled scales and black spotting.

Common Egg-eater *Dasypeltis scabra*

Common Egg-eater, *Dasypeltis scabra*.
Left: Namibia (Bill Branch). **Right:** Carpet viper mimic, Ethiopia (Stephen Spawls). **Right bottom:** Rwanda (Harald Hinkel).

A medium-sized, harmless nocturnal snake, found in trees or on the ground. In all sorts of habitat apart from true desert and montane forest. Occurs virtually throughout sub-Saharan Africa; also in parts of Morocco and Egypt. Recent taxonomic work has split the species (into *Dasypeltis confusa* and *D. sahelensis*), but the rhombic individuals are all similar. Average size 40–90cm. It is something of a mimic, in places resembling small adders, carpet vipers (*Echis*; pp. 83–95), and night adders (*Causus*; pp. 101–11). Note the short head, rhombic pattern and heavily keeled scales; it has a scale-rubbing threat display very similar to that of a carpet viper.

Large-eyed Snake *Telescopus dhara*

A small, rear-fanged nocturnal snake that is largely terrestrial but will climb. Very variable in colour. In moist and dry savanna, semi-desert and vegetated areas in desert. Occurs from northern Tanzania north throughout north-east Africa; similar forms (*Telescopus obtusus* and *T. tripolitanus*) are widespread over much of North Africa. Average size 40–90cm, possibly larger. Might be mistaken for a carpet viper (*Echis*), pp. 83–95. Note the large eyes and large head.

Large-eyed Snake, *Telescopus dhara*.
Above: Kenya (Stephen Spawls).

Tiger Snake *Telescopus semiannulatus*

A medium-sized, rear-fanged nocturnal snake that is largely terrestrial but will climb. In moist and dry savanna and semi-desert. Occurs from Kenya south to South Africa, and west to southern DR Congo, Zambia and Angola; similar species (*Telescopus beetzii* and *T. finkeldeyi*) occur in south-western Africa. Average size 40–90cm, possibly larger. Might be mistaken for a garter snake (*Elapsoidea*), pp. 129–42, or a Coral Shield Cobra (*Aspidelaps lubricus*), p. 187. Note the large eyes and large head.

Tiger Snake, *Telescopus semiannulatus*. **Above:** Tanzania (Stephen Spawls).

Variable Cat Snake (False Carpet Viper) *Telescopus variegatus*

A medium-sized, rear-fanged nocturnal snake that is largely terrestrial but will climb. In moist and dry savanna and semi-desert. Occurs from Senegal east to Chad. Average size 40–80cm. Might be mistaken for a carpet viper (*Echis*), pp. 83–95. Note the large head.

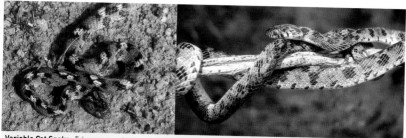

Variable Cat Snake, *Telescopus variegatus*. **Left:** Ghana (Stephen Spawls). **Right:** Nigeria (Gerald Dunger).

White-lipped Snake (Herald Snake/Red-lipped Snake)
Crotaphopeltis hotamboeia

White-lipped Snake, *Crotaphopeltis hotamboeia*. **Left:** Kenya (Stephen Spawls). **Right:** Botswana (Stephen Spawls).

A small, rear-fanged, nocturnal terrestrial snake. In woodland, forest clearings, moist and dry savanna and semi-desert. Occurs virtually throughout sub-Saharan Africa, and common everywhere apart from closed forest. Average size 30–70cm. If seen clearly it does not really resemble any dangerous species, but its habit of flattening the head, hissing and striking fiercely leads people to think it is dangerous; in Setswana it has the same name ('Kake') as the Mozambique Spitting Cobra (*Naja mossambica*), p. 175. Note the white spotting and large head. A very similar species in West Africa, *Crotaphopeltis hippocrepis*, has a black and white head.

Laurent's Green Tree Snake *Dipsadoboa viridis*

Lauren't Gree Tree Snake, *Dipsadoboa viridis*. **Above:** DR Congo (Kate Jackson).

A medium-sized, rear-fanged nocturnal tree snake. In forest and forest clearings. Occurs from western Uganda westwards right through the Central and West African forest to Guinea, and south to Gabon. Average size 0.5–1.1m. Might be mistaken for a green Boomslang (*Dispholidus typus*), p. 226, or a green-coloured mamba (*Dendroaspis angusticeps*, *D. jamesoni* or *D. viridis*), pp. 121–26, but it has vertical pupils. Note the large eyes and large head.

Hook-nosed Snake *Scaphiophis albopunctatus*

Hook-nosed Snake, *Scaphiophis albopunctatus*. **Left:** Kenya (Stephen Spawls). **Right:** threat display, Kenya (Tim Spawls).

A large harmless, diurnal terrestrial snake. In moist and dry savanna. Occurs from Zambia north to Kenya, and west to Ghana (and possibly further west). Average size 0.7–1.4m. Very variable in colour; grey, rufous or pinkish. Large adults might be mistaken for cobras; it has a sinister threat display, elevating the body and opening its black mouth wide. Note the large size and obvious pointed snout.

Smith's Water Snake *Grayia smythii*

Smith's Water Snake, *Grayia smythii*. **Left:** Nigeria (Gerald Dunger). **Right:** Liberia (Bill Branch).

A very large, harmless, diurnal semi-aquatic snake. In forest and moist and dry savanna, but always in or near water. Occurs from Tanzania north to Sudan, and west to Senegal. Average size 0.8–1.6m. Often banded, and dorsal colour very variable; may be black, brown or rufous. Juveniles are strongly banded. Large adults might be mistaken for forest cobras, and juveniles for garter snakes (*Elapsoidea*), pp. 129–42. Note the small eyes and round pupils.

Thollon's Water Snake *Grayia tholloni*

A fairly large, harmless, diurnal semi-aquatic snake. In woodland and moist and dry savanna, but usually in or near water. Occurs from Zambia north to Ethiopia and Sudan, and west to Senegal. Average size 0.6–1.1m. Large adults might be mistaken for forest cobras, and juveniles for garter snakes (*Elapsoidea*), pp. 129–42. Note the small eyes, round pupils and black barring on the lips.

Thollon's Water Snake, *Grayia tholloni*. **Above:** Uganda (Konrad Mebert).

Ornate Water Snake *Grayia ornata*

A fairly large, harmless, diurnal semi-aquatic snake. In forest and woodland, but usually in or near water. In Central Africa, from southern DR Congo to Cameroon. Average size 0.6–1.3m. Large adults might be mistaken for forest cobras. Note the small eyes, round pupils and black barring on the lips.

Ornate Water Snake, *Grayia ornata*.
Above: DR Congo (Konrad Mebert).

Olive Marsh Snake *Natriciteres olivacea*

A small, harmless, diurnal terrestrial snake. In woodland and moist and dry savanna, but usually in or near water. Occurs virtually throughout sub-Saharan Africa. Average size 25–40cm. Not really likely to be mistaken for any dangerous snake if seen clearly, but the black-barred lips might lead to confusion with a small forest cobra. Note the dorsal stripe.

Olive Marsh Snake, *Natriciteres olivacea*.
Above: Rwanda (Max Dehling).

Viperine Snake *Natrix maura*

A fairly small, harmless, diurnal terrestrial snake. In oases, semi-desert, hills and Mediterranean woodland, but usually in or near water. Occurs from Morocco eastwards to western Libya, but north of the desert. Average size 40–80cm. Not really likely to be mistaken for any dangerous snake if seen clearly, but it looks vaguely viper-like. Note the distinctive circular pupil and zigzag dorsal pattern.

Viperine Snake, *Natrix maura*,
Above: Morocco (Tomáš Mazuch).

Snakebite in Africa: the big picture

In the 25 years since Bill and I published the first edition of this book, many things have changed in the African snakebite sphere. Some things have got better, but many have got worse. Some pharmaceutical companies have stopped making antivenom, because it was unprofitable. Some antivenoms have increased dramatically in price. The supply of antivenom to many healthcare facilities has decreased. In some countries healthcare itself has noticeably deteriorated, particularly in those racked by conflict; poverty has increased and more people are at risk. In places, the pharmaceutical standards of antivenom production have slumped (or were never high), and corruption has meant that useless, substandard and inappropriate antivenoms, which have never been formally tested in a clinical trial, are being marketed. In financial terms, antivenom is an easy drug to sell, because 90% of snakebite victims are going to recover anyway, so you're guaranteed 90% success.

It's not all bad news. In many African countries, healthcare is steadily improving, and good antivenom is becoming more widely available. In 2007, snakebite was placed on the World Health Organization (WHO) list of neglected tropical diseases. It was removed, rather inexplicably, from this list in 2013, but reinstated in 2017 as a Category A Neglected Tropical Disease. This has heightened awareness and has led to some positive action. In 2012, an Australian-based charity the Global Snakebite Initiative was founded. This public-spirited body is taking action; read about it on their website. In 2018, Kofi Annan, statesman and former UN Secretary General, recorded a moving podcast on the significance of the snakebite problem and the importance of taking action, making the point that it is largely overlooked. The Lillian Lincoln Foundation has produced an important documentary film in connection with this, *Minutes to Die*, and screening can be arranged via minutestodie.com/the-snakebite-crisis. In 2019, two important events occurred: the Wellcome Trust announced an £80 million, seven-year research programme into snakebite, and WHO launched a well-funded, $130 million, three-phase initiative aimed at halving global snakebite deaths by 50% by 2030. The African Snakebite Research Group of the Liverpool School of Tropical Medicine is working in Kenya and Nigeria to establish self-sustaining regional hubs of snakebite expertise, aiming to reduce snakebite deaths and disability.

At present, the figures suggest that there are between 81,000 and 138,000 snakebite deaths per year, with another 400,000 suffering lasting disabilities; the Global Snakebite Initiative reckon that snakebite affects 4.5 million people globally every year, seriously injures 2.7 million and causes 125,000 deaths, 32,000 of them in Africa. In West Africa, it is estimated that close to 10,000 snakebite-related amputations per year are performed. Whichever statistics are correct (and it is very hard to quantify, for reasons we'll explain shortly), this is a major problem. Kofi Anan called it 'the biggest public health crisis you have never heard of'. Action is needed; it can be simply summed up as: heighten awareness and teach communities to reduce risk by changing behaviour; improve living standards; provide rapid, inexpensive medical treatment for all, including cheaper and more effective antivenom; develop new treatments to reduce tissue damage and retard the onset of life-threatening symptoms following a bite; and teach snakebite protocols in teaching hospitals in the tropical world. But, proverbially, this is easier said than done.

One problem with snakebite is its randomness. You can live all your life in a snake-infested area and hardly see a snake, let alone have one bite you. Although those who are at greatest risk from snakebite are those who are out and about in the countryside, working in fields and plantations, anybody can get bitten, young, old, male or female. There are many different snakes and, as we explain later, the venoms vary not only from species to species but also within the species itself. The circumstances of the bite also vary; every case is different. And, perhaps the most significant factor: it is largely the very poor, living in remote places, who suffer most; their voices are not heard, and politicians can ignore them. In this context, it is worth briefly considering the situation in Australia.

Australia is, unusually, a country where there are more species of venomous snakes than harmless ones. Seventy percent of Australian land snakes are front-fanged venomous elapids and in places they are often common, even abundant. And yet the number of snakebites is relatively small, and this is because most Australians are relatively affluent and medically and snake aware, via the media and public information. They sleep in houses that can be sealed against snakes, wear strong footwear when walking in wild places and use a torch at night. Farming in Australia is largely mechanised; people are not cutting the crops or planting with hand tools. The rugged cane cutters who encountered taipans (*Oxyuranus*) while harvesting with a knife in the fields of northern Australia have passed into history. Snakebites cannot be totally avoided, of course, but the average number of deaths from snakebite in Australia hovers at two to five per year. This is because efficient health systems ensure that antivenom is widely available, information on treatment is widely and quickly shared, most people are literate and have immediate access to fast personal or public transport, and high-level medical care is available to all, rapidly, for the bite and the aftercare. In many places in Africa, few or none of these benefits exist.

One cannot blanket-blame African countries. Most became independent only within the last 70 years, after colonial periods of wildly varying quality. Many are poor, cursed with harsh climates that are not conducive to agriculture and with few natural resources. Many African countries have been at war with their neighbours and among themselves; others have been subjected to ill-advised social experiments, or been taken over by non-elected soldiers or dictators. Talented individuals have risen rapidly to high office and then used that office to enrich themselves and their relatives; altruism may be rare among those who started life with little. Money intended to help the poor is embezzled. Even government medical officers are not immune from taking backhanders. The exploitation of Africa by insiders and outsiders continues unabated. And even genuine patriots may feel that there are somewhat more pressing problems to deal with than snakebite, which does not target any coherent group except the poor.

A number of other harsh factors affect the snakebite problem. Even under the most efficient conditions, antivenom is logistically difficult to produce. Oddly, it is still done by the original method: venom is injected into large animals (usually horses), which develop antibodies against the venom, and the blood from the horses is extracted and purified to produce the antivenom. So if you want to make antivenom, you have to find, catch, transport, identify with certainty and securely maintain relevant deadly snakes. You have to extract venom (a dangerous and highly skilled undertaking), and this then has to be carefully processed. You need to maintain horses or other suitable animals, and you need an efficient, high-tech, clean laboratory with competent and committed technicians, high-maintenance kit, constant electricity, clean water and unobstructed access to reagents to process the drug. These factors make antivenom expensive to produce. And when it is manufactured, it largely has to be transported and stored under refrigeration, the 'cold chain', and you don't know when it's going to be needed.

If someone has been bitten by a dangerous snake and received a lethal dose (and this is a crucial point), the appropriate antivenom is the only thing that will save them. Even if their lives are saved, the aftercare can be prolonged, complicated and horribly expensive. In addition (and this is a problem that has recently reared its ugly head in Africa), antivenom can be most profitably marketed by the unscrupulous. This is because, as previously mentioned, most people (some medical professionals estimate around 90%) could recover from snakebite without treatment, and bite cases rarely receive full clinical monitoring. A crooked medical rep could market ampoules of water, claim they contain antivenom, see them used and claim, convincingly, that 90% of victims treated with the product survived. Until relatively recently, medical centres in Kenya requesting antivenom were given an Indian product that was not raised against the appropriate snakes, had not undergone rigorous clinical testing and was quite inefficient. As Dr Colin Tilbury, eminent South African snakebite expert, has said, people tend to survive their snakebites despite the treatment!

Another problem is the lack of accurate data, largely as a result of victims living in remote places. People suffer snakebites and they live or they die, without the data ever being recorded. In many African countries, governments have no formal reporting systems for births or deaths, let alone individual illnesses, diseases or accidents. The Texan bitten by a rattlesnake hastens to a modern hospital, she is treated by a professional who knows how to deal with snakebite and her misadventure enters government statistics. In contrast, the poor farmer in the remote Sahel bitten by a carpet viper (*Echis*) goes to a local fetish priest, who performs rituals and uses herbs. If she recovers, he takes the credit. If, after a few days, he spots signs of systemic bleeding, he may inform her she has offended the gods and he can do nothing for her. She then sets off for the hospital. There may be no transport and she may have to walk. If she dies on the way, her death is unrecorded. Even if she gets there, no antivenom may be available, or the staff may know little about treating snakebite. No paperwork may be done. If she survives and makes it back, she is likely to have a huge medical bill and may be ostracised by her people, who believe that her snakebite indicates that the gods look on her with disfavour. All over Africa are the graves of snakebite victims who never reached hospital and whose stories died with them. Some seasoned medical professionals have attempted to gather data, extrapolate sensibly and present the big picture, but the fact remains that in many places in Africa (and elsewhere in the tropical world), no one knows with any great accuracy how many people are bitten by dangerous snakes, nor how many die or are crippled, or what species of snake are the main culprits. Snakebite statistics gathered by researchers on the ground often differ in orders of magnitude from those presented by government medical agencies.

Perhaps the biggest difficulty is economic. Big pharma, for good or ill, is largely driven by profit. The costs of producing, testing and marketing a drug are high. It makes financial sense to make a reasonably priced product that many affluent people (or at least people of an affluent nation) need, on a regular basis. Nobody is going to get rich, or keep the shareholders satisfied, with a drug that costs a lot to make, is hard to store and is needed on an erratic basis by the desperate poor who cannot afford it. A recent survey investigating new antivenom production found evidence of 11 registered trials and 36 papers published in medical peer-reviewed journals; only one of the trials and eight of the papers involved African snakes. Research indicates that only 2.5% of those who need antivenom in Nigeria have access to it, and the cost of treatment is well over a month's wages, often far more.

WHAT NEEDS TO BE DONE?

There may not be any profit in treating snakebite, but people, especially the rural poor, need help. Government, health ministries and individuals can do their bit. Obviously, money is needed, and it must be disbursed in a way that eliminates, or at least limits, the potential for corruption or misappropriation. However, this can also be a minefield if that money comes from a high-income country with strings attached; aid that is offered with the donor country exerting powerful control on how it is spent may be labelled neocolonialism.

The first step is to ensure that we have accurate data. We need to know (a) what snakes live where, and (b) how dangerous they are. It is our profound hope that this book goes towards solving that problem in Africa, but no one, least of all us, would suggest that the data is complete. Some indication of what can be done is shown in the intensive surveys by Dr Jean-François Trape and his team in the countries of West and Central Africa, getting local people to preserve snakes; the data they have gathered has been of immense use to us in the preparation of the maps in this book. A small but dedicated group of people made a difference. It is neither difficult nor terribly expensive to do; all you need is a small, expert team with a robust vehicle to get around the villages, talk to people, give some plastic drums of formaldehyde, collect the snakes later, pay the bonuses, top up the drums, curate the specimens, and analyse and publish the results.

Next, we need to know where people are being bitten, what by, the circumstances of bite cases, and the numbers; then sensible action can be taken.

This can be done in health posts. It starts with a training workshop for health professionals, showing them how to identify the common dangerous snakes; they are given posters and asked to record, from snakebite victims, how, when and where they were bitten, and asked if they could identify the snake. The number of bite cases, their locality, the possible culprit, the time of day and the season are recorded. Then you have data you can use for planning. If a regional coordinator (for all of southern Africa, or West Africa, for example) then puts together this data, a picture of who is being bitten by what and where will emerge. This will enable strategic planning; for example, what sort of antivenom is required, in what quantities, and how it should be distributed. From this then comes data on what snakes should be collected, and where, in order to make effective regional antivenom. The clinical effectiveness of this can then be monitored, and the results used to tweak the production process. Professionals who have treated snakebites should be encouraged to publish. Regional cooperation and funds will be needed. It's probably not going to happen, however. At the moment, a handful of pharmaceutical companies make antivenom, of variable quality, with venom sourced from a disparate variety of snakes, with little hard information on exactly where those snakes came from. The antivenom is then sold willy-nilly to whoever wants it.

As the case of Australia, mentioned on p. 280, shows, the really big step is to raise living standards and to prevent snakebite. This costs. But it can start with small steps. It may be beyond the capability of governments in the tropical world to provide closed houses for all their citizens (at present), but it shouldn't be beyond their capability to provide bed nets for all. It shouldn't be that difficult to provide cheap torches (flashlights), either with subsidised batteries or rechargeable batteries, or dynamo torches that light up when you squeeze the handle. No one seems to have thought of the benefits of providing subsidised boots to the rural poor, although a recent attempt to get farmers in India to wear snake-proof shoes was unsuccessful; despite the benefits, people wanted to stick to their comfortable sandals.

Awareness can be heightened, inexpensively. Not enough is being done. Both Bill and I have travelled widely in Africa. With a few honourable exceptions, we have rarely seen, in any public place, school or health centre, in any African country, a good poster showing local dangerous snakes in colour, with a vernacular text, and some pointers on what to do about snakebite. Likewise, good mobile apps with details for the layperson about snakebite need to be widely available. One such is already produced for southern Africa, and is available at africansnakebiteinstitute.com/app. The African Snakebite Institute, in fact, offers a really useful range of materials for the reduction and treatment of snakebite in southern Africa (see their website, africansnakebiteinstitute.com; nothing like this is available elsewhere in Africa). Communications in Africa may be poor, but lots of people have mobile phones, and community centres have noticeboards. Lives can be saved with posters, downloads and apps. In coastal Kenya recently, a snakebite victim arrived at a clinic; it was closed, but she noticed, on the board outside, a poster about snakebite, with a mobile number. She called, and was soon picked up and treated. Such posters are not that difficult to produce and distribute. Likewise, governments need to involve local healers. They are often trusted by their communities; they need to be brought on side. Training courses for local healers (with possibly a small financial inducement for them to attend) would ensure that they stop wasting time when a snakebite victim arrives, do some real first aid if relevant and push the victim to the local hospital, rather than carrying out useless rituals.

What is also needed is a network of clinics – basic health posts that are strategically positioned so that nobody lives more than 30km from one. Such clinics could have solar-powered refrigeration, and a medical technician with a mobile phone. Polyvalent antivenom could be brought in by motorcycle in a cool bag from a handful of regional hospitals that always have stocks. Some specialised medical teams, equipped with four-wheel-drive vehicles and dealing with problems like snakebite (and other medical problems like rabies and scorpion stings), would save much suffering. Countries could also think about producing their own antivenom. In Nigeria, Nigerian snakes are common and easily replaceable. At the very least, local snake farms can produce lyophilised (freeze-dried) venom, which is then sent to a country with an advanced antivenom-producing laboratory. A strategic set of snake farms could be established in Africa.

Improvements are occurring. The WHO strategy, as previously mentioned, intends to try to halve snakebite deaths and has funded studies that look at where the burden of snakebite is greatest: usually in areas where there are several species of deadly snake and medical coverage is poor. A broad band of such regions stretches across tropical Africa. Researchers are looking at enzyme inhibitors, chemicals that restrict the action of tissue-damaging toxins; novel inhibitors may be discovered. Research on the use of anticholinesterases, including their use in actual snakebite cases, indicates the potential for non-antivenom intervention to halt the postsynaptic paralysis induced by snake venom neurotoxins. It may be possible in the not-too-distant future to produce, relatively cheaply, recombinant snakebite antivenom containing relevant mixtures of antibodies that have been raised using cell-based expression systems. Such specific antivenoms are much better tolerated than animal-derived antivenom, which tends to contain a lot of therapeutically irrelevant antibodies. The development of target-specific 'humanised antibodies' for snakebite treatment, including those raised from human snakebite victims as well as test animals, offers exciting possibilities. Possibly in the future, an auto-injector system may be widely available for snakebite that delivers a cocktail of non-allergenic compounds that greatly retard or neutralise the action of the venom, giving time to get the victim to hospital without stimulating any adverse reactions. A cocktail of synthetic nanoparticles that bind to the most dangerous toxins in venom is being developed; it would theoretically be easily usable as first aid by anyone to slow down and partially neutralise the killer components in the venom, giving time to get the victim to hospital. Likewise, an easily usable treatment may be available that prevents the awful tissue damage caused by so many African snakes. The future beckons. It is not too late to hope the new initiative will involve pan-African cooperation, perhaps under regional leaders, who ensure that good data is gathered and regularly published. In addition, as we have mentioned elsewhere, something about snakes and snakebite could be routinely taught in educational establishments.

However, the lack of regional cooperation across Africa to date does not fill one with hope. And in the final analysis, it is not the treatment that is so important – once the patient has reached a decent medical centre, the battle is half over – but the prevention of snakebite. For this to happen, African countries need to put their backs into raising living standards. It's probably not going to happen quickly.

AFRICAN SNAKE VENOMS

Dealing with snakebite is a complicated business, but humanity's understanding of it is gradually improving. We have summarised here some useful information on snakebite and its effects in Africa, but for those whose business involves snakebite treatment in Africa, we suggest reading three main resources:

WHO/Professor David Warrell (2010) *Guidelines for the Prevention and Clinical Management of Snakebite in Africa* (we have extensively made use of in this book, with kind permission), available as a free download at apps.who.int/medicinedocs/documents/s17810en/s17810en.pdf

Dr Colin Tilbury (1996) 'A clinical approach to snakebite and rationale for the use of antivenom', *South African Journal of Critical Care* 9(1): 2–4.

Dr Roger Blaylock (2005) 'The identification and syndromic management of snakebite in South Africa', *South African Family Practice* 47(9): 48–53, available as a free download at ajol.info/index.php/safp/article/view/13271/91641

In addition, a useful paper, Edward R. Chu *et al.* (2010) 'Venom ophthalmia caused by venoms of spitting elapid and other snakes: Report of ten cases with review of epidemiology, clinical features, pathophysiology and management', *Toxicon* 56(3): 259–272, summarises treatment for venom in the eyes after an encounter with a spitting cobra.

Venoms have evolved on numerous occasions throughout the animal kingdom. Sometimes described as evolutionary novelties, their main purpose is to either facilitate, or protect the producing animal from, predation. Defensive venoms, designed to protect the owner (for example, that of bees), usually cause intense immediate localised pain that will distract a predator. However, the purpose of snake venom is to assist the animal with predation (with the exception of venom that is spat by spitting cobras, which is defensive; it deters predators from getting close; see Figure 31). Snake venom is not designed to protect snakes against

Fig 31: Mozambique Spitting Cobra, *Naja mossambica*, spitting (Wolfgang Wuster).

humans. Any snake that is forced to bite a human in perceived self-defence is liable to be killed, so it cannot be selected for in evolutionary terms; in fact, there is evidence that even very dangerous snakes often make strenuous efforts to avoid being in a situation where they are forced to bite in defence.

So why do snakes have venom? It isn't to make life hard for humans, but to rapidly immobilise prey. All snakes are predators. They have no limbs, claws or canine teeth, so they cannot hold, tear or otherwise subdue prey, and yet most eat relatively large animals, which must be killed quickly, before they damage the snake or escape. Non-venomous constricting snakes coil very rapidly around the prey; circulation is stopped instantly. Venomous snakes bite and deliver a relatively large shot of venom; the prey dies very rapidly, either in the snake's jaws or within a short distance, so it may easily be found. The venom may also assist by speeding up digestion. In appearance, snake venom is a modified saliva, a complex cocktail of bioactive compounds; it is usually a sticky white, yellow or colourless fluid.

Treating a snakebite is never routine. The venom varies from species to species, and within the species it may differ according to locality, time of year, what the snake feeds on, and with the size/maturity of the snake. The volume and toxicity of the venom varies. If a snake bites, the depth to which venom is injected and the location (muscle, fat, epidermis or blood vessel) all depend on the circumstances. In addition, the size, age and state of health of the victim, and their physiological reaction to a sudden injection of a venomous cocktail, may all differ. Shock, fear, delay in reaching the hospital and ill-advised local treatment also affect the symptomology. This all complicates the treatment of snakebite, and thus even carefully documented case histories from several examples of the same species can all appear remarkably different. Often the culprit is not known, and there are relatively small numbers of thoroughly clinically documented cases where the snake has been unequivocally identified (often ironically, the best-documented African snakebite cases are those suffered by snake keepers in developed countries, who present rapidly at fully equipped hospitals knowing exactly what bit them). Another problem consists of snakebite cases documented in the 'grey literature' or on the internet, often by medical amateurs and snake keepers, where psychosomosis may play a part. However, although the accuracy of such reports may be unverified, they are often the only literature that is available, and consequently members of the medical profession have to make occasional use of them. So snakebite is very different to a tropical disease like malaria, where many thousands of cases have been carefully documented by doctors and a clear clinical picture of symptomology, causes and treatment exists. In snakebite, the clinical picture is incomplete.

Ongoing research is uncovering more and more details about snake venoms. Venoms are often loosely grouped into neurotoxins ('nerve poisons'), haemotoxins ('blood poisons'), myotoxins ('muscle poisons') and cytotoxins ('cell poisons'), although the latter two terms have fallen out of favour with toxinologists and expert medical professionals. In simple terms, most African snake venoms contain one or more of eight main components: procoagulant enzymes, largely possessed by vipers, which prevent blood from clotting; cytolytic or necrotic toxins, which destroy cell membranes, leading to local swelling and blistering at the site of the bite; peptide toxins, which lower the blood pressure by opening up (dilating) blood vessels, causing shock; haemolytic and myolytic phospholipases A_2, which damage membranes, muscles, nerves and red blood cells; presynaptic neurotoxins, which damage nerve endings, and postsynaptic neurotoxins, which

Fig 32 (left top): Puff Adder, *Bitis arietans*, bite (David Warrell).
Fig 33 (right): Nine-year-old girl, nine months after being bitten by a Nigerian Black-necked Spitting Cobra, *Naja nigricollis* (David Warrell).
Fig 34 (left bottom): Bleeding from mucous membranes following carpet viper bite (David Warrell).

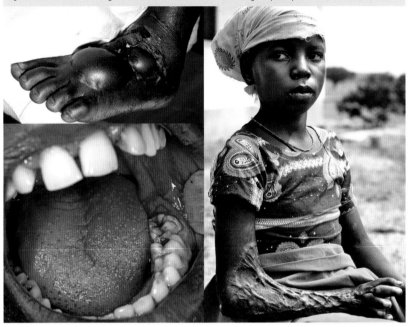

block the physiological transmitter acetylcholine, both leading to paralysis; haemorrhagins, which cause spontaneous systemic bleeding; and biogenic amines, which cause pain and affect permeability at the site of the bite.

Complicated and hideous though they sound, the components have purposes. They ensure that the prey animal, with stunning rapidity, becomes paralysed and shocked, its heart and blood vessels leaking or becoming blocked with blood clots immediately. Intense pain cripples the animal, and the combined effect means the unfortunate victim can only twitch briefly or stagger a short distance before death. The mamba (*Dendroaspis*) that seizes and bites a squirrel in a tree is banking on the animal dying before it can struggle around and bite the snake in the eye; the Puff Adder (*Bitis arietans*) that stabs a passing rat is relying on it dropping dead before it has gone more than a few metres.

In Africa, these eight components can lead to four main types of envenoming. Cytotoxic envenoming is characterised by painful and progressive swelling (see Figure 32), with blood-stained tissue fluid leaking from the bite wound, hypovolaemic shock from loss of plasma or blood into the swollen limb, blistering and bruising. The victim will complain of severe pain at the bite site and throughout the affected limb, and painful and tender enlargement of lymph glands draining the bite site. Irreversible death of tissue (necrosis/gangrene) may occur, leading to horrible scarring (see Figure 33). African snakes that cause this type of envenoming include carpet or saw-scaled vipers, Puff Adders, desert vipers of the genus *Cerastes*, Gaboon and Rhinoceros Vipers (*Bitis gabonica* and *B. rhinoceros*, and *Bitis nasicornis*), and spitting cobras.

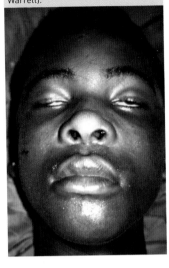

Fig 35: Ptosis following Black Mamba, *Dendroaspis polylepis*, bite (David Warrell).

Haemorrhagic envenoming is characterised by internal and external bleeding, from the gums (see Figure 34), gastrointestinal and genito-urinary tracts, and recent and partly healed wounds. African snakes that cause this include the carpet vipers, desert vipers of the genus *Cerastes*, Gaboon and Rhinoceros Vipers, the Boomslang (*Dispholidus typus*) and vine snakes (*Thelotornis*).

Neurotoxic envenoming, affecting the nervous system, is characterised by moderate or no local swelling, and progressive descending paralysis, starting with drooping eyelids (ptosis) (see Figure 35) and paralysis of eye movements, causing double vision. There may be painful and tender enlargement of lymph glands draining the bite site. The patient may vomit, the saliva may become profuse and stringy, and eventually there may be difficulties with swallowing and breathing. African species involved include mambas and non-spitting cobras. Bites from these snakes are very problematic as they cause what Royjan Taylor, Director of Bio-Ken Snake Farm, described as 'a huge problem in a very short time'. Unusually among African vipers, the venom of the South African Berg Adder (*Bitis atropos*) also has a neurotoxic component. If the victim survives a bite from a neurotoxic snake, however, recovery is usually complete, without tissue destruction.

Finally, myotoxic envenoming from sea snakes is characterised by negligible local swelling, increasing generalised muscle pain and tenderness (myalgia). This is associated with features of neurotoxic envenoming and progressive descending paralysis, culminating in paralysis of breathing. However, only a single species of sea snake, the Yellow-bellied Sea Snake (*Hydrophis platurus*), reaches the African coast. It is very rare and we are not aware of any African bite cases.

Occasionally, mixed types of envenoming may occur. Mixed cytotoxic and neurotoxic bites occur in the case of the Rinkhals (*Hemachatus haemachatus*),

Fig 36: Triangular head of a Green Bush Viper, *Atheris squamigera* (Konrad Mebert).

but not with spitting cobras of the genus *Naja*. There can be mixed haemorrhagic and cytotoxic bites from carpet vipers, North African desert vipers of the genus *Cerastes*, and Puff Adders.

The venom itself is produced and stored in glands that secrete into the snake's mouth, and several gland types are identified. There is esoteric debate about the classification of these glands ('rictal glands', 'Duvernoy's glands', etc.). However, in the really dangerous front-fanged snakes, the nasty stuff comes from the 'venom glands', located behind the eye and above the mouth. When these snakes bite, a large muscle (the *muscularis compressor glandulae*) compresses the glands and venom is pushed under pressure down a duct into a hollow fang, and thus injected into the victim. The venom glands may be large and store a considerable volume; in vipers and some spitting cobras, these glands contribute to the triangular head shape (Figure 36). In some species, notably the night adders (*Causus*) and burrowing asps (*Atractaspis*), the venom glands may extend some distance along the neck.

AVOIDING SNAKEBITE

In Africa, snakebite is a constant deadly hazard faced by all rural and most urban dwellers. As discussed above, reducing snakebite risk lies largely with raising living standards. Many rural dwellers farm using short tools, squat or sit while working, and move around barefoot and bare-limbed, and without lights at night. They sleep on the ground in homes that cannot be sealed. They face a high risk of snakebite all their lives, and only their governments have the ability to deliver large-scale help that will eventually reduce the risk to all.

Many accidental snakebites in Africa are inflicted on the leg, below the knee, when the person treads on the snake. Other major bite sites are the hand or wrist. A sleeping person may be bitten anywhere, and sometimes multiple times. The health worker concerned with preventing snakebite will be aware of the impracticality of telling the rural poor to always use a torch at night, not to use short farming tools, to always wear strong footwear and to seal their sleeping rooms against snakes. However, they can help people lower the incidence of snakebite by offering practical advice.

In general, where feasible, the two lists of precautions (pp. 290–92) will reduce the risk of snakebite, although we are aware that they may not always be practicable or affordable; if you sleep in a brushwood or mud building you are at risk. Do what you can. Note also, there are no effective snake 'repellents' and those who sell them are trying to cheat you. Snakes are eukaryotes, advanced multicellular vertebrates; any chemical strong enough to poison or repel a snake can also poison and repel humans. Burning tyres near your home or spreading used engine oil is not going to deter snakes, only create a toxic environment that may affect the health of you, your family and your stock. Likewise, do not waste your money on the various 'snakebite first aid kits' that are available. Most contain nothing useful, and many are positively dangerous, consisting of, inter alia, blades, rubber tourniquets and devices that supposedly suck out the venom. We do advise familiarising yourself with the dangerous and common snakes of your area, using this book and the maps and checklists in the Appendix (pp. 313–24), as well as regional guides if available, so you know what you are dealing with. You should also find out which of your nearest health facilities has antivenom in stock, and learn how to get there.

And, of great importance, spread your knowledge. Do not keep it to yourself. If you know something about preventing and treating snakebite, or how to identify snakes with certainty, tell your friends and neighbours; teach and show by example. You will save lives and benefit humanity. Education is the way forward. If something about snakes, snake identification and snakebite was routinely taught in schools, there would be fewer snakebites and fewer severe consequences, and more innocuous snakes would survive. But caution is needed. Like many ideas, putting this into practice isn't easy; few people know enough to be able to teach about snakes with confidence, and (as we mentioned in the look-alikes section, pp. 252–78) identification is not always easy and a mistake can have far-reaching consequences; this is a difficult issue.

Preventing snakebite in the home

- If possible, and affordable, try to make your home snake-proof. A hidden snake inside a building is a hazard. Ensure that the doors fit snugly, especially at the bottom; consider fitting a rubber strip (with the lip outside) if there is a large gap under the door. Keep windows closed, especially at night; or fit them with wire gauze (this will also keep out mosquitoes). From time to time check the gauze is unrusted and sound. Put up mosquito nets (impregnated with insecticide if possible) and tuck them under the mattress; elasticated nets won't slip out at night. Never sleep on the floor. Fit a ceiling, so that tree snakes in the roof space cannot descend.
- If your home cannot be sealed (and even if it can, but it has been open during the day), then before you go to sleep, check that there are no snakes inside, under the beds, against the walls or resting up in the rafters. Ensure that all bed nets are tucked in. Likewise, make a quick check for snakes when you get up. If you have to get up in the night, use a torch or lamp, and never put your feet down onto the floor without looking. Shake out your shoes, and the children's shoes, before putting them on. Do the same with clothes, and don't leave clothing on the floor. Get into a safe routine and stick to it.
- Teach your children, as early as possible, to be snake aware and to never touch or play with a snake, even a tiny one. Alert them to always tell an adult, forcefully, if they see a snake.
- If practical, compounds and gardens should be kept free of hiding places for snakes. Such places include piles of stones, bricks, firewood, grass, rubbish tips, pits, etc. Open holes near homes should be blocked (especially any associated with termite mounds, or squirrel warrens). Keep the grass short or the ground cleared around your home.
- Large trees or bushes that touch against houses should be cut back (tree snakes will use them as passageways). Cut away the lower branches of thick bushes, and clear away leaf litter from beneath.
- Rubbish tips and rock piles also attract rats and lizards, which may in turn attract snakes. If practical, clear these from around the home.
- Don't have dripping taps or open water sources; snakes may come to drink in the dry season, or frogs may come for the water and snakes for the frogs.
- Domestic fowl, rabbits and cage birds kept outside or on verandas may attract snakes. Keep your stock away from your home. However, poultry (especially guinea fowl and turkeys) sometimes attack and kill snakes. Likewise, domestic cats will also warn of a snake, and may even kill them. A number of bird species (especially bulbuls, barbets and sunbirds) will make alarm calls if they see a snake, and may mob it. Domestic animals (especially poultry) making a noise at night may indicate a snake is among them. Be very careful if you go to investigate.
- Open food may attract rats, which attract snakes. Keep your silos away from the house and make them rat-proof; in the home, store food in rat-proof containers.
- Be wary when moving things in store rooms or outhouses, especially if they are piled up just off the ground, for example on a pallet. A snake may be hiding underneath. If you lift something from storage or off the ground, roll it towards you, not away, so a snake hiding underneath doesn't rush towards your feet.

- If you have an open veranda or stoep with furniture, look carefully before you sit down, and shift the cushions in case a snake has taken refuge there. A visitor at a Kenyan coast hotel sat in a veranda chair without noticing a Boomslang coiled on the arm.
- If feasible, and affordable, in a particularly snake-infested area you can keep snakes out of the compound by fitting a barrier fence of stout shade cloth or fine-mesh wire netting. Attach it to posts at least a metre or so high, and sink it to a depth of 20–30cm. Ensure the gates fit snugly. This should largely prevent ground-dwelling snakes entering the compound. Some home owners even fit motion sensors to the outside of their house, and cleared the ground between the house and the fence, although some may regard this as excessive.

Preventing snakebite outside and in the field

- When walking, always look carefully where you are going. It isn't always easy to see snakes: look at our pictures of concealed snakes (Figures 37–40). Keep your eyes open! Use a lamp or torch at night, and carry a stick to prod the ground in front of you. Take particular care while walking after a rainstorm, especially after dark. Don't blunder through tall grass, overhanging bushes or thick cover, or up through rock outcrops. Do not put your hands or feet into places you can't see, in particular under objects lying on the ground, or into piles of rock, logs or vegetation. Take care stepping over rocks or logs; it is better to step onto them and then down, rather than right over. Check the ground before sitting down, especially at the base of a tree, and when you sit down, don't lean back into a recess without checking it.

Fig 37 (left top): North-east Africa Carpet Viper, *Echis pyramidum*, hiding in soil crack (Stephen Spawls).
Fig 38 (right top): Katanga Bush Viper, *Atheris katangensis*, concealed in a bush (Frank Willems).
Fig 39 (left bottom): Camouflaged Puff Adder, *Bitis arietans* (Stephen Spawls).
Fig 40 (right bottom): North-east Africa Carpet Viper shuffled down in sand (Stephen Spawls).

- If possible, wear strong footwear that covers the foot and the ankle; long trousers also help. Commercial snake gaiters are available.

Fig 41: Nigerian snake handler – never agree to hold a dangerous snake (David Warrell).

- If you are farming, be very careful if you are digging or harvesting with short tools, and be very careful when you sit or squat. In the savanna and Sahel areas of West and Central Africa, agricultural workers sitting or squatting to make yam mounds, or weeding, are particularly at risk, especially in the early rainy season.
- Don't gather firewood or move material on the ground at night.
- Never tease or play with snakes, or molest them. Don't pick up or play with a supposedly dead snake – some species sham (feign) death, and even if fatally injured, snakes can still bite and kill; in one case, a Puff Adder head gave a venomous bite 30 minutes after it was cut off.

Likewise, be wary of snake charmers. Many are experts, but some are not; audience members have been bitten by charmers' snakes. Stand away from the snakes, and never offer or agree to hold one (Figure 41). If you have to kill a snake, use something long, such as a long stick, hosepipe, panga, machete or whip; or throw a rock or use a gun. Don't play with the body.

- If you meet a snake at close quarters, try to remain calm and stand still. Don't lash out at it or make threatening gestures. Stay calm and move backwards carefully. Snakes do not make unprovoked attacks.
- Never sleep on the ground outside (or inside).
- Avoid deliberately running over a snake on a road. It may not be killed and may pose a hazard to other road users; it may also be injured and become trapped under the vehicle, to emerge later when the vehicle has stopped.
- If working or travelling in remote country, try not to go alone; let people know where you have gone and when you will be back. A snakebite victim on their own is at much greater risk than when a member of a group.
- If you are on safari or an expedition, and sleeping in a tent, it is important to keep the tent zipped shut both when you are inside and when you are away. A snake prowling at ground level may follow the edge of the tent and go inside. If the tent has been left open, check your bedding before you get into it, and check that a snake isn't hiding somewhere else in the tent. When you pack up the tent, be careful that a night snake isn't sheltering underneath. If you go outside at night, be very careful, and use a light. Shake out your clothing and shoes in the morning. Don't plunge your hand into a bag or box that has been left open. Listen to your local guides, who will have advice on snake activity and are skilled at spotting snakes.

Following all these suggestions may seem difficult, but once good habits are developed they are easy to maintain. If you live in an area where there are dangerous snakes, it makes sense to take precautions and stick to them. The personal and financial cost of a snakebite can be enormous. A bite can kill or cripple you for life. Even the affluent are at risk (although less so than the rural poor). An oil geologist in northern Kenya felt it was safe, just after dusk, to walk 10m in open sandals, without a torch, from her cabin to the mess room. It wasn't. She trod on a North-east African Carpet Viper (*Echis pyramidum*) and was bitten. Her subsequent medical evacuation and treatment cost her company in excess of $60,000. Do not step out into the dark, no matter how small the errand.

In a few areas of Africa (notably South Africa, and, to a limited extent, Kenya) there are professional outfits that will come and remove problem snakes, and some of these offer training. Businesses and large educational establishments might consider sending staff on one of these courses. For those who would prefer to capture rather than just kill a problem snake, a large plastic bucket, a long pair of snake tongs (also called a grabstick) and some gaffer or duct tape, plus a pair of goggles for a spitting snake, will enable the removal of the majority of dangerous snakes; the snake is picked up with the tongs, dumped in the bin, and the lid is stuck down (see Figures 42–44); the snake is later released away from humans. Even just a hook will do to remove a large viper.

Fig 42 (top): Safe handling – holding the snake with tongs (Johan Marais).
Fig 43 (centre): The snake is placed in the bin (Johan Marais).
Fig 44 (bottom): A Puff Adder, *Bitis arietans*, can be handled with a hook (Johan Marais).

However, one thing is worth remembering for those who visit Africa's wild places. Snakebite is generally not a significant risk in Africa if you can take basic precautions. Far more people die of diseases like malaria or AIDS, or in vehicle accidents, than are killed by snakes. In South Africa in 2017, over 14,000 people were killed in road accidents, 6,000 died from lung cancer and other diseases related to smoking, 270 were struck and killed by lightning, and fewer than 20 and perhaps as few as 10 died from snakebite. Fear of snakebite should not put off the potential visitor to the African bush. In rural areas, the local people may be at much greater risk (for the reasons detailed above), but, nevertheless, compared to the risk from diseases like malaria, death from snakebite is a much smaller danger.

WHO IS AT RISK, WHEN AND WHERE?

Venomous snakes are like land mines, a deadly hidden danger in the countryside, randomly and indiscriminately putting all at risk. That risk can be reduced. But it cannot be totally removed; venomous snakes cannot be eradicated and in fact their removal would have possibly catastrophic effects on the balance of nature. Many snakes eat rodents, which destroy stored grain; killing all the snakes would cause a dramatic increase in rodents. Snakes are also secretive and thus tend to persist even in well-developed areas. Recently, a relatively large Puff Adder was found in the grounds of a primary school in the southern suburbs of Nairobi; it had been there for a while but had obviously never been seen.

Those at greatest risk, who suffer most from snakebite, are rural dwellers, especially subsistence farmers, often young adults, who live and farm in African lowlands and are the most active, often in agriculture. Many cannot afford closed houses, torches or strong shoes; few can afford the ruinous costs of thorough snakebite treatment, and their nearest medical facility may be more than a day's walk away. The relatively well-off Abuja or Johannesburg resident who is bitten by a snake while on a day out in the countryside will be driven rapidly to a well-equipped hospital and can afford much more than the basic treatment. The outcome to their bite is likely to be much more favourable than that of the poor peasant farmer who is bitten while harvesting in a remote place where the nearest clinic is a long journey away by bicycle or on foot.

Certain seasons and times carry more risk of snakebite than others. In Africa, hazardous seasons are: (a) the rainy season, especially at the start, and particularly after a long dry season; (b) spring in northern and southern Africa; (c) late autumn in southern Africa, when snakes are looking for places to hibernate and for a last meal; and (d) after rainstorms in the desert and Sahel, especially in spring, summer and autumn. Many snakebites also occur when farmers begin to plough and to plant, during harvesting, and during the dry season, when 'bushmeat' hunters burn the bush and hunt small game by checking holes.

The most hazardous time for snakebite is the hour before total darkness and the first two hours after dark, when night snakes emerge and are active on the ground. This, of course, is also the time that people are most active at night. Late at night, fewer snakes will be active, due to falling temperatures. Snakes are also often active just after rainstorms, especially if the rain falls during late morning or early afternoon.

Are there particular habitats in Africa where human–snake encounters, and consequently bites, are more likely to take place? Rigorous data is lacking, but anecdotal evidence indicates that most snakebite accidents occur not, as one might suppose, in the great forests, where snakes are believed to be cunningly waiting everywhere, but in the savanna and semi-desert. This might be a biased view, based on the fact that many of Africa's eminent snakebite doctors have often been based in savanna areas, but an observation by Dr Patrick Malonza, Director of Herpetology at the National Museum in Kenya, may be significant. In a forthcoming book on Kenya's reptiles and amphibians, Malonza notes that in Kenya most snakebites occur not in forested areas, but in dry savanna, and suggests this is because in forest the snakes can live in places that are not easily reached (up in trees, hidden in deep leaf litter or buried) and thus avoid encountering humans. However, in the dry savanna the snakes and humans are in the same place, on the ground, and thus much more likely to encounter one another. This is an idea that invites investigation.

WHAT HAPPENS WHEN A SNAKE BITES AND HOW BAD WILL IT BE?

A frightened snake will try to bite if threatened, injured or restrained. If the snake isn't dangerous, the bite may cause some punctures, cuts and scratches. These may bleed freely and can be severe, especially from a large harmless snake like a python (*Pythonidae*) or a large Mole Snake (*Pseudaspis cana*). If a dangerous snake strikes, it does not always bite; it may miss, by accident or design, or simply bang the intruder with its snout.

If a snake does bite, venom may be injected, but the amount varies, from a much larger than lethal dose to none at all (so-called 'dry bites'). If one or two fangs are embedded in the victim, venom is usually injected simultaneously, emerging from the fang through a small hole near the tip, into the tissue (see Figure 45). Snakes strike quickly, especially the vipers, and on contact the mouth is snapped shut and the fang driven in, injecting the venom. Most vipers have long fangs, so the venom is sent deep into the tissues. Elapid fangs are not as long, so the venom is nearer the surface. Some snakes may actually chew on the victim, to either inject more venom or, in the case of the rear-fanged snakes (which have grooved, not tubular, fangs, set back in the mouth), to engage the fangs and improve the flow of venom into the wound (see Figure 46). Burrowing asps have a bizarre way of biting, pushing the head past the target and sliding a fang out of the mouth, which is then driven in by a backward pull; hence most victims of burrowing asp bites have only a single fang mark.

If venom enters directly into a blood vessel, its effects may be rapid and catastrophic. However, it is usually injected into adipose tissue, or sometimes into muscle fibre; a short-fanged snake may inject only into the dermis. Such rapid injection usually results in a small sac of venom squeezed into a small area of tissue. If the movement of that tissue is greatly reduced, the entry of venom into the bloodstream or lymphatic system may be slowed down, making more time

Fig 45 (left): Puff Adder, *Bitis arietans*, with venom emerging above the fang tip (Stephen Spawls).
Fig 46 (right): Northern Stripe-bellied Sand Snake, *Psammophis sudanensis*, fangs at the rear of the upper jaw; circled (Stephen Spawls).

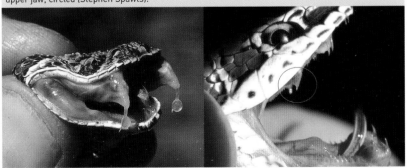

available to get the victim to medical help. Hence immobilisation therapy with a pressure pad (see the 'Snakebite first aid' section, pp. 297–300) may be beneficial, although there is debate about this, as most victims have run or moved rapidly before this is started. It should also be noted that, barring direct injection into a vein or an allergic reaction by the victim to the venom itself, snakebites by large non-spitting elapids are most unlikely to kill in under three hours, and in viper and spitting cobra bites, death in less than 24 hours is most unlikely. Some time is usually available to get the victim to hospital.

The symptoms, their severity and the outcome of a bite by a dangerous snake will depend on a number of factors. Occasionally no venom is injected so nothing will happen, although victims of such bites (and even people who have been bitten by harmless snakes) are known to have experienced not only alarming symptoms of shock, but also appear to display known symptoms of venomous snakebite (in particular, neurotoxic symptoms; it is hard to fake the swelling and discolouration of adder bites). So remember that a bite from a dangerous snake does not necessarily mean a dangerous snakebite.

Perhaps the most significant factor is the volume of venom injected. A bite in defence may involve less venom than a feeding bite. In general, the bigger the snake, the more venom is delivered (there is a strange myth that young snakes deliver more venom than adults; this is incorrect). Whether the snake embedded one fang or two, struck more than once, or hung on and chewed will affect the amount of venom injected. The age, size and state of health of the bitten person are also significant: small children, the very old and those with depressed immune systems are particularly at risk. The ratio between the amount of venom and the victim's mass is significant here; the larger the ratio, the more serious the bite. The site of the bite is also significant; a face or neck bite may lead more rapidly to mechanical obstruction to breathing. A bite over a bone may impede fang penetration and impair or prevent venom injection. An intravenous bite will be rapidly catastrophic. A bite on a toe or finger will not be as serious as one on the trunk or upper limb, but is more likely to cause necrosis and the need for amputation.

Actions taken following the bite may affect the outcome. Although there is debate about first aid (see the 'Snakebite first aid' section, pp. 297–300), sensible support saves lives. The time between the bite and the start of medical treatment is very significant. In Africa, victims may be a long distance from medical help, and may not decide to try to get there until severe systemic symptoms appear, by which time the prognosis is worse. Snakebite victims in Africa often (in some areas, always) visit local healers before going to hospital. No one can blame them; the local healer is usually nearby, and the hospital may be many hours away. Treatments offered by local healers may involve various harmless rituals and applications of potions to the skin – but some may involve incision and the placing of unsterile substances into the wound. Other preparations may be swallowed to induce vomiting or have a laxative effect. These healers have no effective snakebite cures. Their reputation rests with those victims who would have recovered anyway. However, they are often very skilled at spotting systemic symptoms – in West Africa, they watch most carefully for signs of internal haemorrhage. When a patient develops systemic symptoms, he or she is then informed that for some imagined reason their case cannot be treated further. The victim then goes to hospital, meaning that the physician sees the case after a long delay, which considerably complicates the treatment.

SNAKEBITE FIRST AID: DOS AND DON'TS

Dr Roger Blaylock, Zimbabwean snakebite expert, has said that there is no good first aid for snakebites. This is a rather extreme view, but is based on the medical fact that a snakebite victim who has received a lethal dose will survive only if antivenom is administered in suitable quantities. However, this does not mean that nothing should be done other than getting the victim rapidly to hospital. The likelihood of a snakebite victim surviving might depend on action taken before the victim reaches medical help. It is also vitally important that a snakebite is recognised as a medical emergency. If someone has been bitten by a snake, all other plans should be abandoned and safe, rapid transport to hospital should be organised. Do not wait to see if anything is going to happen. Take action immediately.

In the past (and even recently), many recommended first aid procedures not only wasted time but may also have been highly detrimental to the patient's health. So, to start, here are the things you should *not* do after someone has been bitten by a snake. They might be summarised thus: *leave the wound alone.*

- Do not waste time taking the victim to a traditional healer. None have any effective remedies. Do not allow yourself to be persuaded that anyone other than a qualified medical professional can do anything useful about a snakebite.
- Don't make any cuts, either across, along or near the bite. Some venoms cause non-clotting blood, and uncontrollable bleeding may ensue; you may also damage nerves, blood vessels or tendons, and introduce infection.
- Never apply a tourniquet. In adder bites, these greatly increase tissue damage, and they may precipitate a compartment syndrome, kill limbs and increase haemorrhage.
- Don't inject potassium permanganate solution, magnesium sulphate or any random chemicals, or rub potassium permanganate into the wound. These chemicals have no neutralising effect and may cause tissue damage or poisoning of their own.
- Don't pack the wound with ice, or try to keep it cold, and do not rub, massage, heat up or tamper with the wound site; this may encourage absorption of the venom.
- Don't give an electric shock with a stun gun, cattle prod, car plug lead or any other improvised electrical device; it has no effect on the venom and the shock may traumatise or even seriously hurt the victim.
- Don't bother with poultices, herbs, snake stones, etc. They are all useless. However, gently fixing a black stone to the fang punctures with a plaster may reassure the victim.
- Don't give alcohol (except perhaps in very small amounts, for reassurance), and do not give pain relievers containing aspirin, which reduces platelet adhesiveness.
- In general, you should not try to find, catch or kill the snake. Two bites are worse than one. If the snake is still around, warn everyone loudly and get them to move away. If feasible, without taking any risks, take a photograph of the snake. However (you may need to use your judgement here), if the snake is still around and can be killed without further risk to anyone, consider doing so (read the section entitled 'Identifying a snake' in the Introduction, pp. 18–28). When the snake is dead, scoop the body into a hard, closable container, shut

it securely and take it to the hospital. Although snakebites can be treated without the identity of the snake being known, the dead snake, if identifiable, may (a) enable medical staff to decide if the snake is venomous or not, (b) give them an idea of how much time they have, and (c) assist them with symptomatic treatment and strategies for antivenom use.

Fig 47: It won't work – no herbal remedies as yet are of any use (Stephen Spawls).

- Do not start taking unqualified advice. If someone has been bitten by a snake, the only person qualified to give further advice is a medical professional. Do not start looking for help on the internet, particularly on social media forums; you may be offered persuasive and yet idiotic and actively harmful advice and attract the attention of trolls. Do not even consider homeopathic or herbal remedies (see Figure 47). Do not start randomly asking on the internet for an identification. There are a handful of internet forums where you can get good advice on snake identification; see the Appendix (pp. 311–12).

What *should* you do? This section could be summarised: *make the patient comfortable, reassure them, and get them to hospital as quickly and safely as possible.*

- Organise safe, rapid transfer to hospital. Use your phone and any available transport. If feasible, use the flying doctor, or call for an ambulance, if available. Call the hospital, to warn them you are on your way with a snakebite.
- Remove rings and other restrictive jewellery, and loosen tight clothing, boots, etc. If the bitten areas swell, these may cause problems.
- Get the victim to lie down immediately, in the transport (or on a bed if waiting for transport), on their side in the recovery position, and if possible reduce movement of the affected area. Reassure them and try to get them to relax. Quote the statistics on snakebite – most victims recover without treatment, you've got plenty of time, etc. Point out how important it is to stay calm and relaxed. Keep talking, stay calm yourself, and don't leave the victim alone if possible.
- If the bite is from a non-spitting cobra or a mamba, it may help to put on a pressure pad (but not a pressure bandage; recent work indicates that the inexpert application of pressure bandages wastes time and does little good). This is easily done and may be life-saving. A firm pad (folded cloth, rolled bandage or rubber; see Figures 48–50) is held gently over the bite site, and

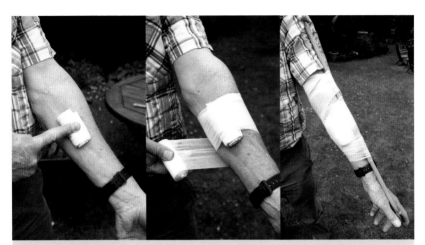

Fig 48: Apply a pad of any available material directly over the bite wound (Laura Spawls).

Fig 49: Secure the pad tightly with an inelastic bandage (Laura Spawls).

Fig 50: Immobilisation: splint the bitten limb to prevent movement at any of its joints (Laura Spawls).

strapped in place using an inelastic bandage. If possible, the limb is then splinted. This reduces the speed with which the venom enters circulation. If the bandage and splint are applied properly, they will be comfortable and may be left on for several hours; they should not be taken off until the victim reaches hospital – their removal may cause venom to move quickly into the bloodstream, so the doctor should have the appropriate drugs ready before the bandage is taken off.

- If possible, without taking any risks, take the dead snake with you (see the final point on p. 297), or take numerous close-up smartphone pictures of the body and head, above and below, that you can show to the doctor treating the bite.
- The victim should be carried or transported by vehicle, with the bitten limb (if the bite was on a limb) immobile. If this is impossible, try to minimise movement of the bitten limb (although obviously if you're alone and have been bitten on the foot, you're going to have to use it).
- If necessary, be prepared to give artificial respiration. In a severe cobra or mamba bite, the lungs stop working but the heart continues to beat, and artificial respiration may keep the victim alive until hospital treatment and a respirator become available (a danger sign of impending respiratory failure is if the victim cannot blow out a match held at arm's length). The best method is by using a bag valve mask (Ambu bag), coupled with an artificial airway; any medical professional who deals with elapid bites should have the appropriate training for this. Manual artificial respiration can be used in an emergency. This is best learnt from a qualified instructor, but important points are: the method used should be mouth-to-mouth or mouth-to-nose, the victim's head should be tipped right back, and the initial rate should be 30 times per minute, and then 15 times per minute (but even four or five good breaths per minute will suffice). Start mouth-to-mouth or mouth-to-nose when and if the patient starts to struggle to breathe. Ensure the airway is not obstructed by a prolapsed tongue or aspirated vomit.

SPITTING SNAKES: FIRST AID FOR VENOM IN THE EYES

The most effective treatment if a spitting cobra or Rinkhals has spat venom into someone's eye is to wash the eye gently but thoroughly with large quantities of water, as soon as possible; this will remove residual venom. Stay calm while organising this; the victim isn't going to die, and as the eyes are essential organs treatment must be gentle and careful. It is better to take a little longer to organise a bucket and some help, rather than quickly squirting a hose in someone's eyes; you don't want to cause further damage. The easiest way to wash out the eye, if the victim can stand, is for them to hold the eye under a gently flowing tap, rotate the eye and move the head around, while holding the eyelid open, so water enters from all directions and reaches all parts of the eyeball. Alternatively, get the victim to put their face underwater in a large bowl or sink (if there is venom in both eyes, this is the best method), and then hold the eyes open and rotate them, at the same time blinking vigorously. If the victim is in severe pain and can't control themselves, lay them down on their back and get someone to hold them while someone else gently pours (not squirts) water into the eyes. Treatment should be started as soon as possible after the accident, but will be beneficial even if started many hours later, and should be continued for at least 20 minutes. In the absence of water, fluids such as beer, soft drinks, cold tea, saliva or even urine can be used, and milk is very suitable and soothing. However, never use potassium permanganate solution or petrol, which could damage the eye.

It may be helpful to thoroughly wash down the face, neck and hands, while the victim keeps their eyes closed, in case any venom is still present and then gets accidentally wiped into the eye (although venom on unbroken skin is not harmful, nor is venom in the mouth, provided it is spat out promptly). Anaesthetic eye drops would be most useful for the pain, but should be applied only once. The victim may find it difficult not to rub the eye, and if it is anaesthetised much damage could be done, so keep their hands away from their eyes. Put a soft cotton wool pad and a soft bandage over the eye, or wear a pair of very dark glasses. Analgesic (painkilling) drugs can be taken. Venom in the eye is not life-threatening, merely very painful; antivenom must not be injected or put into the eye. However, the victim should visit an eye clinic or a hospital, where slit lamp microscopy and/or fluorescein staining may be undertaken to see if there is any corneal damage. Such damage often occurs, and until the cornea regenerates it is prone to secondary bacterial infection. If possible, always use an antibiotic eye ointment for prophylaxis.

TREATMENT OF SNAKEBITE AT A MEDICAL CENTRE: INITIAL PROCEDURES

These two sections are intended for the guidance of the medical professional treating a snakebite case. They are not clinically comprehensive. There is excellent comprehensive literature available on snakebite in Africa, as detailed in the Appendix (pp. 310–11) and at the beginning of this section. Much material is available as free downloads; the medical professional involved in treating snakebite is advised to read and file them, and to examine the pictures showing symptomology in Professor Warrell's monograph for the WHO (referenced on p. 285, at the start of the section on 'African snake venoms'). Much of the material here is taken from this monograph, and gratefully acknowledged.

Obvious though it may seem, a bite case victim should be seen immediately upon arrival at hospital; someone waiting in a queue might die. The health professional must ascertain, straight away: (a) has the victim definitely been bitten by a snake, (b) was the snake venomous and (c) is the patient in danger of dying rapidly? If the victim has definitely been bitten, and received a lethal quantity of venom by a snake for which antivenom is available, then antivenom is going to be needed. However, the use of antivenom involves risks. Thus an asymptomatic, weakly symptomatic or shocked/terrified victim might need initial monitoring, possibly for 24 hours or more, in a bed in full view of medical staff. Those who are treating a suspected snakebite case might also want to consider the following:

- If the victim didn't see what hurt them, and the injury was followed by immediate intense or burning local pain but no bleeding, then it might have been caused by a scorpion or some other venomous invertebrate. Snakebites, even if serious, don't always cause immediate intense local pain; scorpion stings don't bleed but always do cause immediate pain (but bear in mind the possibility of imagined intense pain from a genuine snakebite).
- Even if the bite is serious, systemic symptoms rarely appear in under 30 minutes, except some local pain, discolouration and swelling. So if a patient presents rapidly after a bite with symptoms (and signs) such as pain (tenderness), nausea (vomiting), postural dizziness (hypotension), syncope, inability to breathe (dyspnoea), palpitations (irregular pulse) and anxiety, sweating and dry mouth, these may be a result of fear or shock, and can often be alleviated by reassurance, immobilisation and a warm, sweet drink; a physician might consider a placebo injection, or strapping on a black stone. But remember, a small minority of snakebite victims do react suddenly to venom, and death can follow rapidly.
- Fangs are sharp and may be long; they can tear and puncture the skin, and may break off in the wound, as may solid teeth. Pain akin to being stabbed with blunt or dirty pins may be present and can be quite unpleasant, especially if a fang has snapped off in the wound, but such pain is not connected with the action of the venom.

In general, the most effective medical treatment for a life-threatening snakebite is the intravenous administration of snakebite antivenom. However, in non-life-threatening cases, or for snakes not covered by antivenom (there are a number), symptomatic treatment should be carried out. For life-threatening cases involving an elapid with a neurotoxic venom (large cobra or mamba), ventilation is probably going to be necessary; for life-threatening cases involving a snake with a cytotoxic venom (Puff Adder, Gaboon vipers, Rhinoceros Viper, spitting cobra) or one that causes clotting abnormalities (carpet vipers), blood fluid volumes have to be maintained.

The victim should be asked the following four questions. Devised by Professor David Warrell, these are usefully diagnostic, as detailed following the actual questions.

'In which part of your body have you been bitten?'

Look at where the patient points. There may be evidence that the patient has been bitten by a snake (for example, fang marks). If there are two punctures, the distance between them may give clues to the size of the snake. Look for signs of local envenoming (for example, local swelling, bruising or continuing bleeding from the fang punctures), but also evidence of prehospital treatment (for example, impressions made by a tourniquet or incision marks that may be bleeding, suggesting that the blood is incoagulable). Sometimes the victim may not have been aware that they were bitten; for example, at night during sleep, or in the dark, or in water. In such cases, suspicion of the diagnosis will depend on typical signs such as fang puncture marks, progressive swelling, bleeding gums or descending paralysis.

'When were you bitten?'

Assessment of the severity of envenoming depends on the length of time between the actual bite and when the patient seeks treatment. The patient may seek treatment immediately after the bite, before symptoms and signs of envenoming have developed. Or, the patient may arrive so long after the bite that the only signs are of late consequences of envenoming (for example, gangrene, pneumonia or renal failure).

'Where is the snake that bit you?' or 'What did the snake look like?'

In some parts of Africa, the snake responsible for a bite may be killed and brought to hospital with the victim. If the snake is available, its identification can be extremely helpful but requires an expert who can identify local snakes unerringly. If the snake is obviously a harmless species (or not a snake at all), the patient can be reassured, possibly given an injection of tetanus toxoid and discharged from hospital immediately. Descriptions of the snake by bite victims or onlookers are often unreliable and misleading, but it is worth asking about the snake's size, colouring, markings and behaviour. The surroundings where the bite occurred and the time when it happened can also suggest a particular species; for example, spreading a hood (cobra), hissing loudly (Puff Adder or other large viper), rasping/fizzing sound (carpet viper), snake in a tree (mamba or Boomslang), or small black snake on the ground (burrowing asp). A bite inflicted on someone sleeping in their hut at night is likely to have been caused by a spitting cobra or burrowing asp. Bites in and near rivers, lakes and marshy areas are also most likely to be caused by cobras.

'How are you feeling now?'

The patient's current symptoms can point to what is likely to be the most important effect of envenoming (for example, faintness or dizziness indicating hypotension or shock, or breathlessness indicating incipient respiratory failure). Patients should be asked to describe their symptoms and should then be questioned directly about the extent of local pain, swelling, tenderness; painful, enlarged lymph nodes draining the bite area; bleeding from the bite wounds, at sites of other recent injuries and at sites distant from the bite (gums, nose etc.); motor and sensory symptoms; vomiting; fainting; and abdominal pain. The time after the bite when these symptoms appeared and their progression should be noted. Details of prehospital treatment (tourniquets, ingested and applied herbal remedies, etc.) should also be recorded as these may, themselves, be responsible for some of the symptoms.

SYNDROMES OF ENVENOMING

In many (perhaps most) bite cases, the snake has not been brought along. However, it is often useful if a clinical syndrome of envenoming can be identified, as this will assist with clinical management. There are four major syndromes of envenoming, in snakes that are liable to cause death or major disfigurement, and two minor syndromes, associated with snakes that are unlikely to cause major problems. These six syndromes are outlined below.

Major syndromes

- Marked local swelling with coagulable blood. There is painful, rapidly progressive or extensive local swelling, sometimes with blistering and necrosis, with coagulable blood (detected by the 20WBCT) and an absence of spontaneous systemic or persistent local bleeding. This is suggestive of bites by spitting cobras, especially if the patient was bitten indoors at night, or Puff Adders, especially if the patient is shocked. A rare cause is the Berg Adder (southern Africa only), especially if the patient has paralysis. Treatment: polyspecific antivenom and circulating volume repletion; or supportive only if Berg Adder is proven.

- Marked local swelling with incoagulable blood and/or spontaneous systemic bleeding. There is painful, rapidly progressive or extensive local swelling, sometimes with blistering and necrosis, with incoagulable blood (detected by the 20WBCT) and often spontaneous systemic bleeding (gums, nose and gastrointestinal/genito-urinary tracts) or persistent local bleeding at the bite site. The syndrome is strongly suggestive of bites by carpet or saw-scaled vipers (northern third of Africa only, north of the equator) or Sahara Horned Vipers (*Cerastes cerastes*; Sahara desert only). It may also be caused by some populations of Puff Adders (East and southern African savanna region, but not West Africa). Rare causes may be Gaboon vipers (rainforests) and bush vipers (*Atheris*). Treatment: south of the Sahara and north of the equator, use monospecific *Echis* antivenom; anywhere in Africa, use polyspecific antivenom.

- Progressive paralysis (neurotoxicity). There is negligible, mild or only moderate local swelling with progressive, usually descending, paralysis. This is strongly suggestive of bites by neurotoxic (non-spitting) cobras or mambas. Treatment: polyspecific antivenom; monitor respiratory function carefully, and intubate and assist ventilation if/when necessary.

- Mild or negligible local swelling with incoagulable blood. This is a rare syndrome. There is mild or absent local swelling with incoagulable blood (detected by the 20WBCT) and often spontaneous systemic bleeding. This is suggestive of bites by Boomslangs or, even more rarely, by vine snakes. Treatment: monospecific antivenom for Boomslang; supportive treatment for vine snakes. Note: the likelihood of the average person being bitten by a Boomslang or vine snake is almost non-existent.

Minor syndromes

- Mild swelling alone. There is mild local swelling alone, rarely involving more than half of the bitten limb, with negligible or absent systemic symptoms. This is suggestive of bites by night adders or burrowing asps, and some bush vipers, desert vipers (*Cerastes*) and the small species of the genus *Bitis*. In burrowing asp bites in light-skinned victims, some greenish discolouration often appears around the puncture wound after some hours (Figure 51). Treatment: no antivenom; palliative treatment only.

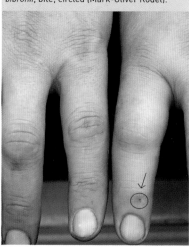

Fig 51: Bibron's Burrowing Asp, *Atractaspis bibronii*, bite; circled (Mark-Oliver Rödel).

- Moderate to marked local swelling associated with neurotoxicity. This syndrome is characterised by moderate to marked local swelling associated with neurotoxicity. Although a rare clinical entity, it is characteristic of Berg Adder bite. This syndrome has also been observed in the bites of some other small adders in the genus *Bitis*, such as Péringuey's Adder (*Bitis peringueyi*) and the Desert Mountain Adder (*B. xeropaga*). Treatment: no antivenom available; palliative treatment only.

When examining the patient, note the following. Local swelling and enlargement and tenderness of regional lymph nodes are often the earliest signs of envenoming (but note swelling may be due to a tourniquet or other restrictive device). Compare the bitten and unbitten limb. Swelling, within two hours, usually follows most cases of significant envenoming by African vipers and spitting cobras, but is usually negligible in the case of some bites by mambas and non-spitting cobras (and also Boomslangs, vine snakes and sea snakes). The only highly venomous snakes with greatly delayed symptomology are the Boomslang and the vine snakes. Persistent bleeding from the fang marks, other recent wounds and venepuncture sites suggest that the blood is incoagulable. Check the gums thoroughly as these are usually the first site of spontaneous systemic bleeding. Shock may occur, shown by a fall in blood pressure, collapse, cold, cyanosed and sweaty skin, and impaired consciousness. The foot of the bed should be raised and an intravenous infusion of isotonic saline, a plasma expander or fresh frozen plasma should be started immediately.

A bite from a mamba or non-spitting cobra may result in neurotoxicity/paralysis, and the onset may be rapid. The earliest symptoms of neurotoxicity after such bites are often blurred vision, a feeling of heaviness of the eyelids and apparent drowsiness. Whether snake venom toxins can exert any direct central effect on the level of consciousness is controversial, but it is unlikely that they cross the blood–brain barrier. The *frontalis* muscle is contracted, raising the eyebrows and puckering the forehead, even before ptosis can be demonstrated. Respiratory muscle paralysis with imminent respiratory failure is suggested by dyspnoea, distress, restlessness, sweating, exaggerated abdominal respiration, central cyanosis and coma. If there is any suggestion of respiratory muscle weakness, objective monitoring should be attempted by measuring peak expiratory flow, forced expiratory volume in one second (FEV1), vital capacity or forced expiratory pressure using the mercury manometer of a sphygmomanometer. Coma is usually the result of respiratory or circulatory failure.

WHEN WILL ANTIVENOM BE USED?

- If there are obvious signs of severe adder envenomation; that is, extensive and increasing swelling, 15cm or more beyond the bite site in less than two hours, or where swelling in the arm extends onto the chest wall, or where swelling from foot bites extends above the knee. Systemic signs such as blood-stained saliva, hypotension or markedly unstable blood pressure, abnormally long coagulation time and external bleeding are absolute indicators for antivenom.
- Initially severe or progressive neuromuscular paralysis demands early administration of antivenom.
- Suspected or confirmed spitting cobra bites with signs of swelling, if they present within six hours of the bite.

Be much readier to give antivenom to children, or anyone with a small body mass. It is almost never too late to try antivenom treatment for persistent systemic envenoming; it has proved effective in reversing coagulopathy 10 days or more after carpet viper bites.

It should be noted that Boomslang and vine snake bites represent a special case. The venom of these snakes causes continual bleeding, due to a coagulation defect, which may not be apparent until several days after the bite. Polyvalent antivenom is not effective against their venoms. Specific Boomslang antivenom (a freeze-dried preparation) can be obtained from the South African Vaccine Producers (SAVP; address details given in the Appendix, in the section entitled 'Current producers of snakebite antivenom useful for Africa', pp. 309–10), who can arrange for it to be flown rapidly anywhere. No antivenom covers the vine snakes, and treatment will consist of whole blood transfusion, infusion of fresh frozen plasma and platelets. However, these two snakes are inoffensive in nature, and bites are most unlikely.

GIVING ANTIVENOM

Snakebite antivenom is best given intravenously, slowly, at room temperature in a saline drip, over 30–60 minutes. Using this method, it usually appears within a few minutes at the area of the bite. Antivenom injected intramuscularly takes too long to appear in the bloodstream. The use of antivenom, however, carries the risk of untoward, unpleasant or even life-threatening reactions. So the use of antivenom 'just in case' following a snakebite (or suspected bite) is totally inadmissible. Antivenom must be used only in cases where envenoming has definitely taken place and specific criteria for its use are fulfilled.

Reactions to antivenom range at around 20% or more depending on the dose and the purification process used in its manufacture. The commonest are mild febrile and/or urticarial skin reactions, but a small percentage will go on to a more severe anaphylaxis, which may involve shock, bronchospasm or swelling of the tongue, gums, throat and face. Serum sickness, seven to 10 days after the injections, presents as itching rashes, slight temperature and joint pains, and may be treated by antihistamines and, if severe, with steroids.

Acute antivenom reaction (anaphylaxis) is much rarer. This is a severe, life-threatening, generalised or systemic hypersensitivity reaction, characterised by rapidly developing life-threatening airway, breathing and/or circulation problems, usually associated with skin and mucosal changes. It presents with a rapid fall in blood pressure and collapse in a few minutes. Persons prone to this reaction are those with a history of previous exposure to horse serum and those with a history of allergy, especially with a history of infantile eczema or asthma. It must be treated with adrenaline. The usual protocols apply: the patient lies flat, the legs are raised and 0.5ml adrenaline 1:1000 is injected intramuscularly; this is repeated after five minutes if there is no improvement.

The antivenom should be added to the drip and allowed to infuse slowly, over at least 20 minutes (30–60 minutes is better, as previously mentioned). The dose will depend on the product used, but in severe cases the minimum dose for an effective antivenom is about 50ml (i.e. five 10ml ampoules). In severe elapid and Gaboon viper bites, the initial dose could be up to 120ml or more. In any case, the situation and clinical response should be reassessed at 4–6 hours to determine the need for further antivenom. The antivenom may counteract the spread of swelling, and abnormal coagulation values should be corrected with an adequate dosage.

In addition to antivenom therapy, central venous pressure should be monitored, and a check kept on urinary output, with a catheter if necessary, as renal failure is a danger in some snakebites; dialysis may be necessary. Blood fluid volumes must be maintained. For adder bites, the limb should be placed in the most comfortable position. Fasciotomy (cutting of tissue to relieve pressure) is no longer recommended in most cases. Impending respiratory paralysis from neurotoxic venoms may be difficult to reverse quickly with antivenom; be prepared to give artificial respiratory support (ventilation) until the antivenom becomes effective, and consider the use of anticholinesterases where relevant. A suction device may be necessary where salivation is pronounced. Painkilling drugs (but NOT morphine or central nervous system depressants) can be given. At a late stage, if tissue damage is severe, necrotic tissue should be surgically excised to prevent secondary infection with anaerobic bacteria, such as tetanus. Every snakebite patient, including those bitten by non-venomous snakes (or non-snakes), should be given a booster dose of tetanus toxoid. The use of antibiotics is not usually indicated in the average snakebite, and when it is, prescription should be guided according to results of culture and sensitivity, if the wound has been incised or if there is frank necrosis.

USE OF ANTIVENOM BY A LAY PERSON

Giving snakebite antivenom is the business of trained medical personnel, for important reasons. To be effective, in a case where the victim is likely to die without antivenom, the antivenom must be given intravenously. However, this may provoke an anaphylactic reaction that can kill.

However, lay persons living, working or travelling in remote areas of Africa may feel the need to carry or store snakebite antivenom, and may need to be able to use it. We offer this advice for desperate cases, where hospital is a long way away, antivenom is available and the patient appears to be dying. It is of the utmost importance that you should know how to use antivenom correctly, and the possible outcomes and risks. Always remember that it is best to try to get a snakebite case to a medical centre. If you have antivenom, and needles or syringes, take them with you.

A snakebite kit should contain at least four 10ml ampoules of polyvalent snakebite antivenom; six ampoules would be better. Polyvalent antivenom acts against a range of snake venoms. A list of antivenoms and suppliers are given in the Appendix (pp. 309–10); in sub-Saharan Africa, the antivenoms from South African Vaccine Producers (SAVP) probably have the best reputation. Make sure you get the right antivenom for your area; in the northern half of Africa the antivenom should cover carpet vipers. The kit should also contain two or three each of 2ml syringes, 5ml syringes and 10ml syringes, with sterile needles to fit them. It should contain two 1ml ampoules of adrenaline 1:1000; these are used to counter allergic shock (anaphylaxis). A couple of ampoules of sterile water are worth having – if you get a few, you can practise breaking them and filling a syringe, a useful skill in a tight situation. You might consider practising with an expired ampoule of antivenom as well. A spill at the wrong time would be catastrophic. Read the instruction leaflet that comes with the antivenom carefully, so you have some ideas in advance.

The antivenom must be kept cold; if it gets hot it may lose its potency. If kept between 2°C and 10°C, it should stay potent for years in excess of the expiry date stamped on the ampoule. Freezing it will not affect its potency but may cause the ampoule to crack. If kept at warmer temperatures, the antivenom will gradually lose its potency; if kept in a hot car boot it can go off quite quickly. Loss of potency is indicated by the antivenom becoming cloudy, and a white condensate may appear. As a general rule, if it is clear it is potent. Check your antivenom periodically; hold it up to the light (don't confuse cloudiness with the condensation that will appear on a cold ampoule in warm air). If you want to take antivenom into the field, pre-cool it, keep it in an ice box and return it to the fridge when you get back to base.

Judging whether or not antivenom is needed after a bite may be difficult, especially since a snakebite victim in a remote place, knowing help is far away, is likely to become distressed and exhibit alarming symptoms of fear and shock, become hysterical or even faint. Such behaviour must be rapidly controlled. Point out that the more agitated the victim becomes, the quicker the venom will reach the bloodstream. A subcutaneous injection of saline, sterile water or tetanus toxoid in the guise of antivenom will do no harm and may well be calming.

You will have to use your judgement, but in general, antivenom should be used only if there are obvious signs of systemic neurotoxic envenoming (drooping eyelids, difficulty with speech, difficulty in breathing – probable culprit mamba or non-spitting cobra) and medical help is still more than 30 or so minutes away. Adder bites are far slower acting; it is almost unheard of for adder bites to kill in less than six hours, but antivenom use might be justified if the victim was bitten by a Gaboon viper or a large Puff Adder, swelling has become gross, the saliva is blood-stained and medical help is more than six hours away.

If you are going to inject antivenom, you must question the victim beforehand about whether they have a history of allergy – in a field team operating in remote areas, find this out in advance. Ask if they have had infantile eczema, or suffer from asthma. Ask also if they have had antivenom or similar products before. Such persons are prone to anaphylactic shock. Recent research indicates that for all snakebite cases where antivenom is going to be injected, a syringe of the appropriate dose of adrenaline 1:1000 (0.5ml for adults and children over 12, 0.3ml for children of 6–12 years, and 0.15ml for a child under six) should be prepared, and pre-treatment with subcutaneous low-dose adrenaline (0.25ml) is safe and significantly reduces severe anaphylactic reactions, by 75% or more.

Antivenom is often in glass ampoules. Check carefully (preferably beforehand) how to open the ampoule; some unscrew, some snap off, in some you put the needle through the lid, and in others you need to file a little on one side before pushing the neck off. Check you have the bit of metal for filing. It is best to draw up the liquid from the neck of the ampoule with the syringe and then put the needle on. If your hands are shaky, it might be easier to suck it up through the needle, but antivenom is glutinous stuff and this may take a while. When the syringe is full, hold it with the needle vertically upwards, tap it with your fingernail to make any air bubbles rise, and express a drop through the needle. Now you are ready to inject. Injections should be made intramuscularly. The best site is in the middle third of the thigh, at the upper outer edge of the *vastus lateralis* muscle. Several small injections should be made, slowly, and firmly massaged after each injection to speed up absorption. Watch very carefully as you inject for signs of a severe anaphylactic reaction – signs will be (not necessarily in this order) difficulty in breathing, wheezing, swelling of the face and neck, itchy rashes and red blotches on the face and body, swelling and excessive pain at the site of the injection, and faintness (indicating a fall in blood pressure). If these appear, discontinue injecting antivenom and inject adrenaline intramuscularly. If the signs of anaphylaxis then abate, continue with a little more antivenom; if the signs come back, stop giving antivenom – the victim may survive the snakebite, but has little chance of surviving a severe anaphylactic reaction in the bush.

If antivenom has to be given, a competent paramedic may consider giving the injections intravenously. However, remember that the antivenom must be given slowly, that those prone to anaphylaxis may experience a rapid and severe reaction to an intravenous injection of antivenom, and that adrenaline to counter this must be given intravenously.

Finally, remember that in most cases you won't have to give antivenom; it will be sufficient to take it with you to hospital. Even if you do have to give antivenom, the likelihood of having to deal with an anaphylactic reaction is small.

A list of antivenom producers is given in the Appendix (pp. 309–10).

Appendix

CURRENT PRODUCERS OF SNAKEBITE ANTIVENOM USEFUL FOR AFRICA

The list below covers commercially produced antivenoms that can be purchased. Their efficiency varies considerably, and some are efficient for one species and not for another; some have no evidence of being clinically tested. For those interested, we recommend reading the 2017 paper by Harrison *et al.* on preclinical testing of the efficiency of African antivenoms ('Preclinical antivenom-efficacy testing reveals potentially disturbing deficiencies of snakebite treatment capability in East Africa', *PLOS Neglected Tropical Diseases* 11(10): e0005969), available as a free download here: ncbi.nlm.nih.gov/pmc/articles/PMC5646754

In addition to the list here, some other producers may be found via the Munich AntiVenom Index (MAVIN) at toxinfo.org/antivenoms/Index_Product.html

South African Vaccine Producers (SAVP), South Africa
Modderfontein Road, Sandringham, Johannesburg, South Africa; PO Box 28999, Sandringham 2131, Johannesburg, South Africa
Tel: +27(11) 3866063/2, fax: (011) 386-6016, website: savp.co.za
Produce a polyvalent antivenom active against venom from *Dendroaspis polylepis*, *Dendroaspis angusticeps*, *Dendroaspis jamesoni*, *Naja nivea*, *Naja annulifera*, *Naja haje*, the (*Naja melanoleuca*) complex, *Bitis gabonica*, *Naja mossambica*, *Bitis arietans* and *Hemachatus haemachatus*; and two monovalent antivenoms, against the venom of the *Dispholidus typus* and the *Echis ocellatus* group.

MicroPharm Ltd, United Kingdom
Station Road Industrial Estate, Newcastle Emlyn, Carmarthenshire, SA38 9BY, United Kingdom
Tel/fax: +44(0) 1239710529, email: enquiries@micropharm.co.uk, website: micropharm.co.uk
Produce EchiTAbG antivenom for use against the bite of *Echis ocellatus/ Echis romani.*

Instituto Clodomiro Picado, Costa Rica
Next to the Sports Park, Dulce Nombre de Coronado, San Jose, Costa Rica
Tel: +506 25117888, email: icp@ucr.ac.cr, website: icp.ucr.ac.cr
Produce EchiTAb-plus-ICP antivenom, active against the venom of *Echis ocellatus*, *Bitis arietans* and *Naja nigricollis*; and working on a new product, EchiTAb+, against the venom of *Echis ocellatus, Bitis arietans, Naja nigricollis, N. mossambica, N. annulifera, Dendroaspis polylepis* and *Hemachatus haemachatus.*

King Fahad National Guard Hospital, Saudi Arabia Riyadh
Saudi Arabia; National Antivenom and Vaccine Production Centre: PO Box 22490, Riyadh 11426, Kingdom of Saudi Arabia
Tel: +966 11 80 45626, email: navpc@ngha.med.sa, website: antivenom-center.com

Produce a polyvalent antivenom against the venom of the following Arabian-sourced snakes: *Bitis arietans, Echis coloratus, E. carinatus, Naja arabica, Cerastes cerastes* and *Walterinnesia aegyptia.* They also produce a bivalent antivenom, active against the venom of *Naja arabica* and *Walterinnesia aegyptia;* this is said to also neutralise the venom of *Naja haje, N. melanoleuca, N. nigricollis* and *N. nivea.*

Premium Serums and Vaccines PVT Ltd, India
S.No 354-1,354-2A/1 At and Post Narayangaon, Near Champagne India, Tal-Junnar, Dist- Pune-410504, Maharashtra, India
Tel: +91 9021983818/+91 8275754803, email: sales@premiumserums.com, website: premiumserums.com
Produce polyvalent Snake Venom Antiserum (Pan Africa). Effective against *Bitis arietans, B. gabonica, B. g. rhinoceros, B. nasicornis, Echis ocellatus, E. leucogaster, E. carinatus, Naje haje, N. melanoleuca, N. nigricollis, N. mossambica, Dendroaspis polylepis, D. viridis, D. jamesoni* and *D. angusticeps.*

USEFUL LITERATURE

A wide range of material on African snakes is available and can be found relatively easily through the internet, and from specialised booksellers such as the NHBS (nhbs.com). For anyone interested in snake taxonomy, a most useful starting point is The Reptile Database, available at reptile-database.org. We salute Peter Uetz and Jiri Hošek, who maintain this essential resource.

There is (as yet) no complete guide to all the snakes of Africa, although Tony Phelps's *Old World Vipers: A Natural History of the Azemiopinae and Viperinae* (Frankfurt: Edition Chimaira, 2010) covers all the African vipers, and has excellent pictures and maps for all species. However, a handful of well-illustrated popular books on the snakes of various regions of Africa are available; these are detailed below.

Southern Africa

Bill Branch (1988) *Field Guide to the Snakes and other Reptiles of Southern Africa.* Cape Town: Struik. Covers all southern African reptiles; slightly dated; pictures bound in a block at the end.

Johan Marais (2004) *A Complete Guide to the Snakes of Southern Africa.* Cape Town: Struik Nature. Excellent classic field guide with large, clear pictures, maps and text on the same page, of all the southern African (everywhere south of 15°S) snakes.

Also useful is Johan Marais's *Snakes and Snakebite in Southern Africa* (Cape Town: Struik Nature, 2014). Clear pictures and distribution maps of all dangerous species of southern Africa plus relevant 'look-alikes', and some useful material on avoiding snakebite and on first aid, etc.

East Africa

Stephen Spawls, Kim Howell, Harald Hinkel and Michele Menegon (2018) *Field Guide to East African Reptiles.* London: Bloomsbury. Virtually all East African snakes illustrated; maps for all species.

West and Central Africa

Jean-François Trape and Youssouph Mane (2006) *Guide des Serpents d'Afrique Occidentale: Savane et Désert*. Paris: IRD Editions. In French; large, clear illustrations of all the snakes of the western Sahel; maps for all. In the not too distant future, Trape should be publishing his *Guide des Serpents d'Afrique Occidentale et Centrale, Afrique du Nord*, with comprehensive coverage of all the snakes of North, West and Central Africa.

Jean-Philippe Chippaux and Kate Jackson (2019) *Snakes of Central and Western Africa*. Baltimore: Johns Hopkins University Press. Pictures of nearly all the snakes of West and Central Africa; good maps for all species.

North Africa

Philippe Geniez (2018) *Snakes of Europe, North Africa and the Middle East: A Photographic Guide*. Princeton: Princeton University Press. Excellent pictures, maps and text, covering all African snakes from north of the Sahara.

USEFUL WEBSITES AND FORUMS

Several Facebook forums are involved with African herpetology, and some offer identification; we list a selection below. As always on the internet, some are more active than others. Although several of the forums below (particularly the South African and East African ones) are pretty rapid, largely troll-free and well policed, often with expertise beyond their geographical area, there is still a tendency by some posters to confidently misidentify snakes. Use your judgement.

East African Snakes and other reptiles/East African Herpetofauna: facebook.com/ groups/662521540444058

Herpetology of the Horn of Africa: facebook.com/ groups/147119015300743

Reptiles and amphibians of Central Africa: facebook.com/ groups/324342077985776

Reptiles and amphibians of the Middle East and North Africa: facebook.com/groups/Arabianherps

Snakes and Creepy Crawlies of Malawi: facebook.com/ groups/158862031304054

Snakes of Eswatini (Swaziland): facebook.com/ groups/367549667045931

Snakes of Namibia: facebook.com/ groups/187156948143224

Snakes of South Africa: facebook.com/groups/snakesofsouthafrica

West African Reptiles/Reptiles de l'Afrique de l'Ouest: facebook.com/ groups/196041673924460

Zambian Snakes and other crawlies: facebook.com/groups/Zamsnakes/?fref=nf

There are also several internet forums/websites where herpetological information (particularly distributions) can be published and accessed; they include inaturalist. org, herpnet.org, vertnet.org and the virtual museum at the Animal Demography Unit (ADU) of the University of Cape Town, at vmus.adu.org.za. Many extensive African snake collections in museums can also be accessed via the internet; they include the Field Museum in Chicago (fieldmuseum.org/science/research/area/ amphibians-reptiles) and the California Academy of Sciences (calacademy.org/ scientists/herpetology-collection).

A number of academic journals publish material on snakes and many medical journals publish papers on snakebite; search online for 'African snakebite research'. Major science journals that regularly publish African herpetological material include *Zootaxa* and the *African Journal of Ecology*. The only organisation solely devoted to African herpetology is the Herpetological Association of Africa, which produces a scholarly journal and a regular newsletter. Details of all these can be sourced from the internet, and many welcome contributions.

An immense set of resources (posters, literature, snake-handling equipment, apps, training courses, first aid packs) are available at the African Snakebite Institute; their website is africansnakebiteinstitute.com

Stephen Spawls is always delighted to receive communications concerning African herpetology. He can be reached via Facebook or at stevespawls@ hotmail.com

CHECKLISTS OF DANGEROUS SNAKES FROM THE REGIONS AND COUNTRIES OF AFRICA

NORTH AFRICA		Western Sahara	Morocco	Algeria	Tunisia	Libya	Egypt
Puff Adder	*Bitis arietans*	x	x				
Burton's Carpet Viper	*Echis coloratus*						x
North-east African Carpet Viper	*Echis pyramidum*					x	x
White-bellied Carpet Viper	*Echis leucogaster*	x	x	x	x	x	
Sahara Horned Viper	*Cerastes cerastes*	x	x	x	x	x	x
Sahara Sand Viper	*Cerastes vipera*	x	x	x	x	x	x
Böhme's Horned Viper	*Cerastes boehmei*				x		
Western False Horned Viper	*Pseudocerastes fieldi*						x
Moorish Viper	*Daboia mauritanica*	x	x	x	x	x	
North African Blunt-nosed Viper	*Macrovipera lebatina*			x	x		
Lataste's Viper	*Vipera latastei*		x	x	x		
Egyptian Cobra	*Naja haje*	x	x	x	x	x	x
Nubian Spitting Cobra	*Naja nubiae*						x
Desert Black Snake	*Walterinnesia aegyptia*						x
Palestine Burrowing Asp	*Atractaspis engaddensis*						x
Western Montpellier Snake	*Malpolon monspessulanus*	x	x	x			
Eastern Montpellier Snake	*Malpolon insignitus*			x	x	x	x
Moila Snake	*Malpolon moilensis*	x	x	x	x	x	x

SAHEL		Mauritania	Mali	Burkina Faso	Niger	Chad	Sudan
Puff Adder	Bitis arietans	x	x	x	x	x	x
Burton's Carpet Viper	Echis coloratus						x
North-east African Carpet Viper	Echis pyramidum						x
White-belled Carpet Viper	Echis leucogaster	x	x	x	x	x	
West African Carpet Viper	Echis ocellatus	x	x	x	x		
Roman's Carpet Viper	Echis romani				x	x	x
Mali Carpet Viper	Echis jogeri	x	x				
Sahara Horned Viper	Cerastes cerastes	x	x		x	x	x
Sahara Sand Viper	Cerastes vipera	x	x		x	x	x
West African Night Adder	Causus maculatus	x	x	x	x	x	x
Velvety-green Night Adder	Causus resimus					x	x
Rhombic Night Adder	Causus rhombeatus					x	x
Black Mamba	Dendroaspis polylepis		x	x			x
Central African Garter Snake	Elapsoidea laticincta					x	x
Half-banded Garter Snake	Elapsoidea semiannulata	x	x	x	x	x	x
Trape's Garter Snake	Elapsoidea trapei	x	x				
Egyptian Cobra	Naja haje	x(?)	x	x	x	x	x
Senegal Cobra	Naja senegalensis		x	x	x		
Eastern Forest Cobra	Naja subfulva					x	
Banded Forest Cobra	Naja savannula		x	x	x	x	
Black-necked Spitting Cobra	Naja nigricollis	x	x	x	x	x	x
West African Brown Spitting Cobra	Naja katiensis	x	x	x	x		
Nubian Spitting Cobra	Naja nubiaea				x	x	x
Slender Burrowing Asp	Atractaspis aterrima		x	x		x	
Dahomey Burrowing Asp	Atractaspis dahomeyensis		x	x		x	
Western Small-scaled Burrowing Asp	Atractaspis microlepidota	x					
Sahelian Burrowing Asp	Atractaspis micropholis	x	x	x	x	x	
Watson's Burrowing Asp	Atractaspis watsoni	x	x	x	x	x	

SAHEL		Mauritania	Mali	Burkina Faso	Niger	Chad	Sudan
Magretti's Burrowing Asp	*Atractaspis magrettii*						x
Sudan Burrowing Asp	*Atractaspis phillipsi*						x
Central African Rock Python	*Python sebae*	x	x	x	x	x	x
Boomslang	*Dispholidus typus*	x	x	x	x	x	x
Olive Sand Snake	*Psammophis mossambicus*	x	x	x	x	x	x
Moila Snake	*Malpolon moilensis*	x	x		x	x	x

WEST AFRICA		Senegal	Gambia	Guinea-Bissau	Guinea	Sierra Leone	Liberia	Ivory Coast	Ghana	Togo	Benin	Nigeria
Western Bush Viper	*Atheris chlorechis*				x	x	x	x	x	x		x
Hairy Bush Viper	*Atheris hirsuta*					x	x					
Green Bush Viper	*Atheris squamigera*											x
Puff Adder	*Bitis arietans*	x	x	x	x	x		x	x	x	x	x
Rhinoceros Viper	*Bitis nasicornis*				x	x	x	x	x	x		x
Eastern Gaboon Viper	*Bitis gabonica*											x
Western Gaboon Viper	*Bitis rhinoceros*				x	x	x	x	x	x		
White-belled Carpet Viper	*Echis leucogaster*	x	x	x(?)	x							x
West African Carpet Viper	*Echis ocellatus*	x			x			x	x	x	x	x
Roman's Carpet Viper	*Echis romani*											x
Mali Carpet Viper	*Echis jogeri*	x	x	x	x							
Forest Night Adder	*Causus lichtensteini*				x	x	x	x	x			x
West African Night Adder	*Causus maculatus*	x	x	x	x	x	x	x	x	x	x	x
Velvety-green Night Adder	*Causus resimus*											x
Rhombic Night Adder	*Causus rhombeatus*											x
Jameson's Mamba	*Dendroaspis jamesoni*								x	x	x	x
West African Green Mamba	*Dendroaspis viridis*	x	x	x	x	x	x	x	x	x	x	
Black Mamba	*Dendroaspis polylepis*	x	x (?)	x	x			x				
Half-banded Garter Snake	*Elapsoidea semiannulata*	x	x	x	x			x	x	x	x	x
Trape's Garter Snake	*Elapsoidea trapei*	x										
Egyptian Cobra	*Naja haje*										x	x

WEST AFRICA		Senegal	Gambia	Guinea-Bissau	Guinea	Sierra Leone	Liberia	Ivory Coast	Ghana	Togo	Benin	Nigeria
Senegal Cobra	*Naja senegalensis*	x	x	x	x				x	x(?)	x	x
Eastern Forest Cobra	*Naja subfulva*											x
Central African Forest Cobra	*Naja melanoleuca*											x
Banded Forest Cobra	*Naja savannula*	x	x	x	x			x	x	x	x	x
West African Forest Cobra	*Naja guineensis*			x	x	x	x	x	x	x		
Black-necked Spitting Cobra	*Naja nigricollis*	x	x	x	x	x	x	x	x	x	x	x
West African Brown Spitting Cobra	*Naja katiensis*	x	x		x			x	x	x	x	x
Gold's Tree Cobra	*Pseudohaje goldii*							x	x			x
Black Tree Cobra	*Pseudohaje nigra*			x	x	x	x	x	x	x		x
Slender Burrowing Asp	*Atractaspis aterrima*	x	x	x	x	x		x	x	x	x	x
Bill Branch's Burrowing Asp	*Atractaspis branchi*				x		x					
Fat Burrowing Asp	*Atractaspis corpulenta*				x	x	x	x	x			x
Dahomey Burrowing Asp	*Atractaspis dahomeyensis*	x	x	x	x	x	x	x	x	x	x	x
Variable Burrowing Asp	*Atractaspis irregularis*				x		x	x	x	x	x	x
Reticulate Burrowing Asp	*Atractaspis reticulata*								x(?)			x
Western Small-scaled Burrowing Asp	*Atractaspis microlepidota*	x	x									
Sahelian Burrowing Asp	*Atractaspis micropholis*	x	x								x	x
Watson's Burrowing Asp	*Atractaspis watsoni*	x									x	x
Central African Rock Python	*Python sebae*	x	x	x	x	x	x	x	x	x	x	x
Boomslang	*Dispholidus typus*	x	x	x	x	x		x	x	x	x	x
Forest Vine Snake	*Thelotornis kirtlandii*			x	x	x	x	x	x	x	x	x
Blanding's Tree Snake	*Toxicodryas blandingii*	x		x	x	x	x	x	x	x	x	x
Powdered Tree Snake	*Toxicodryas pulverulenta*				x	x	x	x	x	x	x	x
Olive Sand Snake	*Psammophis mossambicus*	x	x	x	x	x		x	x	x	x	x

CENTRAL AFRICA

Common Name	Scientific Name	Cameroon	Central African Republic	Bioko Island	Equatorial Guinea (mainland)	São Tomé	Gabon	Republic of the Congo	Democratic Republic of the Congo
Broadley's Bush Viper	*Atheris broadleyi*	x	x		x		x	x	
Rough-scaled Bush Viper	*Atheris hispida*								x
Shaba Bush Viper	*Atheris katangensis*								x
Great Lakes Bush Viper	*Atheris nitschei*								x
Green Bush Viper	*Atheris squamigera*	x	x	x	x		x	x	x
Cameroon Bush Viper	*Atheris subocularis*	x							
Puff Adder	*Bitis arietans*	x	x				x	x	x
Rhinoceros Viper	*Bitis nasicornis*	x	x	x	x		x	x	x
Eastern Gaboon Viper	*Bitis gabonica*	x	x		x		x	x	x
Northeast African Carpet Viper	*Echis pyramidum*		x						
Roman's Carpet Viper	*Echis romani*	x	x						
Two-striped Night Adder	*Causus bilineatus*								x
Forest Night Adder	*Causus lichtensteini*	x	x		x		x	x	x
West African Night Adder	*Causus maculatus*	x	x		x		x	x	x
Velvety-green Night Adder	*Causus resimus*	x	x						x
Rhombic Night Adder	*Causus rhombeatus*	x	x					x	x
Rasmussen's Night Adder	*Causus rasmusseni*								x
Jameson's Mamba	*Dendroaspis jamesoni*	x	x	x	x		x	x	x
Black Mamba	*Dendroaspis polylepis*	x	x						x
Boulenger's Garter Snake	*Elapsoidea boulengeri*								x
Gunther's Garter Snake	*Elapsoidea guentheri*							x	x
Central African Garter Snake	*Elapsoidea laticincta*	x	x						x
East African Garter Snake	*Elapsoidea loveridgei*								x
Half-banded Garter Snake	*Elapsoidea semiannulata*	x	x				x	x	x
Egyptian Cobra	*Naja haje*	x	x						x
Anchieta's Cobra	*Naja anchietae*								x
Burrowing Cobra	*Naja multifasciata*	x	x		x		x	x	x
Banded Water Cobra	*Naja annulata*	x	x		x		x	x	x

CENTRAL AFRICA		Cameroon	Central African Republic	Bioko Island	Equatorial Guinea (mainland)	São Tomé	Gabon	Republic of the Congo	Democratic Republic of the Congo
Congo Water Cobra	*Naja christyi*								x
Eastern Forest Cobra	*Naja subfulva*	x	x				x	x	x
Central African Forest Cobra	*Naja melanoleuca*	x	x	x	x		x	x	x
Banded Forest Cobra	*Naja savannula*	x	x						
São Tomé Forest Cobra	*Naja peroescobari*					x			
Black-necked Spitting Cobra	*Naja nigricollis*	x	x				x	x	x
West African Brown Spitting Cobra	*Naja katiensis*	x							
Gold's Tree Cobra	*Pseudohaje goldii*	x	x	x	x		x	x	x
Slender Burrowing Asp	*Atractaspis aterrima*	x	x					x	x
Bibron's Burrowing Asp	*Atractaspis bibronii*								x
Central African Burrowing Asp	*Atractaspis boulengeri*	x	x		x		x	x	x
Congo Burrowing Asp	*Atractaspis congica*	x						x	x
Fat Burrowing Asp	*Atractaspis corpulenta*	x	x		x		x	x	x
Dahomey Burrowing Asp	*Atractaspis dahomeyensis*	x							
Variable Burrowing Asp	*Atractaspis irregularis*	x	x		x		x	x	x
Reticulate Burrowing Asp	*Atractaspis reticulata*	x	x		x		x	x	x
Sahelian Burrowing Asp	*Atractaspis micropholis*	x							
Watson's Burrowing Asp	*Atractaspis watsoni*	x	x						
Southern African Rock Python	*Python natalensis*								x
Central African Rock Python	*Python sebae*	x	x		x		x	x	x
Boomslang	*Dispholidus typus*	x	x					x	x
Forest Vine Snake	*Thelotornis kirtlandii*	x	x		x		x	x	x
Southern Vine Snake	*Thelotornis capensis*								x
Blanding's Tree Snake	*Toxicodryas blandingii*	x	x		x		x	x	x
Powdered Tree Snake	*Toxicodryas pulverulenta*	x	x				x	x	x
Jackson's Tree Snake	*Thrasops jacksoni*	x	x				x	x	x
Olive Sand Snake	*Psammophis mossambicus*	x	x				x	x	x

EAST AFRICA		Djibouti	Ethiopia	Eritrea	South Sudan	Somalia	Kenya	Tanzania	Uganda	Rwanda	Burundi
Barbour's Short-headed Viper	Atheris barbouri							x			
Horned Bush Viper	Atheris ceratophora							x			
Mount Kenya Bush Viper	Atheris desaixi						x				
Rough-scaled Bush Viper	Atheris hispida						x	x	x		
Acuminate Bush Viper	Atheris acuminata								x		
Matilda's Bush Viper	Atheris matildae							x			
Great Lakes Bush Viper	Atheris nitschei								x	x	x
Mount Rungwe Bush Viper	Atheris rungweensis							x			
Green Bush Viper	Atheris squamigera				x		x	x	x		
Kenya Montane Viper	Montatheris hindii						x				
Floodplain Viper	Proatheris superciliaris							x			
Puff Adder	Bitis arietans	x	x	x	x	x	x	x	x	x	x
Rhinoceros Viper	Bitis nasicornis				x		x	x	x	x	x
Eastern Gaboon Viper	Bitis gabonica				x		x	x	x	x	x
Ethiopian Mountain Viper	Bitis parviocula		x								
Bale Mountain Viper	Bitis harenna		x								
Kenya Horned Viper	Bitis worthingtoni						x				
North-east African Carpet Viper	Echis pyramidum	x	x	x	x	x	x				
Somali Carpet Viper	Echis hughesi					x					
Two-striped Night Adder	Causus bilineatus							x		x	
Snouted Night Adder	Causus defilippii						x	x			
Forest Night Adder	Causus lichtensteini				x		x		x		
West African Night Adder	Causus maculatus		x		x				x		
Velvety-green Night Adder	Causus resimus		x		x	x	x	x	x	x	x
Rhombic Night Adder	Causus rhombeatus		x		x		x	x	x	x	x
Rasmussen's Night Adder	Causus rasmusseni							x			
Eastern Green Mamba	Dendroaspis angusticeps						x	x			

EAST AFRICA		Djibouti	Ethiopia	Eritrea	South Sudan	Somalia	Kenya	Tanzania	Uganda	Rwanda	Burundi
Jameson's Mamba	Dendroaspis jamesoni				x		x	x	x	x	x
Black Mamba	Dendroaspis polylepis		x	x	x	x	x	x	x	x	
Boulenger's Garter Snake	Elapsoidea boulengeri								x		x
Broadley's Garter Snake	Elapsoidea broadleyi					x					
Southern Somali Garter Snake	Elapsoidea chelazziorum					x					
Gunther's Garter Snake	Elapsoidea guentheri							x			
Central African Garter Snake	Elapsoidea laticincta				x				x		
East African Garter Snake	Elapsoidea loveridgei		x		x		x	x	x	x	x
Usambara Garter Snake	Elapsoidea nigra						x	x			
Half-banded Garter Snake	Elapsoidea semiannulata				x						
Egyptian Cobra	Naja haje	x(?)	x	x	x	x	x	x	x		
Banded Water Cobra	Naja annulata							x			x
Eastern Forest Cobra	Naja subfulva		x		x	x	x	x	x	x	x
Central African Forest Cobra	Naja melanoleuca							x	x	x	x
Black-necked Spitting Cobra	Naja nigricollis		x		x		x	x	x	x	x
Ashe's Spitting Cobra	Naja ashei		x		x	x	x	x	x		
Mozambique Spitting Cobra	Naja mossambica							x			
Red Spitting Cobra	Naja pallida	x	x		x	x	x	x			
Nubian Spitting Cobra	Naja nubiae	x		x							
Gold's Tree Cobra	Pseudohaje goldii						x		x		
Yellow-bellied Sea Snake	Hydrophis platura	x				x	x	x			
Slender Burrowing Asp	Atractaspis aterrima				x			x	x		
Bibron's Burrowing Asp	Atractaspis bibronii					x	x	x		x	x
Engdahl's Burrowing Asp	Atractaspis engdahli					x	x				
Variable Burrowing Asp	Atractaspis irregularis		x	x	x		x	x	x	x	x
Ogaden Burrowing Asp	Atractaspis leucomelas	x	x			x					
Somali Burrowing Asp	Atractaspis scorteccii		x			x					

EAST AFRICA		Djibouti	Ethiopia	Eritrea	South Sudan	Somalia	Kenya	Tanzania	Uganda	Rwanda	Burundi
Reticulate Burrowing Asp	*Atractaspis reticulata*							x			
Magretti's Burrowing Asp	*Atractaspis magrettii*			x		x					
Sudan Burrowing Asp	*Atractaspis phillipsi*		x		x						
Eastern Small-scaled Burrowing Asp	*Atractaspis fallax*	x(?)	x			x	x	x			
Southern African Rock Python	*Python natalensis*							x	x		x
Central African Rock Python	*Python sebae*		x	x	x	x	x	x	x	x	
Boomslang	*Dispholidus typus*	x	x	x	x	x	x	x	x	x	x
Forest Vine Snake	*Thelotornis kirtlandii*				x			x	x	x	x
Eastern Vine Snake	*Thelotornis mossambicanus*					x(?)	x(?)	x			x
Usambara Vine Snake	*Thelotornis usambaricus*							x	x		
Dagger-tooth Vine Snake	*Xyelodontophis uluguruensis*							x			
Blanding's Tree Snake	*Toxicodryas blandingii*				x		x	x	x		
Powdered Tree Snake	*Toxicodryas pulverulenta*				x		x		x		
Jackson's Tree Snake	*Thrasops jacksoni*				x		x	x	x	x	x
Olive Sand Snake	*Psammophis mossambicus*	x	x	x	x	x	x	x	x	x	x
Moila Snake	*Malpolon moilensis*			x							

SOUTH-CENTRAL AFRICA		Angola	Zambia	Malawi	Northern Mozambique
Shaba Bush Viper	*Atheris katangensis*		x		
Mount Mabu Bush Viper	*Atheris mabuensis*				x
Mount Rungwe Bush Viper	*Atheris rungweensis*		x	x	
Green Bush Viper	*Atheris squamigera*	x			
Floodplain Viper	*Proatheris supciliaris*			x	x
Puff Adder	*Bitis arietans*	x	x	x	x
Rhinoceros viper	*Bitis nasicornis*	x			
Eastern Gaboon Viper	*Bitis gabonica*	x	x	x	x
Horned Adder	*Bitis caudalis*	x			
Peringuey's Adder	*Bitis peringueyi*	x			
Angolan Adder	*Bitis heraldica*	x			
Two-striped Night Adder	*Causus bilineatus*	x	x		
Snouted Night Adder	*Causus defilippii*		x	x	x
Forest Night Adder	*Causus lichtensteini*	x	x		
West African Night Adder	*Causus maculatus*	x			
Velvety-green Night Adder	*Causus resimus*	x			
Rhombic Night Adder	*Causus rhombeatus*	x	x	x	x
Rasmussen's Night Adder	*Causus rasmusseni*	x	x		
Eastern Green Mamba	*Dendroaspis angusticeps*			x	x
Jameson's Mamba	*Dendroaspis jamesoni*	x			
Black Mamba	*Dendroaspis polylepis*	x	x	x	x
Boulenger's Garter Snake	*Elapsoidea boulengeri*		x	x	x
Gunther's Garter Snake	*Elapsoidea guentheri*	x	x		x
Half-banded Garter Snake	*Elapsoidea semiannulata*	x	x		
Snouted Cobra	*Naja annulifera*		x	x	x
Anchieta's Cobra	*Naja anchietae*	x	x		
Banded Water Cobra	*Naja annulata*	x	x		
Eastern Forest Cobra	*Naja subfulva*	x	x	x	x
Central African Forest Cobra	*Naja melanoleuca*	x			
Black-necked Spitting Cobra	*Naja nigricollis*	x	x	x	
Zebra Cobra	*Naja nigricincta nigricincta*	x			
Mozambique Spitting Cobra	*Naja mossambica*	x	x	x	x
Gold's Tree Cobra	*Pseudohaje goldii*	x			
Coral Shield Cobra	*Aspidelaps lubricus*	x			
Yellow-bellied Sea Snake	*Hydrophis platurus*				x

SOUTH-CENTRAL AFRICA		Angola	Zambia	Malawi	Northern Mozambique
Bibron's Burrowing Asp	Atractaspis bibronii	x	x	x	x
Congo Burrowing Asp	Atractaspis congica	x	x		
Variable Burrowing Asp	Atractaspis irregularis	x			
Reticulate Burrowing Asp	Atractaspis reticulata	x			
Southern African Rock Python	Python natalensis	x	x	x	x
Central African Rock Python	Python sebae	x			
Boomslang	Dispholidus typus	x	x	x	x
Forest Vine Snake	Thelotornis kirtlandii	x	x		
Eastern Vine Snake	Thelotornis mossambicanus		x	x	x
Southern Vine Snake	Thelotornis capensis	x	x	x	x
Usambara Vine Snake	Thelotornis usambaricus				x
Blanding's Tree Snake	Toxicodryas blandingii	x	x		
Powdered Tree Snake	Toxicodryas pulverulenta	x			
Jackson's Tree Snake	Thrasops jacksoni	x	x		
Olive Sand Snake	Psammophis mossambicus	x	x	x	x

SOUTHERN AFRICA		Namibia	Botswana	Zimbabwe	Southern Mozambique	South Africa	Eswatini (Swaziland)	Lesotho
Floodplain Viper	Proatheris superciliaris				x			
Puff Adder	Bitis arietans	x	x	x	x	x	x	x
Eastern Gaboon Viper	Bitis gabonica			x	x	x		
Horned Adder	Bitis caudalis	x	x	x		x		
Namaqua Dwarf Adder	Bitis schneideri	x				x		
Peringuey's Adder	Bitis peringueyi	x						
Berg Adder	Bitis atropos			x		x	x	x
Many-horned Adder	Bitis cornuta	x				x		
Albany Adder	Bitis albanica					x		
Southern Adder	Bitis armata					x		
Red Adder	Bitis rubida					x		
Plain Mountain Adder	Bitis inornata					x		
Desert Mountain Adder	Bitis xeropaga	x				x		
Snouted Night Adder	Causus defilippii			x	x	x	x	
Rhombic Night Adder	Causus rhombeatus	x	x	x	x	x	x	x
Eastern Green Mamba	Dendroaspis angusticeps			x	x	x		

SOUTHERN AFRICA		Namibia	Botswana	Zimbabwe	Southern Mozambique	South Africa	Eswatini (Swaziland)	Lesotho
Black Mamba	Dendroaspis polylepis	x	x	x	x	x	x	
Boulenger's Garter Snake	Elapsoidea boulengeri	x	x	x	x	x	x	
Gunther's Garter Snake	Elapsoidea guentheri			x	x			
Half-banded Garter Snake	Elapsoidea semiannulata	x	x					
Sundevall's Garter Snake	Elapsoidea sundevalli	x	x	x	x	x	x	
Snouted Cobra	Naja annulifera			x	x	x	x	
Anchieta's Cobra	Naja anchietae	x	x	x				
Cape Cobra	Naja nivea	x	x			x		x
Eastern Forest Cobra	Naja subfulva			x	x	x		
Black-necked Spitting Cobra	Naja nigricollis		x					
Zebra Cobra	Naja nigricincta nigricincta	x						
Southwestern Black Spitting Cobra	Naja nigricincta woodi	x				x		
Mozambique Spitting Cobra	Naja mossambica	x	x	x	x	x	x	
Coral Shield Cobra	Aspidelaps lubricus	x				x		
Speckled Shield Cobra	Aspidelaps scutatus	x	x	x	x	x	x	
Rinkhals	Hemachatus haemachatus			x		x	x	x
Yellow-bellied Sea Snake	Hydrophis platurus	x (vagrant)			x	x		
Bibron's Burrowing Asp	Atractaspis bibronii	x	x	x	x	x	x	
Congo Burrowing Asp	Atractaspis congica	x						
Duerden's Burrowing Asp	Atractaspis duerdeni	x	x			x		
Spotted Harlequin Snake	Homoroselaps lacteus					x	x	
Striped Harlequin Snake	Homoroselaps dorsalis					x	x	
Southern African Rock Python	Python natalensis	x	x	x	x	x	x	
Boomslang	Dispholidus typus	x	x	x	x	x	x	
Eastern Vine Snake	Thelotornis mossambicanus			x	x			
Southern Vine Snake	Thelotornis capensis	x	x	x	x	x	x	
Olive Sand Snake	Psammophis mossambicus	x	x	x	x	x	x	
Kwazulu-Natal Black Snake	Macrelaps microlepidotus					x		
Many Spotted Snake	Amplorhinus multimaculatus			x		x		

LIST OF MEDICAL AND SNAKEBITE TERMS

Where there is more than one meaning, the words/terms here are defined in the context of herpetology and the medical treatment of snakebite.

20WBCT: 20-minute whole blood clotting test. A few millilitres of blood taken by venepuncture is placed in a new, clean, dry glass vessel, left undisturbed at room temperature for 20 minutes, then tipped once to see if the blood has clotted or not.

Acetylcholine: one of the most common neurotransmitters; released by motor neurons of the nervous system, it functions at the neuromuscular junction by binding to receptors, thus activating muscles.

Acute kidney injury: failure or damage to the kidneys, causing them to stop working.

Adipose tissue: connective tissue beneath the skin that stores fat.

Anaphylactic shock: an extreme, often life-threatening reaction to an antigen, in which the blood pressure falls precipitously. This may result from allergy caused by the body becoming hypersensitive (e.g. bee and wasp sting anaphylaxis). However, although early reactions to antivenoms are commonly anaphylactic, they are only rarely the result of acquired hypersensitivity.

Anaphylaxis: see anaphylactic shock.

Angina: central chest pain usually due to lack of oxygenated blood flow to the heart, caused by blockage or narrowing of the coronary arteries.

Angioedema: the swelling of the deeper layers of the skin, especially of the tongue, lips, gums, face and upper airway, due to a build-up of fluid as part of a severe anaphylactic reaction.

Anoxic: lacking oxygen.

Antigen: any molecule or part of a molecule, such as a venom toxin or other foreign substance, that induces an immune response in the body, especially the production of antibodies.

Antivenom: a medication containing antibodies directed against venom toxins that is used to treat venomous bites or stings.

Arrhythmia: irregular heartbeat, including bradycardia (slow), tachycardia (fast) and atrial fibrillation (irregular).

Atrioventricular block: a disruption in heart rhythm, adversely affecting the conduction between the atria and ventricles of the heart.

Autonomic nervous system: an efferent branch of the vertebrate peripheral nervous system that regulates the internal environment; it controls the function of smooth muscle, blood vessels and glands, and thus affects the function of internal organs.

Bivalent: an antivenom raised using, and active against, the venom of two species of snake.

Blepharospasm: an involuntary tight closing of the eyelids.

Bronchospasm: spasm (sudden involuntary contraction) of bronchial smooth muscle, producing narrowing of the bronchi (the main air-transporting tubes in the lungs). May be part of a severe anaphylactic reaction.

Cardiorespiratory arrest: loss of blood flow and breathing resulting from the failure of the heart to pump effectively.

Cardiotoxic: see cardiotoxin.

Cardiotoxin: one of a group of phospholipase-based or three-finger polypeptide toxins, often in snake venoms, that cause cellular injury and may affect the heart, causing irregular heartbeat.

Cardiovascular: relating to the heart and blood vessels.

Clotting abnormalities: a disorder that affects the normal mechanism by which the body controls bleeding from damaged blood vessels by activating the blood clotting cascade.

Coagulation profile: a screening test to detect any abnormalities in blood clotting.

Coagulopathic toxins: toxins that cause activation and breakdown ('consumption') of blood clotting factors, resulting in incoagulable blood and bleeding.

Coagulopathy: a condition in which the blood's ability to clot is impaired.

Compartment syndrome: occurs when there is excessive

pressure within a body compartment (usually within the muscles of a limb) as a result of swelling; it may be due to venom and can lead to insufficient supply of blood to tissue in the affected compartment.

Consumptive coagulopathy: a derangement of the coagulation system as a consequence of massive and systemic activation.

Contraindicated: of a medical condition or circumstance, suggesting or indicating that a particular technique or drug should not be used in the case in question.

Cranial: to do with the head or the skull.

Cyanosed skin: bluish colour of the skin resulting from lack of oxygen.

Cytolytic: refers to a toxin that destroys or extensively damages cells.

Cytotoxic: see cytotoxin.

Cytotoxin: a substance that is toxic to living cells.

Dendrotoxin: neurotoxins found in the venom of mambas (*Dendroaspis*); they enhance acetylcholine release, which then causes muscular convulsions and hyperexcitability.

Descending paralysis: acute diffuse weakness and paralysis that begins with the cranial nerves, affecting the muscular functions of the head, and moves gradually downwards.

Disseminated intravascular coagulation (DIC): a disorder in which small blood clots are formed in the entire

bloodstream; this tends to block small blood vessels, thus depleting the number of platelets and associated clotting factors that would normally control bleeding, leading to excessive bleeding.

Dry bite: a bite from a venomous snake in which no venom is injected.

Dyspnoea: difficult or laboured breathing.

ECG (EKG): electrocardiogram; a test that monitors the heart's electrical activity with the aim of detecting abnormalities.

Electrocardiographic: pertaining to the electrical activity of the heart.

Endothelin: a peptide that constricts blood vessels and raises blood pressure.

Enzyme: a biological catalyst, usually a globular protein, that accelerates biochemical reactions.

Factor X: a blood clotting factor and enzyme, or biological catalyst, that is involved in the coagulation cascade, whereby the blood forms clots.

Fasciculations (myokymia): contractions of groups of muscle fibres innervated by the same spinal motor neuron, producing a rippling contraction under the skin that can be confused with shivering.

Fasciotomy: the cutting of the fascia (the connective tissue that encloses and bundles muscles) to decrease damaging pressure in a muscle cluster.

Febrile: relating to a fever.

Fixed dilated pupils: pupils of the eye open wide and showing no response to light.

Forced expiratory pressure: the pressure created by the air leaving the lungs in the middle part of a forced expiration.

Forced expiratory volume in one second (FEV$_1$): the volume of air a patient can expel from their lungs in one second.

Ganglion: a small, solid mass of nervous tissue along a nerve containing numerous cell bodies.

Genito-urinary: relating to the reproductive organs and the urinary system (the kidneys, ureter, bladder, etc.).

Haematemesis: vomiting blood.

Haematoma: localised collection of blood, usually manifesting as dark discolouration under the skin (bruising).

Haematuria: the presence of blood in the urine.

Haemolytic: refers to the destruction of red blood cells.

Haemolytic phospholipases A$_2$: enzymes that break down the membranes of red blood cells.

Haemorrhagic: produced by haemorrhage.

Haemorrhaging: losing blood.

Haemorrhagins: toxins that cause degeneration and lysis (disintegration of a cell by rupture of the cell membrane) of endothelial cells lining capillaries and small blood vessels, thereby resulting in numerous small haemorrhages in the tissues.

Haemostatic dysfunction (antihaemostasis): a syndrome usually associated with bleeding caused by toxins that activate and deplete blood clotting, and cause platelet dysfunction and damage to blood vessel walls.

Haemotoxins (haematotoxins): a somewhat outmoded term describing toxins that destroy red blood cells (haemolysis), disrupt blood clotting and/or damage blood vessels; little used nowadays as not only the blood is affected.

Hyperaesthesia: exaggerated physical sensitivity, particularly of the skin.

Hypertension: high blood pressure.

Hypofibrinogenaemia: decrease in circulating concentrations of fibrinogen, a blood clotting factor, caused by its activation or consumption by the action of venom procoagulant enzymes, leading to consumptive coagulopathy or disseminated intravascular coagulation (DIC) associated with an acute hemorrhagic state.

Hyponatraemia: low sodium concentration in the blood.

Hypotension: low blood pressure.

Hypovolaemia: a state of decreased circulating plasma or blood volume, often resulting from leakage of blood or plasma from blood vessels into the tissues of an envenomed limb, or sometimes generalised.

Hypovolaemic shock: a medical emergency, whereby severe blood or fluid loss means that the heart cannot pump enough blood to the body; it is a form of shock and results in organs ceasing to function.

Incoagulable: incapable of coagulating (usually referring to blood, meaning it will not form a clot).

Intravenous: administered directly into a vein.

IV toxicity: the effect of a toxin administered straight into a vein.

Lesions: areas of an organ or tissue that have been damaged by injury or disease; examples include ulcers, tumours and abscesses.

Leucocytosis: an increase in the number of white cells (white blood corpuscles) in the blood, usually due to inflammation caused by infection or envenoming.

Lymphadenitis: painful, tender enlargement in one or more lymph nodes, due to passage of venom toxins or infection.

Lymphadenopathy: inflammation of the lymph nodes, making them abnormal in size or consistency.

Lymphangitis: an inflammation of the lymph channels, usually due to an infection or envenoming of tissue drained by those channels.

Melaena: production of sticky, foul-smelling jet-black faeces containing digested blood as a result of internal bleeding in the upper part of the gastrointestinal tract.

Menorrhagia: heavy menstrual bleeding.

Microangiopathic haemolysis: destruction of red blood cells secondary to alteration of the lining endothelium of small blood vessels.

Monovalent (monospecific): an antivenom raised using, and

active against, the venom of a single species of snake.

Myalgia: pain in a muscle or a group of muscles.

Mydriasis: dilation of the pupil of the eye.

Myocardium: the muscular tissue of the heart.

Myolytic phospholipases A_2: enzymes that break down the membranes of muscle cells.

Myotoxic: see myotoxin.

Myotoxin: a toxin that causes muscle necrosis, notably venom phospholipases A_2.

Nasal congestion: a nose that is either runny or blocked by sticky mucus.

Nausea: an unpleasant sensation of being about to retch or vomit.

Necrosis: the death of cells in an organ or tissue due to envenoming, disease, injury, or failure of the blood supply.

Neurotoxic: see neurotoxin.

Neurotoxin: a toxin that produces an adverse effect on the structure or function of the central and/or peripheral nervous system. In neurotoxic envenoming, the usual effect of neurotoxins is to block or abnormally excite transmission of nerve impulses at peripheral neuromuscular junctions.

Neutrophil leucocytosis: an abnormally high number of neutrophils (a type of white cell) in the blood in response to envenoming or infection.

Ocular: related to the eye.

Oedema: a condition characterised by an excess of

watery fluid collecting in the cavities or tissues of the body, manifested as swelling.

Ophthalmia: inflammation or disease of the eye.

Ophthalmoplegia: paralysis of the muscles within or surrounding the eye.

Palliative: relieving pain or other symptoms without dealing with the causes of the condition.

Paraesthesia: an odd sensation felt by the skin, manifesting in several ways – such as pricking, pins and needles, numbness, chilling or burning – sometimes without an obvious physical cause.

Paraspecific: having curative actions or properties in addition to the specific one considered medically useful.

Peak expiratory flow: maximum speed of expiration of air from the lungs, and thus an indicator of obstruction of the airways.

Peptide: a compound consisting of a short chain of amino acids, not long enough to be classed as a protein.

Polyvalent (polyspecific): an antivenom raised using, and active against, the venoms of more than one species of snake.

Presynaptic: relating to a nerve cell that releases a transmitter substance into a synapse during transmission of an impulse.

Procoagulant: relating to or denoting substances that promote the conversion in the blood of the inactive precursor protein (clotting factor) into an active clotting enzyme;

for example, prothrombin to thrombin by prothrombin activators.

Proteinuria: the presence of abnormal quantities of protein in the urine, usually an indicator of damage to the kidneys.

Prothrombin: a plasma protein produced by the liver; precursor of thrombin, involved in the blood clotting cascade.

Psychosomosis: a disorder where mental factors are significant in the development of the affliction.

Ptosis (usually bilateral – both sides): paralysis of the upper eyelids, resulting in drooping of the upper eyelids and failure to elevate the lids during upward gaze. The important earliest sign of snake venom neurotoxicity.

Renal: to do with the kidneys.

Replacement therapy: treatment aimed at making up a deficit of a substance normally present in the body, often whole blood or components of whole blood.

Respiratory depression: decreased drive to breathe, manifested by slow and ineffective breathing.

Sarafotoxins: toxins present in the venom of the Palestine Burrowing Asp (*Atractaspis engaddensis*) and some other members of the genus, whose actions resemble those of endothelins, endogenous peptides that have powerful effects on blood vessels.

Skip lesion: an area of necrotic skin caused by envenoming that is separated from another site of necrosis by an area of apparently intact skin.

Subcutaneous: applied or situated under the skin.

Sutures: stitches holding together the edges of a wound.

Syncope: loss of consciousness, informally called 'fainting' or 'passing out', caused by a drop in blood pressure.

Systemic: affecting the entire body.

Thrombocytopenia: having a low blood platelet count.

Thrombosis: local coagulation or clotting of the blood in a part of the circulatory system.

Urticaria: a rash with raised itchy, circumscribed red or white bumps, which may also burn or sting.

Venepuncture: the puncture of a vein as a medical procedure, to take a blood sample out or inject material in.

Vertigo: a form of dizziness, when a person feels that objects around them are moving, or that they themselves are moving when they are not. Often associated with nausea.

Vital capacity: the maximum volume of air that a human can expel from their lungs after breathing in as much as they possibly can.

Glossary

Note: for terms relating to snakebite and its medical treatment, see the 'List of medical and snakebite terms' (pp. 325–28).

Allopatric: occurring in separate regions, not overlapping.

Buccal: to do with the mouth.

Caniculate: with grooves or channels on the surface.

Caudal lure/caudal luring: use of tail movements by a predator to attract potential prey, which approach because they mistake the tail for a food item.

Clade: a group of organisms believed to comprise all the evolutionary descendants of a common ancestor.

Cline: in zoology, a measurable change in a character across the geographical range of a species.

Cloaca: in lower vertebrates, including reptiles, the common chamber into which the urinary, intestinal and genital tracts open.

Colubridae (colubrids): a family of 'typical' snakes, which is now divided and much reduced; in its older sense, a snake without front fangs and with broad ventral scales that span the width of the belly.

Colubroidea: a superfamily of snakes; essentially all snakes apart from the pythons, boas, typhlopids and a few other small groups.

Conspecific: belonging to the same species.

Convergent evolution: the independent evolution of similar features in two species that are not closely related.

Crepuscular: active at twilight.

Diurnal: active by day.

Dorsal: pertaining to the back or upper surface of a snake; the part of the body that doesn't touch the ground.

Ecotype: a distinct form or race of a species of animal, occupying a particular habitat.

Elapidae (elapids): a family of snakes with short, fixed fangs at the front of the upper jaw, notably cobras and mambas (*Dendroaspis*).

Endemic: native to and restricted to a certain place (usually a political country, but sometimes an altitudinal or vegetation zone).

Evolutionary species: a single lineage of populations of an organism, which is distinct from and remains separate from other such lineages.

Extralimital: coming from outside of a particular area.

Fang: a tooth modified to transport venom.

Festoon: a garland or chain of flowers hung in a curve as a decoration; in snakes, a pattern consisting of such a shape.

Fynbos: natural heathland or shrubland of southern Africa.

Gravid: carrying eggs or young; pregnant.

Gular: relating to the throat.

Holotype: a single type specimen on which the description of a new species is based.

Intergrade: a form intermediate between two separate species or subspecies, or the transition from one taxon to another.

Interorbital: between the eyes.

Karoo: semi-desert inland region of South Africa characterised by widely spaced, drought-resistant vegetation and subdued topography.

Keeled: of a scale, having a ridge down the centre.

Lamprophiidae: a family of harmless, small to medium-sized African, Madagascan and Asian snakes.

Lepidosis: the pattern, disposition and numbers of scales.

Loreal: a scale between the eye and the nostril (but not touching either) of a snake; its presence or absence is a useful diagnostic tool.

Monophyletic: of a group of organisms that are descended from a single common ancestor, not shared with any other group.

Monotypic: having only one type of representative; in this case usually a genus containing only a single species.

Morph: different forms of the same species, usually (in snakes) identifiable by sight; for example, yellow and black colour varieties of the Egyptian Cobra (*Naja haje*), or green and grey Boomslangs (*Dispholidus typus*).

Morphology: the form, shape or structure of a living organism.

Neonate: a newborn animal.

Nocturnal: active by night.

Nominate subspecies: the subspecies that was first described as the species itself (e.g. *Naja nigricollis nigricollis*).

Ocellate: shaped like an eye.

Ontogenic: relating to the development of an organism.

Orbit: the bony socket of the eye, or the skin bordering the eye of a reptile.

Postocular: the scale behind the eye.

Prehensile: capable of grasping.

Relict population: a population of organisms that was originally more widespread but now occurs in a restricted area.

Rostral: related to the nose; also a reptile scale at the front of the head.

Sahel: broad, dry east–west region of Africa between the Sahara Desert and the Sudan savanna.

Savanna: a mixed woodland/ grassland ecosystem, where the tops of the trees do not overlap.

Shamming or feigning death: pretending to be dead to avoid the attentions of a predator.

Sidewinding: a method of movement used by desert snakes for moving quickly across the sand, in which a loop of the body is lifted and moved forward, and the rest of the body follows the loop as a fresh loop is formed; this type of movement means that little of the belly is in contact with the ground, preventing the snake from overheating.

Sister species: the species most closely related to the organism being considered, and thus having a single common ancestor.

Sloughing: shedding a layer of dead skin.

Species complex: a cluster of organisms, closely related, that are so similar in morphology that it is difficult to distinguish between them.

Stridulation: producing sound by rubbing body parts together; in snakes this refers to the rubbing together of scales by forming C-shaped coils, typical of carpet vipers, bush vipers and egg-eating snakes.

Subcaudals: scales under the tail; usually refers to these relatively broad scales in snakes.

Sudan savanna: a wide band of tropical savanna that crosses Africa; it extends from Mauritania and Senegal eastwards to the western lowlands of Ethiopia, between the Sahel to the north and the more densely wooded Guinea savanna to the south.

Supranasals: scales above the nose.

Supraoculars: scales above the eyes.

Suprarostral: above the rostral scale.

Sympatric: of two or more species, occurring in the same geographical areas or regions.

Synonymy: a set of (usually scientific) names that have been used for a particular species, only one of which is valid.

Taxonomy: the branch of science concerned with the classification of organisms.

Temporals: the scales behind the postocular scales.

Type (specimen): the specimen(s) on which the original description of a species is based.

Venomous: possessing venom.

Venter/ventral: referring to the underside of the snake.

Ventrals: ventral scales; the broad scales on the underside of a snake.

Vermiculation: a marking resembling the track of a worm.

Viperidae (vipers): a family of snakes with relatively long, folding fangs.

Wadi: an Arabic term meaning a dry ravine that carries water in the wet season.

Zoogeographic: relating to the science of the distribution of animals.

Index